Soft Condensed Matter Physics in Molecular and Cell Biology

Scottish Graduate Series

Soft Condensed Matter Physics in Molecular and Cell Biology

W C K Poon
School of Physics
Edinburgh University, Scotland, UK

D Andelman
School of Physics and Astronomy
Tel Aviv University, Israel

CRC Press
Taylor & Francis Group
Boca Raton London New York

CRC Press is an imprint of the
Taylor & Francis Group, an **informa** business

CRC Press
Taylor & Francis Group
6000 Broken Sound Parkway NW, Suite 300
Boca Raton, FL 33487-2742

First issued in paperback 2019

ISBN-13: 978-0-7503-1023-9 (hbk)
ISBN-13: 978-0-367-39136-2 (pbk)

Library of Congress Cataloging-in-Publication Data

Catalog record is available from the Library of Congress

Visit the Taylor & Francis Web site at
http://www.taylorandfrancis.com

and the CRC Press Web site at
http://www.crcpress.com

SUSSP Proceedings

Lecturers

David Andelman	Tel Aviv University
David Bensimon	École Normale Supérieure, Paris
Stefan Egelhaaf	The University of Edinburgh
Ron Elber	Cornell University, Ithaca
Daan Frenkel	Institute of Atomic & Molecular Physics, Amsterdam
Jean-François Joanny	Curie Institute, Paris
Michael Kozlov	Tel Aviv University
Fred MacKintosh	Free University, Amsterdam
Tom McLeish	Leeds University
Peter Olmsted	Leeds University
Rudi Podgornik	University of Ljubljana
Wilson Poon	The University of Edinburgh
Matthias Rief	The Technical University, Munich
Christoph Schmidt	Free University, Amsterdam
Claus Seidel	Max Planck Institute for Biophysical Chemistry, Göttingen
Jeremy Smith	Ruprecht Karls University, Heidelberg
Patrick Warren	Unilever Research, Wirral

Executive Committee

Wilson Poon	The University of Edinburgh	Director
David Dryden	The University of Edinburgh	Secretary
Stefan Egelhaaf	The University of Edinburgh	Treasurer
Daniel Berry	The University of Edinburgh	Steward

International Committee

Wilson Poon	The University of Edinburgh, Edinburgh, Scotland
David Andelman	Tel Aviv University, Tel Aviv, Israel
Fred MacKintosh	Free University, Amsterdam, The Netherlands
Tom McLeish	University of Leeds, Leeds, UK

Preface

The chapters in this book originated as lectures in the NATO Advanced Science Institute (ASI) and Scottish Universities Summer Schools in Physics (SUSSP) 59, entitled *Soft Condensed Matter Physics in Molecular and Cell Biology* and held in Edinburgh from 29 March to 8 April 2004. All but one of the lecture courses are represented in this volume.

The rationale for this ASI/SUSSP can be simply stated. Soft condensed matter physics is concerned with the study of colloids, polymers and surfactants (or surface-active molecules). Biology is 'soft matter come alive'. Thus, the lecture courses were aimed at introducing participants to the basic principles of contemporary soft condensed matter physics and showing how these principles could be used to elucidate the operation of biomolecules and cells.

As directors of the ASI/SUSSP, we want to thank all the lecturers for the effort they put into preparing for the school: the questionnaires returned by participants tell us that they very much appreciated what the lecturers have done for them. As editors of this volume, we thank all the contributors for easing our task by writing such well-presented chapters to a tight schedule.

The ASI/SUSSP itself could not have taken place except for Daniel Berry's efficient and cheerful local organisation. Afterwards, Drs. Eirini Theofanidou and Helen Sedgwick provided vital editorial assistance. In particular, the former valiantly converted a number of Word files into LATEX while the latter engaged in the minutiae of proof editing and constructed the Index. We are most grateful to all three for their help.

This ASI/SUSSP was part of the six-month *Statistical Mechanics of Biomolecules and Cells* programme at the Isaac Newton Institute for Mathematical Sciences, Cambridge, UK. Many of the lecturers benefitted from stimulating periods of residence at the Institute, who also paid for the lecturers' local expenses at the Edinburgh school. The main sources of funding for the school, however, came from NATO and SUSSP, while the UK's Engineering and Physical Sciences Research Council (EPSRC) and Biotechnology and Biological Sciences Research Council (BBSRC) provided significant 'top-ups'. We thank all of these organisations for their support.

Wilson Poon (Edinburgh) and David Andelman (Tel Aviv), April 2005.

Contents

Introduction: Coarse graining in biological soft matter

Wilson C K Poon

School of Physics, The University of Edinburgh
Mayfield Road, Edinburgh EH9 3JZ, UK

wckp@ph.ed.ac.uk

1 Introduction

Soft condensed matter physics is concerned with the study of complex fluids: liquids in which there is an intermediate, or mesoscopic, length scale between the atomic ($\sim$ 0.1 − 1 nm) and the macroscopic ($\sim$ 1 mm or more). It became a recognisable, separate branch of physics relatively recently (Poon and Warren 2001). The key idea in modern soft condensed matter physics is that complex fluids possess features that are *independent of chemical details*. All colloids undergo Brownian motion. Polymers of all kinds share features arising out of intramolecular connectivity. Aspects of the self-assembly of any surfactant (or surface-active molecule) can be understood by treating it as a truncated cone of suitable shape. (For detailed introductions to soft condensed matter physics, see the chapters by Frenkel, Warren and Olmsted in this volume.)

Biology is 'soft matter come alive'. DNA, RNA and proteins are essentially random[1] polymers (for DNA as polymer, see the chapters by Warren and Bensimon in this volume). The lipids that make up various biological membranes are surfactants (see the chapter by Olmsted in this volume), and the membranes themselves can be viewed as soft elastic sheets (see the chapter by Kozlov in this volume). Vesicles, inclusion bodies and even globular proteins can be viewed as colloids. At first sight, therefore, nothing can be more obvious than the claim, embodied in the title of this school, that soft condensed matter physics should find extensive application in biology. But matters are not so straightforward.

---

[1]The adjective 'random' here needs qualification. The sequence of any biological macromolecule has evolved to enable it to perform certain functions, and from that point of view, there is little randomness!

The modern era in the life sciences began in 1953 when Crick, Watson, Wilkins and Franklin discovered that DNA had a double-helix structure. Not long afterwards, the 'Central Dogma' was established, viz., that biological 'information' is transcribed from DNA into RNA, which is then translated into proteins. The latter are the most abundant macromolecules of life. Proteins make up the bulk of the structural components of cells and are key players in intra- and inter-cellular signalling. After working out much of the molecular details of the Central Dogma (see Alberts *et al.* 2002), biologists then turned to sequencing genomes. By the end of the twentieth century, the genomic challenge has been met for *Homo sapiens* and many 'model organisms'. The next target is 'proteomics' — a complete catalogue of the proteins involved in cellular biochemistry and their mutual interactions. A key component is 'structural proteomics' — solving structures to atomic resolution (first done by Perutz for haemoglobin in 1959).

This brief review of the history of modern biology (Morange 1998) reminds us of the subject's essentially *atomistic* paradigm. Chemical details are everything. Ultimately, no two proteins are alike. A 'point mutation' changing one 'residue' (= amino acid) in a 100-residue protein can make the difference between life and death. Such specificity contrasts starkly with the instinct of the soft condensed matter physicist to focus on generic features by *coarse graining* — ignoring enough details until similarities begin to emerge. The key issue in the applicability of soft condensed matter physics to biology is then this: will coarse graining always throw the baby out with the bath water, or are there situations in which *judicious* coarse graining can give biologically relevant insights?

The contributions to this volume suggest that the latter situation prevails: there are indeed a number of (perhaps even many!) biological problems in which the coarse-grained approach of soft condensed matter physics can usefully be applied. Doing so sometimes answers questions that biologists themselves have been asking. Often the soft matter approach casts these questions in new light. Every now and then, it may even suggest wholly new questions that have not yet been asked within the atomistic paradigm.

In this introductory chapter, I give a brief introduction to coarse graining by reviewing the different levels of description that can be applied to globular proteins and the kinds of questions that one may hope to answer at each level. I do not believe that such an exercise has been attempted systematically before. So, while there are no new pieces of information in what follows, I hope that the juxtaposition of material may offer some insights for soft condensed matter physicists seeking to contribute to biology, and for biologists who want to understand what makes some of their physics colleagues 'tick'.

2 The atomistic description of globular proteins: the tertiary structure

During the research that eventually led A. Fleming to the discovery of the antibiotic penicillin, he found (in 1922) that a substance in his own nasal mucus (he was having a cold at the time!) also killed bacteria. This substance is the protein *lysozyme*. It catalyses the cleavage of a particular bond in the cell wall of 'gram-negative' bacteria (a large class that includes *E. coli*), leading to the swelling and bursting ('lysis', hence *lysis* en*zyme*: lysozyme) of cells. Hen egg white lysozyme has the distinction of being the first enzyme

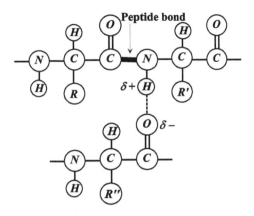

Figure 1. *A schematic representation of amino acid residues in a polypeptide chain. Single (double) lines represent single (double) bonds. The bold line is the peptide bond linking two residues. Residues are distinguished by their side chains,* R, R', *etc. A polypeptide chain is directed, and conventionally runs from the* N- *to* C-*termini. Two residues in spatial proximity can form a hydrogen bond (dotted line).*

to have its structure determined to atomic resolution, by D. Phillips and his team in the 1960s. (For a treatise on lysozymes, see Jollès 1996.)

Proteins are polypeptides, or random polymers made up of amino acid monomers (or, in biological parlance, 'residues'); see Figure 1. Different amino acids are distinguished by their 'side chains' (R, R', etc., in Figure 1); 20 occur naturally. Hen egg white lysozyme has 129 residues and a molecular weight of $M_w \approx 14400$. The sequence of residues in lysozyme (the protein's 'primary structure') and its three-dimensional structure obtained by Phillips and his team (the molecule's 'tertiary structure') can be found in the Protein Data Bank (http://www.rcsb.org) by typing in its 'PDB code', 6LYZ. (For introductions to physical aspects of proteins, see Creighton 1993 and Finkelstein and Ptitsyn 2002.) The atomic structure of lysozyme is shown in Figure 2(a).[2]

The tertiary structure of proteins form the basis of the modern biological discussion of their functions. In the case of lysozyme, the interest lies in how it cleaves the relevant type of bonds in gram-negative bacterial cell walls. There is no doubt that *for this purpose*, a detailed atomistic description is an essential part of the answer. It is a matter of particular atoms on particular residues in the enzyme contacting particular atoms in the target molecules (the 'substrate'). But even for the purpose of obtaining such atomistic understanding, coarse graining can come in handy. The gross shape of lysozyme, Figure 2(a), suggests (correctly!) that the catalytic site resides in the 'cleft'.[3]

---

[2]The figure is generated using the freeware Rasmol (http://www.umass.edu/microbio/rasmol).

[3]Of course, this 'guess' reflects the very old 'lock and key' hypothesis of enzyme function: that substrates geometrically fit into enzymes like keys into locks.

(a)   **(b)**

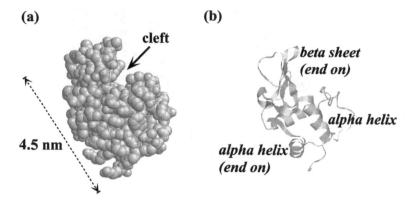

Figure 2. *(a) Space-filling model of the atomic structure of hen egg white lysozyme. The 'cleft' in the molecule contains the catalytic site responsible for digesting bacterial cell walls. (b) Ribbon diagram of exactly the same view of lysozyme showing the secondary structure of α helices, β sheets and coils (wiggly lines).*

3   Coarse-graining level 1: Secondary structure

The primary and tertiary structures of a protein — its sequence of amino acids and the co-ordinates of all of its atoms — constitute the most fine-grained description of a protein.[4] The interesting thing about the complex tertiary structure of proteins, such as that of lysozyme shown in Figure 2(a), is that it is made up of certain smaller structural motifs. These motifs constitute the protein's 'secondary structure'.

The most common secondary structural motifs are α helices and β sheets. The two motifs are about equally common, each accounting for just under one third of residues in all known globular proteins (Creighton 1993). Both α helices and β sheets are held together by hydrogen bonds of the type shown schematically in Figure 1. Certain sequences of residues are known to have propensities for forming helices or sheets, so that given a sequence of residues, reasonably accurate secondary structure predictions can be made (e.g., Garnier *et al.* 1996). From a coarse-grained perspective, we can think of both helices and sheets as formed by a tape (representing the 'backbone' of the chain of residues) whose two sides want to 'stick' to each other. Helices and sheets are just two ways in which opposite sides of such a tape can be brought into contact with each other in an orderly fashion (Figure 3).

A protein's secondary structure is not at all obvious from a fully atomistic representation (Figure 2(a)). To render the secondary structure, biologists use 'ribbon diagrams'. The one for lysozyme is shown in Figure 2(b), where the 129 residues are classified into α helices, β sheets and those that do not fall into either category (wiggly lines).

The ribbon diagram is one of the most important inventions in protein research, and it is a coarse-grained description! It makes clear the folding topology of a protein. Looking at such diagrams makes it clear that there are *families* of proteins. Within each family, members with widely different primary structures (i.e., amino acid sequences) neverthe-

---

[4]This is true for 'statics'. For the importance of *dynamics*, see Smith's chapter in this volume.

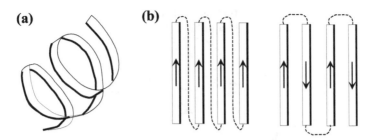

Figure 3. *Schematic coarse-grained representation of secondary structure motifs as tapes with two distinguishable sides (bold and thick lines respectively) that want to stick to each other. (a) α helix; (b) β sheets. The chain has direction, so that neighbouring tapes can be parallel (left) or antiparallel (right), with intervening portions of the chain adopting other conformations (dashed lines). Beta sheets have 'sticky edges'; if exposed, these can cause aggregation. Such aggregation may play a key role in amyloidosis.*

less adopt the same generic tertiary structure made up of very similar secondary structural motifs packed together in nearly identical fashions in three dimensions (see Elber's chapter in this volume for details). The existence of secondary structural motifs also has important consequences for the kinetics of folding — how a randomly structured chain of amino acids searches a very high dimensional space (see McLeish's chapter in this volume) to find the 'native' tertiary structure (for one possibility, see McLeish 2005).

Alpha helices and β sheets can be characterised by certain coarse-grained parameters — from the 'stickiness' of the sides (e.g., so many $k_B T$ of bonding energy per unit length) to various mechanical moduli (e.g., the bending rigidity of a rod). I will now briefly describe one example to illustrate how combining such coarse-grained descriptions with appropriate atomistic details may yield biological insight. The schematic representation of β sheets in Figure 3 shows that it is a potentially dangerous structure: a flat sheet, however extended, will have extensive 'sticky edges'. If exposed, such sticky edges may cause trouble by initiating intermolecular aggregation. Such aggregation probably plays a key part in generating 'amyloids' — insoluble fibrilar structures, often with well-defined diameters, found in tissues of patients suffering from diseases such as Alzheimer's and CJD. There are two common strategies for avoiding exposed sticky edges from β sheets: stick them together to form a 'β barrel', or bury them in the interior.

Wild-type human lysozyme does *not* aggregate to form amyloid fibres. However, a single mutation of residue 67 from aspartic acid to histidine generates a mutant that *does* participate in amyloidosis. Comparison of the atomic-resolution structure of the wild type and the 'Asp67His' mutant (Booth *et al.* 1997) shows that there is significant change to the accessibility of a β sheet (the one labelled as such in Figure 2(b)). This may be enough to 'tip' the mutant over to become amyloidic. But how do exposed sticky edges generate fibres with seemingly well-defined diameters? Here, coarse-graining may help. A. Semenov and coworkers have suggested a theory to describe how sticky chiral objects may give rise to fibrilar aggregates with rather well-defined diameters; the experiments of Aggeli and coworkers using model β sheet forming peptides are consistent with this theory (Aggeli *et al.* 2001). If this or a similar theory is right, then there may well be a generic, coarse-grained mechanism for amyloidosis, but to understand the genesis of

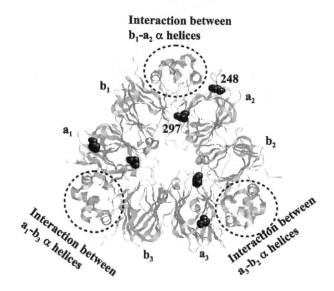

Figure 4. *The structure of proglycinin shown as a ribbon diagram. This protein is a 'homo-trimer', consisting of three identical subunits ('monomers') non-covalently bound together (chiefly via the intertwined α helices, enclosed in dashed circles). Each subunit is made up of an acidic domain and a basic domain ($a_i, b_i; i = 1, 2, 3$). (Compare lysozyme, which is smaller than a single domain in a proglycinin monomer.) There are five main disordered regions on each monomer. The largest of these disordered regions begins and ends at residues 248 and 297, respectively (black space-filling atoms).*

'exposed stickiness' in any particular protein requires atomistic description.

About two thirds of the residues in known globular proteins order into α helices or β sheets. The other one third of residues are accounted for in three ways. Some form small loops linking helices and sheets. Others form extended 'coil' regions (e.g., the wiggly lines in Figure 2(b)), the nomenclature simply reflecting the fact that they are not ordered into helices or sheets. Yet other residues are too disordered even in the crystalline state to be resolved by crystallography.[5] Lysozyme has no such disordered regions, but they are in fact rather common, and show up as gaps in PDB files. There are a number of examples in the structure of proglycinin (Adachi *et al.* 2001; PDB code 1FXZ), a precursor to one of the storage proteins in soya bean. Figure 4 shows where the largest disordered region (consisting of nearly 50 residues) begins and ends in this protein.

Partly because of their absence in PDB files, dynamically disordered regions are typically ignored in discussions of biological function. However, 'disordered' may *not* be synonymous with 'useless'. Thus, for example, a disordered region can act as a 'lid'. In proglycinin, the largest disordered region (residues 248-297) 'covers up' a binding site. Immediately after synthesis, proglycinin aggregates into trimers. When these trimers arrive at the organelle for protein deposition in the developing seed, the large disor-

---

[5]Note that disordered domains may still contain some secondary structure; it is just that somehow the atomic positions cannot be resolved crystallographically.

dered loops are removed to allow two trimers to aggregate into a hexamer (now known as glycinin). Disordered regions can also be used to cover and uncover active sites *reversibly*, without permanent cleavage. Other functions can be envisaged. But the key point is that coarse-grained polymer physics (see Warren's chapter in this volume) can be applied rather directly to disordered regions. This should be a fruitful area for soft condensed matter physicists to explore (for reviews, see Bright *et al.* 2001 and Dunker *et al.* 2002).

4 Coarse-graining level 2: Domains

To introduce the next two levels of coarse graining, I will start by asking what appears at first sight to be a rather strange question: is there any physics hiding in the molecular weight of lysozyme ($M_w \approx 14400$)? Or, given that all the 20 amino acids relevant to biology have molecular weights in the region of 100, the question can be rephrased into this form: is there anything significant about the fact that lysozyme has the order 10^2 residues? Or do we just have to take these numbers as 'mere accidents of evolution'?

To answer these questions, we go back to the history of protein science (Tanford and Reynolds 2001). One of the startling early discoveries, from chemical experiments, was the apparently very large molecular weights of proteins. T. Svedberg confirmed these chemical measurements using a physical method, ultracentrifugation. In the process, he noticed that the molecular weights of all the proteins that he had studied were approximately 'quantised', in units of about 160 residues (Svedberg 1929). Lysozyme, with its 129 residues, is a single 'quantum' on the light side. We now know that larger proteins tend to consist of 'domains', each containing $\approx 100 - 200$ residues — Svedberg's 'quantum' of 160 residues fits nicely in the middle of this range. Figure 4 shows an example of domain structure: each proglycinin monomer is made up of an acidic domain with $M_w \approx 30000$ and a basic domain with $M_w \approx 20000$. The existence of multiple domains in some proteins can also be inferred when single molecules are unfolded by pulling: the force-extension curve clearly shows individual domains 'popping open' successively (see the chapter by Rief in this volume).

But why are there domains of $\approx 160 \pm 50$ residues? A definitive answer cannot yet be given. One intriguing suggestion is that this domain size reflects coarse-grained DNA physics in the primitive environment where DNA-directed protein synthesis first evolved (reviewed in Trifonov and Berezovsky 2003). Let us assume that since the beginning of this synthetic pathway, three base pairs specify one residue. Under physiological conditions (~ 0.1 M salt and pH ~ 7 at ~ 300 K), DNA is a semi-flexible polymer with persistence length $l_p \approx 50$ nm, or about 150 base pairs (see the chapters by Warren, Mackintosh and Bensimon in this volume). The optimal contour length for the cyclisation (or ring closure) of a linear polymer is $\approx 3.5 l_p$, or about 500 base pairs (for references, see Trifonov and Berezovsky 2003). If DNA fragments that could optimally form closed loops were favoured in the primitive environment, e.g., because of their extra stability against degradation (no exposed ends in the cyclised state), then we expect that primitive proteins may preferentially have a size of $\sim 500/3 \approx 160$ residues. These primitive units can then be bolted together for proteins of higher complexity.

In a multi-domain protein, the different domains undergo thermal motion relative to one another. Binding of substrates may significantly affect the spectrum of these low-frequency vibrations, and therefore the ability of the protein to undergo further binding. Models of such 'entropic allostery'[6] have been suggested in which the domains are just treated as featureless 'blobs' (Cooper and Dryden 1984; Hawkins and McLeish 2004).

To make the transition to the next level of coarse graining, I should point out that many proteins are 'multimeric'. (For a review, see D'Allessio 1999.) A multimeric protein is made up of a number of whole proteins (each is called a 'monomer' in this context) non-covalently bound together.[7] An example is the proglycinin trimer (Figure 4). Here the monomers are identical (thus, proglycinin is a 'homo-trimer'), but that need not be the case ('hetero-trimer', etc.). It is tempting to treat each monomer in a multimeric protein as a single particle. We have arrived at the protein as a colloid.

5   Coarse-graining level 3: Proteins as colloids

Globular proteins in their native state are rather compact objects: their interiors approach close packing densities. This compactness is a reflection of the high degree of internal ordering (into secondary structural motifs). The radius of gyration of a typical globular protein is considerably smaller than that of a synthetic polymer of comparable molecular weight but lacking internal structure. Thus, it is possible to think of a globular protein as a hard colloidal particle.[8] (Nowadays, one may be tempted to call proteins 'nanocolloids'!)

5.1   Dilute protein solutions

Colloid scientists today have an impressive array of experimental tools for characterising particles. Even given the atomic details that x-ray crystallography (and nuclear magnetic resonance spectroscopy) can reveal, the tools of colloids science can still be used profitably to study dilute protein solutions. For instance, a careful combination of data from dynamic light scattering (which measures the protein's translational diffusion coefficient in solution) and ultracentrifugation showed, before a crystal structure was available, that the protein ocr (PDB code 1S7Z) could be modelled as a prolate ellipsoid with long and short axes of 10.4 nm and 2.6 nm, respectively (Blackstock *et al.* 2001). (For an introduction to light scattering, see the chapter by Egelhaaf in this volume.) This rather unusual, elongated shape is consistent with the protein's known function: as a DNA mimic used by bacteriophage T7 to 'fool' the defence mechanism of its host.[9] Subsequent crystallography (Walkinshaw *et al.* 2002) confirms this finding.

---

[6] Allostery is the effect of binding at one site of a protein on binding at a distant site.

[7] The multimeric constitution of a protein is known as its 'quaternary structure'.

[8] The founders of protein science in the nineteenth century would not have chosen to describe proteins as colloids. At that time, 'colloid' was taken by the majority of practitioners of colloid science (also founded in the nineteenth century) to mean more or less ill-defined aggregates of small molecules. Indeed, nineteenth century colloid scientists typically believed that proteins were just such aggregates, rather than *bona finde* macro*molecules*. See Tanford and Reynolds 2001 for details.

[9] Ocr is the first protein synthesised by bacteriophage T7 using the host cell's machinery as it injects its DNA into *E. coli*. Ocr mimics DNA, so that bacterial 'restriction' enzymes (ocr = 'overcome classical restriction'), which otherwise would recognise and destroy the phage DNA, find and bind to ocr instead.

(a) 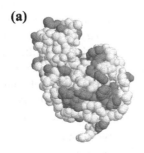 **(b)**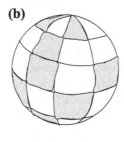

Figure 5. *(a) Hen egg white lysozyme showing hydrophilic (white) and hydrophobic (grey) residues. Note the 'patchiness' of the surface. Hydrophobic patches on neighbouring molecules may stick to each other. (b) A protein as a sphere with sticky patches.*

5.2 Concentrated protein solutions

Coarse-grained colloid physics can be used to even greater effect when it comes to understanding certain aspects of the behaviour of concentrated protein solutions. Consider first crystallising proteins *in vitro*, a crucial step in structure determination. This is often done by adding salt, which induces a short-range attraction between proteins. The strength of this attraction can be quantified by measuring the second virial coefficient, B_2, by static light scattering. B_2 is related to the orientationally-averaged (and therefore isotropic) interparticle interaction, $U(r)$ (where r is the centre-to-centre distance), by

$$B_2 = 2\pi \int_0^\infty \left(1 - e^{-U(r)/k_B T}\right) r^2 dr. \tag{1}$$

Interestingly, the value of B_2 at crystallisation as a function of the protein volume fraction (ϕ), $B_2^{\text{cryst}}(\phi)$, is almost independent of the particular protein, provided that it is crystallisable in the first place (George and Wilson 1994). Significantly, the crystallisation of colloids with a short-range attraction maps onto the same universal boundary (Poon 1997). Thus, as far as the equilibrium crystallisation boundary is concerned, proteins behave as coarse-grained sticky colloids.

This is all very well for those proteins that *do* crystallise. But it is well known that proteins in general are rather recalcitrant to the crystalliser's art. It is still not wholly clear why this should be so, but the story so far has involved an intriguing combination of coarse-grained and atomistic descriptions. A way into this problem comes from asking the seemingly rather pointless question: what space groups[10] do proteins adopt when they do crystallise? Small non-polar organic molecules crystallise into space groups that allow them to maximise *packing* (Wright 1995). It appears that a different principle may be at work in proteins. While there is no overwhelming favourite, over a third of crystallisable proteins adopt the orthorhombic space group $P2_1 2_1 2_1$. Wukovitz and Yeates (1995) have investigated this question, and suggested that proteins crystallise to maximise intermolecular contact.

---

[10]There are only 230 ways the symmetry of individual objects, proteins in our case, can be combined with the symmetry of regular lattices; these symmetry combinations are described by 230 'space groups'.

To understand intermolecular contact, one has to look at the *heterogeneity* of protein surfaces (Figure 5(a)). It is known that the amino acid lysine is the most common surface residue on proteins. On the other hand, it is the *least* common residue at sites for intermolecular contacts (e.g., where two monomers touch in a multimeric protein). Z. Derewenda and colleagues have investigated the significance of these pieces of information for protein crystallisation (Longenecker *et al.* 2001). They took human guanine nucleotide dissociation inhibitor (RhoGDI) and a number of its mutants, and replaced some of the surface lysines with alanine. In every case, the crystallisability improved.

Now, of the 20 biologically relevant amino acids, lysine has the second-most extended side chain, and it terminates in a positively charged group ($R = $ —$(CH_2)_4NH_3^+$, Figure 1), whereas alanine has the smallest side chain ($R = CH_3$).[11] To the colloid physicist's coarse-graining mind, lysine therefore provides 'local' steric and charge stabilisation, dis-favouring contact; replacing it with alanine removes both stabilising effects. Importantly, Derewenda and co-workers found that lysine-to-alanine replacement did not prevent their proteins from folding into native or near-native structures (with native or near-native enzymatic function). So, perhaps one could think of lysine[12] as one of nature's 'surface modification agents' for proteins. When these modifiers are absent, specific 'sticky patches' on the protein surface then permit crystallisation. The equilibrium statistical mechanics of particles with sticky patches has been studied, either for random patches (Asherie *et al.* 2002, Figure 5(b)), or regular patches (Sear 1999). A detailed comparison with protein crystals is yet to be made.[13] Note that 'sticky patches' imply an anisotropic interparticle interaction, $U(\mathbf{r})$.

Before leaving the subject of crystallisation, I should point out that proteins do occasionally crystallise *in vivo*.[14] One example is the Woronin body in filamentous fungi (*Neurospora crassa* and other *Euacomycetes*): single-crystalline hexagonal platelets about 0.5 μm across that serve to stop cell content leakage in the event of filaments being severed. It seems that crystallinity is somehow essential to this fungal organelle, at least to its synthesis in the cell if not to its function (Yuan *et al.* 2003). Some insect viruses synthesise crystalline protein shells for their dispersal (Smith 1976). A transgenic example is the occurrence of crystals in grains of wheat incorporating glycinin genes (Stöger *et al.* 2001). Physicists are yet to pay serious attention to *in vivo* crystallisation.

When conditions are right for proteins to crystallise, various forms of amorphous aggregation are never far away.[15] Often, very small change in parameters can lead to significant changes in the morphology of the aggregates (Sedgwick *et al.* 2005). One of the morphologies observed in *in vitro* experiments, micron-sized amorphous protein beads (Figure 6) is reminiscent of proteinaceous 'inclusion bodies' in bacteria, formed when, for example, certain proteins are over-expressed. Many features of the aggregation of proteins when there is little or no disruption of the tertiary structure of the individual molecules[16] can be understood by borrowing concepts from recent work on glass and gel formation in colloids with uniform interparticle attraction (Sedgwick *et al.* 2004, 2005).

---

[11] Arginine is longer by one bond, $R = $ —$(CH_2)_3NHC(NH_2)_2^+$; glycine has no side chain, i.e., $R = H$.

[12] And perhaps some of the other charged residues as well.

[13] It would also be interesting to use such models to study the solution physics of multimeric proteins.

[14] Here I discuss only 3D crystals; there are also examples of *in vivo* 2D crystalline ordering.

[15] E.g. precipitation (for purification purposes) by adding salt such as $(NH_4)_2SO_4$, known as 'salting out'.

[16] This is in contrast to, e.g., the case when a protein partially unfolds to expose sticky β sheet edges which then leads to aggregation.

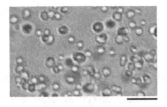

Figure 6. *Salt-induced aggregation of lysozyme (pH = 4, room temperature) into amorphous beads. The scale bar is 10 μm. See Sedgwick et al. (2005) for details.*

The effect of surface heterogeneities, however, has not been considered to date.

Finally, it is clear that ordinarily proteins *in vivo* should neither crystallise nor aggregate, leaving strong interaction to just (and only just) those species that have evolved specifically to bind (such as an enzyme and its substrates). It is therefore legitimate to talk of evolution subjecting the whole *proteome* (the entire collection of proteins in, say, a cell) to *negative design*[17] to ensure stability against general phase separation or aggregation of any kind. The physics behind this question has begun to be addressed (Braun 2004, Sear 2004, Doye *et al.* 2004).

Before leaving the subject of 'proteins as colloids' altogether, I ought briefly to mention electrostatics. (For a comprehensive and advanced survey, see Holm *et al.* 2001. Cohn and Edsall (1943) is still useful.) Proteins are charged objects. Traditionally, charge effects in soft matter are understood within a mean-field framework (see the chapter by Andelman in this volume). However, it is increasingly realised that there are conditions (including biologically relevant ones, such as the presence of multi-valent ions) under which the mean-field framework fails, and fails badly. For example, multi-valent ions can induce an attraction between like-charged surfaces due to correlation effects neglected in the mean-field treatment. Such effects have been extensively thought about for DNA (partly because of the relevance for the 'packaging' of this macromolecule; see the chapter by Podgornik in this volume). They should be equally intriguing for proteins.

6 Further coarse-graining

Considering proteins as particles, perhaps with chemically heterogeneous surfaces, does not represent the highest possible level of coarse-graining. In this section, I briefly mention two further levels of 'blurring the details' in protein physics.

(a) *Rigid rods*: Actin (M_w = 43000) and tubulin (α and β varieties, M_w = 50000) are two proteins that are widely utilised for mechanical and locomotory functions. (For physically and biologically motivated treatments, see Boal 2002 and Bray 2001, respectively.) The details differ, but in both cases, individual globular proteins (called 'monomers' in this context) self-assemble into 'polymers' with very long persistence lengths (17 μm for actin and $\sim 1 - 5$ mm for tubulin). A coarse-grained approach in which these long-persistence-length polymers are simply treated as rods with certain mechanical properties turns out to be adequate for understanding some of their biological functions. For exam-

---

[17]To evolve *not* to do certain things, e.g., proteins not to bind to the 'wrong' partners.

ple, F. Mackintosh's discussion of the rheology of concentrated actin solutions in this volume should be highly relevant for elucidating the mechanics of the cytoskeleton. (See also Schmidt's chapter in this volume for an introduction to the use of optical tweezers to study biopolymer rheology.) Such understanding eventually feeds into coarse-grained models for the movement of whole cells (see Joanny's chapter in this volume).

(b) *Computational elements*: Many enzymes show allostery — binding at one site may affect the activity of other binding sites on the same protein. The activity of proteins can also be modulated by reactions such as phosphorylation, acetylation, methylation, etc. The products of certain enzymatic reactions can be 'fed back' to modify the activity of the enzyme itself. D. Bray suggests that proteins are 'computational elements' (Bray 1995). In this highly coarse-grained description, a protein is a 'black box' that takes certain inputs (substrate concentrations) and computes an output (product concentration), taking into account various control parameters (e.g., phosphorylation states).

7   Conclusion

In this whirl wind tour of globular proteins, I have started from the most atomistic description of the tertiary structure and progressively shed details and coarse-grained. The description at each level, especially when judiciously combined with descriptions at lower levels, can illuminate particular kinds of biological issues. The highest level of coarse-grained description, which I reviewed only very briefly, treats protein molecules as computational elements. This suggests a comparison with computer science. (Hartwell *et al.* 1999 discusses how and what biology can learn from computer science.)

The scientific description of computers can be divided into a number of levels (Figure 7). On the most detailed level, there is the physics of atomic energy levels and how these combine to form energy bands in crystalline solids. Lying at the interface between physics and electrical engineering is the study of semiconductor devices such as transistors. Here, we deal in current-voltage characteristic curves controlled by device properties (conductivity, etc.). These are ultimately derivable from lower-level physics. But once we know their values, we can study devices without further recourse to atoms.

Moving into electrical engineering proper, whole *circuits* appear on the scene. Collections of circuits can be 'coarse grained' into 'black boxes' called *logic gates*, and so on, until we arrive at central processing units, memories, etc. Now we are dealing with *computer architecture*, which straddles electrical engineering and computer science.

Computer science starts with systems architecture, but rapidly moves away from hardware to *algorithms* and the modelling of *processes* using Petri nets or others forms of 'process calculi'. At its most abstract, computer science deals with *computation* — e.g., whether certain propositions are 'decidable' by a Turing machine in finite time.

This brief analysis of the study of computers and computation provides a fruitful analogy on two levels. First, the schema in Figure 7 is quite similar to the sequence of coarse-graining applied to proteins in this chapter. The same sort of analysis can be repeated for other classes of biological macromolecules: there are a number of quasi-independent levels of description, with 'leakage' between levels.

But secondly, and perhaps more interestingly, the computer/computation analogy can

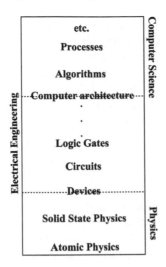

Figure 7. *A schematic summary of the different levels at which computers can be studied. Going from bottom to top, each level is more coarse-grained than the previous. The dashed lines represent where 'leakage' between levels of description occurs. Otherwise, the three 'boxes' can, and do, function more or less independently.*

be applied to the whole of biology: the science of life. On this broader canvas, the ground covered by, say, *Molecular Biology of the Cell* (Alberts *et al.* 2002) would correspond to the box marked 'Physics' in Figure 7. This is the most detailed level of description relevant to the subject. The science of genetics perhaps corresponds to the interface occupied by 'devices' in Figure 7. At this level, we can talk about a coarse-grained entity, the 'gene' occurring as different 'alleles', but molecular-level details do 'leak through'.[18]

What corresponds to the 'interfacial' level of 'architecture' and to the top box in Figure 7 depends on what kind of biological system we are talking about. For the whole biosphere, the individual organism lies at the interface, and sciences such as population genetics and ecology lie in the 'top box'. These top-level descriptions are well established. The challenge is to see how lower-level details fit in. On the other hand, for the biology of single organisms (from a single *E. coli* bacterium to a whale), matters are less clear. Perhaps the interface is occupied by physiological description in terms of either organs (for, say, mammals) or organelles (for single cells). But what now corresponds to the top box proper is unclear. We do not yet have a 'theory of life'!

It may be thought that current efforts in 'systems biology' may eventually yield descriptions at the 'top level'. But much of current systems biology seems to be directed towards obtaining *complete* descriptions, rather than necessarily *higher-level* descriptions. The limitations of this approach are revealed in the current run of '-omics':[19] genomics, proteomics, metabolomics, and then what? At or slightly beyond the metabolic pathway, we seem to run out of coarse-graining descriptors.

To see what is needed, note that every coarse-graining category in Figure 7 from

---

[18]Of course, when Mendel invented genetics by enunciating his laws of inheritance, and for some time afterwards, genetics was a fully coarse-grained science, with no lower-level details.

[19]BLAHomics means a complete description of all the BLAHs in, say, a cell.

the level of 'logic gates' upwards is there because of our understanding of the theory of computation. Someone looking at a circuit diagram of a computer for the first time could invent the category of 'logic gates' if he or she has some inkling of what the abstract idea of 'computation' is all about. Otherwise, less fruitful (or downright useless) coarse-graining categories may have been invented. Now, the important point is that historically, engineers designing and analysing circuit diagrams *did* have, from the beginning, a good idea of abstract ideas of computation.

The application to biology is twofold. First, coarse-graining is only fruitful (certainly in the long run) if it is consonant with higher-level theories. Lacking a 'theory of life', any kind of coarse graining we do in biology at present will remain provisional. For example, we do not know if it is fundamentally fruitful to introduce computational categories, whether in neurobiology or in Bray's proteins. Secondly, it may be fruitful to proceed to seek general theories of life without worrying too much about the lower-level details, at least at the beginning. It is true that, right now, the very concept of a 'theory of life' appears so vague that it does not even sound 'scientific'. But if we had not had Turing, would we have had a similar feeling of unease about a 'theory of computation'?

Even after we have learnt to give 'top-level' description to whole organisms, another question remains: does this sort of schema actually work for biology? In computer science, there are higher levels of description where lower-level details do not have to feature *at all*. But currently, it is hard to find any biological discussion at or below the level of single organisms that does not contain significant reference to molecules. Perhaps there is something special about the biology of a living organism so that, of necessity, 'leakage' from the bottom level occurs all the way. If that is the case, then we should try to understand why this is so. In physics, a system in which every level of detail matters is close to a critical point.

To me, these are the fundamental questions concerning coarse-graining in biology.

Acknowledgements

David Dryden (Edinburgh) introduced me to ocr, while Clare Mills (Institute of Food Research) and Rod Casey (John Innes Centre) educated me on seed proteins. I learnt much about *in vivo* crystallisation from Jon Doye and Aard Louis (both Cambridge). Daan Frenkel (Amsterdam) told me about Trifinov and Berezovsky. Peter Olmsted (Leeds) first made me think about protein crystal space groups. The Isaac Newton Institute, Cambridge, provided a most congenial atmosphere for thinking about these things for six months in 2004. Mike Cates and David Dryden's comments improved the manuscript.

References

Adachi M, Takenaka Y, Gidamis A B and Utsumi B, 2001, *J Mol Biol* **305** 291.
Aggeli A, Nyrkova I A, Bell M, Harding R, Carrick L, McLeish T C B, Semenov A N and Boden N, 2001, *Proc Natl Acad Sci USA* **98** 11857.
Alberts B, Johnson A, Lewis J, Raff M, Roberts K and Walter P, 2002, *Molecular Biology of the Cell*, fourth edition (Garland Science, New York).
Asherie N, Lomakin A and Benedek G B, 2002, *Proc Natl Acad Sci* **82** 1537.

Blackstock J J, Egelhaaf S U, Atanasiu A, Dryden D T F and Poon W C K, 2001, *Biochem* **40** 9944.

Boal D, 2002, *Mechanics of the Cell* (Cambridge University Press, Cambridge).

Booth D R, Sunde M, Bellotti V, Robinson C V, Hutchinson W L, Fraser P E, Hawkins P N, Dobson C M, Radford S E, Blake C C F and Pepys M P, 1997, *Nature* **385** 787.

Braun F N, 2004, *Phys Rev E* **69** article 011903.

Bray D, 1995, *Nature* **376** 307.

Bray D, 2001, *Cell Movements: from Molecules to Motility*, second edition (Garland Publishing, New York).

Bright J N, Woolf T B and Hoh J H, 2001, *Prog Biophys Mol Biol* **76** 131.

Cohn E J and Edsall J T, eds, 1943, *Proteins, Amino Acids and Peptides as Ions and Dipolar Ions* (Reinhold, New York).

Cooper A and Dryden D F T, 1984, *Eur Biophys J* **11** 103.

Creighton T E, 1993, *Proteins: Structures and Molecular Properties*, second edition (W H Freeman and Co., New York).

D'Alessio G, 1999, *Prog Biophys Mol Biol* **72** 271.

Doye J P K, Louis A A and Vendruscolo M, 2004, *Phys Biol* **1** P1.

Dunker A K, Brown C J, Lawson D, Iakoucheva L M and Obradović Z, 2002, *Biochem* **41** 6573.

Finkelstein A V and Ptitsyn O B, 2002, *Protein Physics: A Course of Lectures* (Academic Press, London and San Diego).

Garnier J, Gibrat J-F and Robson B, 1996, *Methods Enzymol* **266** 540.

George A and Wilson W, 1994, *Acta Cryst D* **50** 361.

Hartwell L H, Hopfield J J, Leibler S and Murray A W, 1999, *Nature* **402** C47.

Hawkins R J and McLeish T C B, 2004, *Phys Rev Lett* **93** art. no. 098104.

Holm C, Kekicheff P and Podgornik R, eds, 2001, *Electrostatic Effects in Soft Matter and Biophysics* (Kluwer, Dordrecht).

Jollès P, ed, 1996, *Lysozymes: Model Enzymes in Biochemisry and Biology* (Birkhauser Verlag, Basel).

Longenecker K L, Garrad S M, Sheffield P J and Derewenda Z S, 2001, *Acta Cryst D* **57** 679.

McLeish T C B, 2005, *Biophys J* **88** 172.

Morange M, 1998, *A History of Molecular Biology* (Harvard University Press, Cambridge, Mass. and London).

Poon W C K, 1997, *Phys Rev E* **55** 3762.

Poon W C K and Warren P B, 2001, *Contemp Phys* **42** 179.

Sear R P, 1999, *J Chem Phys* **111** 4800.

Sear R P, 2004, *J Chem Phys* **120** 998.

Sedgwick H, Egelhaaf S U and Poon W C K, 2004, *J Phys Condens Matter* **16** S4913.

Sedgwick H, Kroy K, Salonen A, Robertson M B, Egelhaaf S U and Poon W C K, 2005, *Eur Phys J E* **16** 77.

Smith K M, 1976, *Virus-Insect Relationships* (Longman, London).

Stöger E, Parker M, Christou P and Casey R, 2001, *Plant Physiol* **125** 1732.

Svedberg T, 1929, *Nature* **123** 871.

Tanford C and Reynolds J, 2001, *Nature's Robots: a History of Proteins* (Oxford University Press, Oxford).

Trifinov E N and Berezovsky I N, 2003, *Curr Opinion Struct Biol* **13** 10.

Walkinshaw M, Taylor P, Sturrock S S, Atanasiu C, Berge T, Henderson RM, Edwardson J M and Dryden D T F, 2002, *Mol Cell* **9** 187.

Wright J D, 1995, *Molecular Crystals*, second edition (Cambridge University Press, Cambridge).

Wurkovitz S W and Yeates T O, 1995, *Nature Struct Biol* **2** 1062.

Yuan P, Jedd G, Kumaran D, Swaminathan S, Shio H, Hewitt D, Chua N-H, Swaminathan K, 2003, *Nature Struct Biol* **10** 264.

Part I

Soft Matter Background

Introduction to colloidal systems

Daan Frenkel

FOM Institute for Atomic and Molecular Physics
Kruislaan 407, 1098 SJ Amsterdam, The Netherlands
frenkel@amolf.nl

1   Introduction

Colloid physics has been a meeting place of physics and biology long before the word 'biophysics' emerged. In this context, the discovery of Brownian motion (by Robert Brown in 1827), is not a good example because the crucial aspect of his observation was that pollen particles appeared to move, even though they were *not* motile. Hence, certainly at the time of Brown, the biological relevance of Brownian motion was not clear. But things became different with the discovery of osmotic pressure. Osmotic pressure is very important in biology — it is responsible for the turgor of plant cells, and animal cells usually do not survive in a medium with the wrong osmotic pressure. The systematic study of osmotic pressure, started by a biologist (Hugo de Vries), was taken over by physiologists (Hamburger and Donders) and was given a theoretical foundation by a chemical physicist (van 't Hoff), who showed that the osmotic pressure of macromolecules in solution depends on their concentration in the same way that the pressure of a noble gas depends on the number density of the constituent atoms. This is a surprising finding, because the law holds for macromolecules of *any* size. In particular, it applies to the class of macromolecules called 'colloids'.

In fact, colloids behave like giant atoms in more than one way, and quite a bit of the physics of colloids can be understood by making use of this analogy. However, much of the interesting behaviour of colloids is related to the fact that they are, in many respects, *not* like atoms. In this lecture, I shall start from the picture of colloids as oversized atoms or molecules, and I shall then selectively discuss some features of colloids that are different. My presentation of the subject will be a bit strange because I am a computer simulator rather than a colloid scientist. Colloids are the computer simulator's dream, because many of them can be represented quite well by models — such as the hard-sphere and Yukawa models — that are far too simple to represent molecular systems. On the other hand, colloids are also the simulator's nightmare, or at least challenge, because if we look more closely, simple models do not work: this is sometimes true for the static properties of colloids (e.g., in the case of charged colloids) and even more often, in the case of colloid dynamics.

What are colloids? Usually, we refer to a substance as a 'colloidal suspension' if it is a dispersion of more-or-less compact particles with sizes within a certain range (typically 1 nm – 1μm). However, it would be more logical to classify colloids according to some physical criterion. To this end, we should compare colloidal particles with their 'neighbours': small molecules on one end of the scale and bricks on the other. What distinguishes colloids from small molecules? I would propose that the important difference is that for the description of colloids, a detailed knowledge of the 'internal' degrees of freedom is not needed — in particular, the discrete, atomic nature of matter should be irrelevant. That is *not* to say that the chemical nature of the constituent atoms or molecules is irrelevant — simply that in order to describe a colloid, we do not need to know the detailed microscopic arrangement of these constituents. This definition has the advantage that it allows for the fact that particles may behave like colloids in some respects, and like 'molecules' in others. For instance, we cannot hope to understand the biological function of proteins if we do not know their atomic structure. However, we can understand a lot about the phase behaviour of proteins without such knowledge (see further Poon's chapter on coarse graining in this volume). This ambiguous nature of macromolecules may persist even at length scales that are usually considered colloidal. For instance, for the biological function of the tobacco mosaic virus, the precise sequence of its genetic material is important. But its tendency to form colloidal liquid crystals depends only on 'coarse-grained' properties, such as shape, flexibility and charge.

Let us next consider the other side of the scale. What is the difference between a colloidal particle and a brick? The behaviour of colloids is governed by the laws of statistical mechanics. In equilibrium, colloidal suspensions occur in the phase with the lowest free energy, and the dynamics of colloids in equilibrium is due to thermal ('Brownian') motion. In principle, this should also be true for bricks. But in practice, it is not. In order for bricks to behave like colloids, they should be able to evolve due to Brownian motion. There are two reasons why bricks do not. First of all, on earth, all particles are subject to gravity. The probability to find a particle of mass m at a height h above the surface of the earth is given by the barometric height distribution:

$$P(h) = \exp(-mgh/k_{\mathrm{B}}T), \qquad (1)$$

where m is the effective mass of the colloidal particle (i.e., the mass minus the mass of the displaced solvent), T is the temperature and k_{B} is Boltzmann's constant. The average height of the colloid above the surface is equal to $< h > = kT/(mg)$. For a 1 kg brick at room temperature, $< h > = O(10^{-20})$ cm. This tells us something that we all know: bricks do not float around due to thermal motion. One way to delimit the colloidal regime is to require that $< h >$ be larger than the particle diameter. Suppose we have a spherical particle with diameter σ and (excess) mass density ρ, then our criterion implies

$$\frac{\pi}{6} g \rho \sigma^4 = k_{\mathrm{B}}T. \qquad (2)$$

For a particle with an excess density of 1 g/cm^3, this equality is satisfied on earth for $\sigma \approx 1\ \mu m$. In the microgravity environment that prevails in space, much larger particles would behave like colloids (not bricks though, because it is virtually impossible to reduce all accelerations to less than $10^{-20}g$). Another way to make large particles behave like colloids on earth is to match the density of the solvent to that of the particle. Yet, even if we would succeed in doing all this for a brick, it would still not behave like a colloid. Colloidal particles should be able to move due to diffusion (i.e., thermal motion). How long does it take for a particle to move a distance equal to its own diameter? In a time t, a particle typically diffuses a distance $\sqrt{2Dt}$. For a spherical particle, the diffusion constant is given by the Stokes-Einstein relation $D = k_B T/(3\pi\eta\sigma)$, where η is the viscosity of the solution. Hence, a particle diffuses a distance comparable to its own diameter in a time

$$\tau = O(\eta\sigma^3/kT). \tag{3}$$

For a 1μm colloid in water, this time is of the order of one second. For a brick, it is of the order of ten million years. Hence, even though bricks in zero-gravity may behave like colloids, they will not do so on a human timescale. Clearly, what we define as a colloid also depends on the observation time. Again, 1 micron comes out as a natural upper limit to the colloidal domain.

In summary, a colloid is defined by its behaviour. For practical purposes, the colloidal regime is between 1 nanometre and 1 micrometre. But these boundaries are not sharp. And the lower boundary is ambiguous: a particle may behave like a colloid in some respects, but not in others.

Now consider biological systems. The 'colloidal domain' spans an important range of sizes in biological systems. Compact proteins typically have sizes in the nanometre range. Viruses have sizes in the range from tens to hundreds of nanometres (except in some very special cases). Bacteria (or, more generally, prokaryotic cells) have sizes in the range of hundreds of nanometres to a few microns. They are all colloids. Eukaryotic cells have sizes that are sufficiently large (tens of microns) to make them non-colloidal. If you see a neutrophil chase a bacteria, the bacteria may move by Brownian motion, but the cell can only keep up because it is motile.

1.1 Forces between colloids

Most colloidal suspensions are solutions of relatively large particles in a simple molecular solvent. Yet, the description of the static properties of such a solution resembles that of a system of atoms in vacuum — somehow, the solvent does not appear explicitly. At first sight, this seems like a gross omission. However, as pointed out by Onsager (Onsager 1949), we can eliminate the degrees of freedom of the solvent in a colloidal dispersion. What results is the description that only involves the colloidal particle, interacting through some *effective* potential (the 'potential of mean force') that accounts for all solvent effects. This is the mysterious simplification that was already noted by van 't Hoff. In the following, I briefly sketch how this works. Consider a system of N_c colloids in a volume V at temperature T. The solvent is held at constant chemical potential μ_s, but the number of solvent molecules N_s is fluctuating. The 'semi-grand' partition function of such a system is given by

$$\Xi(N_c, \mu_s, V, T) \equiv \sum_{N_s=0}^{\infty} \exp(\beta\mu_s N_s) Q(N_c, N_s, V, T), \tag{4}$$

where $\beta = 1/k_B T$. The canonical partition function $Q(N_c, N_s, V, T)$ is given by

$$Q(N_c, N_s, V, T) = \frac{q_{id,c}(T)^{N_c} q_{id,s}(T)^{N_s}}{N_c! N_s!} \int dr^{N_c} dr^{N_s} \exp[-\beta U(r^{N_c}, r^{N_s})]. \quad (5)$$

where $q_{id,\alpha}$ is the kinetic and intra-molecular part of the partition function of a particle of species α. These terms are assumed to depend only on temperature, and not on the inter-molecular interactions (sometimes this is not true, e.g., in the case of polymers — I shall come back to that point later). In what follows, I shall usually drop the factors $q_{id,\alpha}$ (more precisely, I shall account for them in the definition of the chemical potential: i.e., $\mu_\alpha \Rightarrow \mu_\alpha + kT \ln q_{id,\alpha}$). The interaction potential $U(r^{N_c}, r^{N_s})$ can always be written as $U_{cc} + U_{ss} + U_{sc}$, where U_{cc} is the direct colloid-colloid interaction (i.e., $U(r^{N_c}, r^{N_s})$ for $N_s = 0$), U_{ss} is the solvent-solvent interaction (i.e., $U(r^{N_c}, r^{N_s})$ for $N_c = 0$), and U_{sc} is the solvent-colloid interaction $U(r^{N_c}, r^{N_s}) - U_{cc}(r^{N_c}) - U_{ss}(r^{N_s})$. With these definitions, we can write

$$Q(N_c, N_s, V, T) = \frac{1}{N_c!} \int dr^{N_c} \exp[-\beta U_{cc}] \left\{ \frac{1}{N_s!} \int dr^{N_s} \exp[-\beta(U_{ss} + U_{sc})] \right\},$$
$$(6)$$

and hence

$$\Xi(N_c, \mu_s, V, T) = \frac{1}{N_c!} \int dr^{N_c} \exp[-\beta U_{cc}]$$
$$\times \left\{ \sum_{N=0}^{\infty} \frac{\exp(\beta \mu_s N_s)}{N_s!} \int dr^{N_s} \exp[-\beta(U_{ss} + U_{sc})] \right\}. \quad (7)$$

We can rewrite this in a slightly more suggestive form by defining

$$Q_s(N_s, V, T) \equiv \frac{1}{N_s!} \int dr^{N_s} \exp[-\beta U_{ss}] \quad (8)$$

and

$$\Xi(\mu_s, V, T) \equiv \sum_{N_s=0}^{\infty} \exp(\beta \mu_s N_s) Q_s(N_s, V, T). \quad (9)$$

Then

$$\Xi(N_c, \mu_s, V, T) = \frac{1}{N_c!} \int dr^{N_c} \exp[-\beta U_{cc}]$$
$$\times \left\{ \sum_{N_s=0}^{\infty} \exp(\beta \mu_s N_s) Q_s(N_s, V, T) \langle \exp[-\beta U_{sc}] \rangle_{N_c, N_s, V, T} \right\}$$
$$= \frac{\Xi(\mu_s, V, T)}{N_c!} \int dr^{N_c} \exp[-\beta U_{cc}] \langle \exp[-\beta U_{sc}] \rangle_{\mu_s, T}, \quad (10)$$

where

$$\langle \exp[-\beta U_{sc}] \rangle_{\mu_s, V, T} \equiv \frac{\sum_{N_s=0}^{\infty} \exp(\beta \mu_s N_s) Q_s(N_s, V, T) \langle \exp[-\beta U_{sc}] \rangle_{N_c, N_s, V, T}}{\Xi(\mu_s, V, T)}.$$
$$(11)$$

We now define the *effective* colloid-colloid interaction as

$$U_{cc}^{\text{eff}}(r^{N_c}) \equiv U_{cc}(r^{N_c}) - k_B T \ln \left\langle \exp[-\beta U_{sc}(r^{N_c})] \right\rangle_{\mu_s, V, T}. \tag{12}$$

We refer to $U_{cc}^{\text{eff}}(r^{N_c})$ as the *potential of mean force*. Note that the potential of mean force depends explicitly on the temperature and on the chemical potential of the solvent. In case we study colloidal suspensions in *mixed* solvents, the potential of mean force depends on the chemical potential of all components in the solvent (an important example is a colloid dispersed in a polymer solution).

At first sight, it looks as if the potential of mean force is totally intractable. For instance, even when the colloid-solvent and solvent-solvent interactions are pairwise additive,[1] the potential of mean force is not. However, we should bear in mind that even the 'normal' potential energy function that we all think we know and love is also not pairwise additive — that is why we can hardly ever use the pair potentials that describe the intermolecular interactions in the gas phase to model simple liquids. In fact, in many cases, we can make very reasonable estimates of the potential of mean force. It also turns out that the dependence of the potential of mean force on the chemical potential of the solvent molecules is a great advantage: it will allow us to *tune* the effective forces between colloids *simply by changing the composition of the solvent.*[2] In contrast, in order to change the forces between atoms in the gas phase, we would have to change Planck's constant or the mass or charge of the electron. Hence, colloids are not simply giant atoms, they are *tunable* giant atoms.

After this general introduction, let me briefly review the nature of inter-colloidal interactions. It will turn out that almost all colloid-colloid interactions depend on the nature of the solvent and are, therefore, potentials of mean force.

1.1.1 Hard-core repulsion

Colloidal particles tend to have a well-defined size and shape. They behave like solid bodies — in fact, many colloidal particles *are* fairly solid (e.g., the colloids that Perrin used to determine Avogadro's number were small rubber balls, silica colloids are small glass spheres, and PMMA colloids are made out of plastic). Solid bodies cannot inter-penetrate. This property can be related to the fact that, at short range, the interaction between (non-reactive) atoms is harshly repulsive. This is due to the Pauli exclusion principle. This hard-core repulsion is about the only colloid-colloid interaction that is essentially independent of the solvent. In fact, colloidal crystals can be dried and studied in the electron microscope because the Pauli exclusion principle works just as well in vacuum as in solution. However, there are also other mechanisms that lead to repulsive interactions between colloids. For instance, short-ranged Coulomb repulsion between like-charged colloids, or entropic repulsion between colloids that have a polymer 'fur', or even solvent-induce repulsion effects. All these repulsion mechanisms are sensitive to the nature of the solvent. We shall come back to them later.

---

[1] In fact, we have, thus far, not even assumed this.

[2] We all know this: simply add some vinegar to milk and the colloidal fat globules start to aggregate.

1.1.2 Coulomb interaction

Coulomb interaction would seem to be the prototype of a simple, pairwise additive interaction. In fact, it is. However, for every charge carried by the colloidal particles, there is a
compensating charge in the solvent. These counter charges 'screen' the direct Coulomb
repulsion between the colloids. I put the word 'screen' between quotation marks, because it is too passive a word to describe what the counter-ions do: even in the presence
of counter-ions and added salt ions, the direct, long-ranged Coulomb repulsion between
the colloids exists — but it is almost completely compensated by a net attractive interaction due to the counter-ions. And the net result is an *effective* interaction between the
colloids that is short-ranged (i.e., that decays asymptotically as $\exp(-\kappa r)/r$, where κ is
the inverse screening length that appears in the Debye-Hückel theory of electrolytes).

$$\kappa = \sqrt{\frac{4\pi}{\epsilon k_B T} \sum \rho_i q_i^2},$$

where ϵ is the dielectric constant of the solvent and ρ_i is the number density of ionic
species i with charge q_i. The first expression for the effective electrostatic interaction between two charged colloids was proposed by Derjaguin, Landau, Verweij and Overbeek
(DLVO) (Verwey and Overbeek 1948; see also Andelman's chapter in this volume):

$$V_{Coulomb} = \left(\frac{Q \exp(\kappa R)}{1 + \kappa R} \right)^2 \frac{\exp(-\kappa r)}{\epsilon r}, \tag{13}$$

where r is the distance between the two charged colloids, Q is the (bare) charge of the
colloid and R is its 'hard-core' radius. Ever since, there have been attempts to improve on
the DLVO theory. However, the theory of the effective electrostatic interaction between
colloids is subtle and full of pitfalls. For a discussion of interactions between charged
colloids, I refer the reader to the chapter by Podgornik in this volume.

1.1.3 Dispersion forces

Dispersion forces are due to the correlated zero-point fluctuations of the dipole moments
on atoms or molecules. As colloids consist of many atoms, dispersion forces act between
colloids. However, it would be wrong to conclude that the solvent has no effect on the
dispersion forces acting between colloids. After all, there are also dispersion forces acting between the colloids and the solvent, and between the solvent molecules themselves.
In fact, for a pair of polarisable molecules, the dispersion interaction depends on the
polarisabilities (α_1 and α_2) of the individual particles

$$u_{disp}(r) \approx -\frac{3\alpha_1 \alpha_2 h \sqrt{\nu_1 \nu_2}}{4\pi r^6} \equiv -\frac{C_{disp}(12)}{r^6}, \tag{14}$$

where ν_i is a characteristic frequency associated with the optical transition responsible
for the dipole fluctuations in molecule i (in what follows, we shall assume this frequency
to be the same for all molecules). The net dispersion force between colloidal particles
in suspension depends on the difference in polarisability per unit volume of the solvent
and the colloid. The reason is easy to understand: if we insert two colloidal particles
in a polarisable solvent, we replace solvent with polarisability density $\rho_s \alpha_s$ by colloid

with polarisability density $\rho_c \alpha_c$. If the two colloidal particles are far apart, each colloid contributes a constant amount proportional to $-\rho_s \alpha_s (\rho_c \alpha_c - \rho_s \alpha_s)$ to the dispersion energy. However, at short inter-colloidal distances there is an additional *effective* colloid-colloid interaction that is proportional to $-(\rho_c \alpha_c - \rho_s \alpha_s)^2 / r_{cc}^6$. This is an *attractive interaction* irrespective of whether the polarisability density of the colloids is higher or lower than that of the solvent. However, in a colloid mixture, the dispersion force need not be attractive: if the polarisability density of one colloid is *higher* than that of the solvent and the polarisability density of the other is *lower*, then the effective dispersion forces between these two colloids are *repulsive*. The polarisability density of bulk phases is directly related to the refractive index. For instance, the Clausius-Mosotti expression for the refractive index is

$$\frac{n^2 - 1}{n^2 + 2} = \frac{4\pi \rho \alpha}{3}. \tag{15}$$

Hence, if the refractive index of the solvent is equal to that of the colloidal particles, then the effective dispersion forces vanish! This procedure to switch off the effective dispersion forces is called *refractive index matching*. In light-scattering experiments on dense colloidal suspensions, it is common to match the refractive indices of solvent and colloid in order to reduce multiple scattering. Hence, precisely the conditions that minimise the dispersion forces are optimal for light-scattering experiments.[3]

Colloids are not point particles; hence, Equation 14 has to be integrated over the volumes of the interacting colloids to yield the total dispersion interaction

$$V_{disp}(r) = -\frac{A}{6} \left\{ \frac{2R^2}{r^2 - 4R^2} + \frac{2R^2}{r^2} + \ln \frac{r^2 - 4R^2}{r^2} \right\}, \tag{16}$$

where A is the so-called *Hamaker constant*. In the simple picture sketched here, A would be proportional to $(\rho_c \alpha_c - \rho_s \alpha_s)^2$. However, in a more sophisticated theoretical description of the dispersion forces between macroscopic bodies (Israelachvili 1992), the Hamaker constant can be related explicitly to the (frequency-dependent) dielectric constants of the colloidal particles and the solvent. This analysis affects the value of the constant A but not, to a first approximation, the functional form of Equation 16.

1.1.4 DLVO Potential

Combining Equation 13 and Equation 16, we obtain the DLVO potential that describes the interaction between charged colloids (see further Andelman's chapter in this volume):

$$V_{DLVO}(r) = \left(\frac{Q \exp(\kappa R)}{1 + \kappa R} \right)^2 \frac{\exp(-\kappa r)}{\epsilon r} - \frac{A}{6} \left\{ \frac{2R^2}{r^2 - 4R^2} + \frac{2R^2}{r^2} + \ln \frac{r^2 - 4R^2}{r^2} \right\}. \tag{17}$$

Note that, at short distances, the dispersion forces always win. This would suggest that the dispersion interaction will always lead to colloidal aggregation. However, the electrostatic repulsion usually prevents colloids from getting close enough to fall into the

---

[3]In addition, it is also possible to match the density of the solvent to that of the colloid. This has an little effect on the interaction between colloids. But, as far as gravity is concerned, density-matched colloidal particles are neutrally buoyant - that is they behave as if they have a very small (ideally zero) positive or negative effective mass. This is the mass that enters into Equation 1. Hence, by density-matching, we can study bulk suspensions of colloids that would otherwise quickly settle on the bottom of the container.

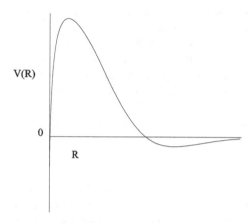

Figure 1. *The DLVO potential has a deep minimum at short distances. At larger distances, the Coulomb repulsion dominates. This gives a local maximum in the curve. At still larger distances, the dispersion interaction may lead to a secondary minimum.*

primary minimum of the DLVO potential (Figure 1). The height of this stabilising barrier depends (through κ) on the salt concentration. Adding more salt will lower the barrier and, eventually, the colloids will be able to cross the barrier and aggregate.

1.1.5 Depletion interaction

One of the most surprising effects of the solvent on the interaction between colloids is the so-called depletion interaction. Unlike the forces that we discussed up to this point, the depletion force is not a solvent-induced modification of some pre-existing force between the colloids. It is a pure solvent effect. It is a consequence of the fact that the colloidal particles exclude space to the solvent molecules. To understand it, return to Equation 12

$$U_{cc}^{\text{eff}}(r^{N_c}) \equiv U_{cc}(r^{N_c}) - k_B T \ln \left\langle \exp[-\beta U_{sc}(r^{N_c})] \right\rangle_{\mu_s, V, T}.$$

Let us consider a system of hard particles with no longer-ranged attractive or repulsive interaction. In that case, all longer-ranged contributions to the *effective* potential in Equation 12 are depletion interactions. These interactions can be attractive, even though all direct interactions in the system are repulsive.

To illustrate this, consider a trivial model-system, namely a d-dimensional cubic lattice with at most one particle allowed per square (Frenkel and Louis 1992).

Apart from the fact that no two particles can occupy the same square face, there is no interaction between the particles. For a lattice of N sites, the grand canonical partition function is:

"Solvent"

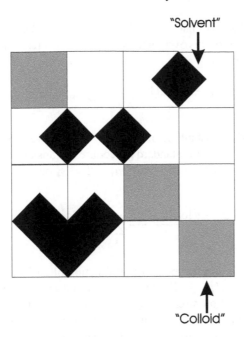

"Colloid"

Figure 2. *Two-dimensional lattice model of a hard-core mixture of 'large' colloidal particles (black squares) and 'small' solvent particles (white squares). Averaging over the solvent degrees of freedom results in a net attractive interaction (depletion interaction) between the 'colloids'.*

$$\Xi = \sum_{\{n_i\}} \exp[\beta\mu_c \sum_i n_i]. \tag{18}$$

The sum is over all allowed sets of occupation numbers $\{n_i\}$, and μ_c is the chemical potential of the 'colloidal' particles. Next, we include small 'solvent' particles that are allowed to sit on the links of the lattice (see Figure 2). These small particles are excluded from the edges of a cube that is occupied by a large particle. For a given configuration $\{n_i\}$ of the large particles, one can then exactly calculate the grand canonical partition function of the small particles. Let $M = M(\{n_i\})$ be the number of free spaces accessible to the small particles. Then clearly:

$$\Xi_{small}(\{n_i\}) = \sum_{l=0}^{M} \frac{M! z_s^l}{l!(M-l)!} = (1 + z_s)^{M(\{n_i\})}, \tag{19}$$

where $z_s \equiv \exp(\beta\mu_s)$ is the activity of the small particles. M can be written as

$$M(\{n_i\}) = dN - 2d\sum_i n_i + \sum_{<ij>} n_i n_j, \tag{20}$$

where dN is the number of links on the lattice, and the second sum is over nearest-neighbour pairs and comes from the fact that when two large particles touch, the number of sites excluded for the small particles is $4d - 1$, not $4d$. Whenever two large particles touch, we have to correct for this overcounting of excluded sites. The total grand partition function for the 'mixture' is:

$$\Xi_{mixture} = \sum_{\{n_i\}} \exp[((\beta\mu_c - 2d\log(1 + z_s))\sum_i n_i + [\log(1 + z_s)]\sum_{<ij>} n_i n_j], \quad (21)$$

where we have omitted a constant factor $(1 + z_s)^{dN}$. Now we can bring this equation in a more familiar form by using the standard procedure to translate a lattice-gas model into a spin model. We define spins s_i such that $2n_i - 1 = s_i$ or $n_i = (s_i + 1)/2$. Then we can write Equation 21 as

$$\Xi_{mixture} = \sum_{\{n_i\}} \exp[\frac{\beta\mu_c - d\log(1 + z_s)}{2}\sum_i s_i + \frac{\log(1 + z_s)}{4}\sum_{<ij>} s_i s_j + \text{Const.}].$$

$$(22)$$

This is simply the expression for the partition function of an Ising model in a magnetic field with strength $H = (\mu_c - d\log(1 + z_s)/\beta)$ and an effective nearest neighbour attraction with an interaction strength $\log(1 + z_s)/(4\beta)$. There is hardly any model in physics that has been studied more than the Ising model. As is well known, this lattice model can again be transformed to a 2-D Ising spin model that can be computed analytically in the zero field case (Onsager 1944). In the language of our mixture model, no external magnetic field means:

$$(1 + z_s)^d = z_c, \quad (23)$$

where $z_c = \exp\beta\mu$, the large particle activity.

Several points should be noted. First of all, in this simple lattice model, summing over all 'solvent' degrees of freedom resulted in effective *attractive* nearest neighbour interaction between the hard-core 'colloids'. Secondly, below its critical temperature, the Ising model exhibits spontaneous magnetisation. In the mixture model, this means that, above a critical value of the activity of the solvent, there will be a phase transition in which a phase with low $< n_c >$ (a dilute colloidal suspension) coexists with a phase with high $< n_c >$ (concentrated suspension). Hence, this model system with a purely repulsive hard-core interaction can undergo a demixing transition. This demixing is purely entropic.

1.1.6 Depletion Flocculation

Let us next consider a slightly more realistic example of an entropy-driven phase separation in a binary mixture, namely polymer-induced flocculation of colloids. Experimentally, it is well known that the addition of a small amount of free, non-adsorbing polymer to a colloidal suspension induces an effective attraction between the colloidal particles and may even lead to coagulation. This effect has been studied extensively and is theoretically well understood (Asakura and Oosawa 1958, Vrij 1976, Gast *et al.* 1983, Lekkerkerker *et al.* 1992, Meijer and Frenkel 1991, Meijer and Frenkel 1994, Meijer and Frenkel 1995). As in the example discussed earlier, the polymer-induced attraction between colloids is an *entropic* effect: when the colloidal particles are close together, the total number of accessible polymer conformations is larger than when the colloidal particles are far apart.

To understand the depletion interaction due to polymers, let us again consider a system of hard-core colloids. To this system, we add a number of ideal polymers. 'Ideal' in this case means that, in the absence of the colloids, the polymers behave like an ideal gas. The configurational integral of a single polymer contains a translational part (V) and an intramolecular part, Z_{int}, which, for an ideal (non-interacting) polymer, is simply the sum over all distinct polymer configurations. In the presence of hard colloidal particles, only part of the volume of the system is accessible to the polymer. How much depends on the conformational state of the polymer. This fact complicates the description of the polymer-colloid mixture, although numerically the problem is tractable (Meijer and Frenkel 1991, Meijer and Frenkel 1994, Meijer and Frenkel 1995). To simplify matters, Asakura and Oosawa (Asakura and Oosawa 1958) introduced the assumption that, as far as the polymer-colloid interaction is concerned, the polymer behaves like a hard sphere with radius R_g. What this means is that as the polymer-colloid distance becomes less than R_g, most polymer conformations will result in an overlap with the colloid, but when the polymer-colloid distance is larger, most polymer conformations are permitted (this assumption has been tested numerically (Meijer and Frenkel 1991, Meijer and Frenkel 1994, Meijer and Frenkel 1995) and turns out to be quite good). As the polymers are assumed to be ideal, it is straightforward to write down the expression for the configurational integral of N_p polymers in the presence of N_c colloids at fixed positions $\{r_c^{N_c}\}$:

$$\int dr^{N_p} \exp[-\beta(U_{ss} + U_{sc})] = \left\{ \int dr_p \exp[-\beta U_{sc}(\mathbf{r}_c^{N_c}; r_p)] \right\}^{N_p} = V_{\text{eff}}^{N_p}(\mathbf{r}_c^{N_c}),$$

where V_{eff} is the effective volume that is available to the polymers. Equation 10 then becomes

$$\Xi(N_c, \mu_s, V, T) = \frac{1}{N_c!} \int dr^{N_c} \exp[-\beta U_{cc}(\mathbf{r}_c^{N_c})] \sum_{N_p=0}^{\infty} \exp(\beta \mu_p N_p) \frac{V_{\text{eff}}^{N_p}(\mathbf{r}_c^{N_c})}{N_p!}$$

$$= \frac{1}{N_c!} \int dr^{N_c} \exp[-\beta U_{cc}(\mathbf{r}_c^{N_c})] \exp(z_p V_{\text{eff}}(\mathbf{r}_c^{N_c})), \qquad (24)$$

where $z_p \equiv \exp(\beta \mu_p)$. Clearly, the effective colloid-colloid potential is now

$$U_{\text{eff}}(\mathbf{r}_c^{N_c}) = U_{cc}(\mathbf{r}_c^{N_c}) - \beta^{-1} z_p V_{\text{eff}}(\mathbf{r}_c^{N_c}). \qquad (25)$$

This equation shows that the correction to the colloid-colloid interaction is due to the fact that the volume available to the polymers depends on the configuration of the colloids. The reason why this should be so is easy to understand. Consider two colloids of radius R at distance $r_1 \gg 2(R + R_G)$. In that case, every colloid excluded a volume with radius $R + R_G$ to the polymers (see Figure 3).

Equation 25 shows that the depletion attraction increases with the polymer activity or, what amounts to the same thing, with the osmotic pressure of the polymers in solution. The more polymer we add to the suspension, the stronger the attraction. The range of the attraction depends on the radius of gyration of the polymers. The larger R_G, the longer the range of attraction. If we model polymers as mutually interpenetrable spheres with

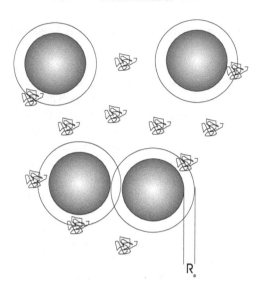

Figure 3. *Hard-core colloids exclude a shell with thickness R_G to the ideal polymers in the solutions. When the colloids are far apart, the excluded volumes simply add up. At shorter distances, the excluded volumes overlap and the total volume available to the polymers increases.*

radius R_G, then the explicit expression for the depletion interaction between a pair of colloids is

$$V_{dep}(r) = -\frac{4\pi(R + R_G)^3 z_p kT}{3}\left\{1 - \frac{3r}{4(R + R_G)} + \frac{1}{16}\left(\frac{r}{R + R_G}\right)^3\right\} \quad (26)$$

[for $2R < r < 2(R + R_G)$],

where we have subtracted a constant term from the potential (namely the contribution of two colloids at a distance $r \gg 2(R + R_G)$). Equation 26 shows clearly that, by changing the size of the added polymers and their concentration, we can change both the range and the strength of the attractive interaction between the colloids. In Section 2, I shall discuss the effect of this tunable attraction on the phase behaviour of polymer-colloid mixtures.

One final comment is in place: the true depletion interaction is *not* pairwise additive. This is clear if we consider three colloidal spheres: if the three exclusion zones overlap, the total excluded volume is larger than would be estimated on basis of the pair-terms alone. Hence, three-body forces yield a *repulsive* correction to the depletion interaction. Note that three-body forces are only important if R_G/R is large enough to get the three exclusion zones to overlap. This holds *a fortiori* for the four-body forces (that are, again, attractive), etc. This feature of the depletion interaction does not depend on the details of the Asakura-Oosawa model. In fact, direct simulations of hard colloids and (lattice) polymers (Meijer and Frenkel 1991, Meijer and Frenkel 1994, Meijer and Frenkel 1995) show exactly the same effect.

1.1.7 Why colloidal materials are soft

Let me return to the picture of colloids as giant atoms. We now know that this is an oversimplification — the origins of the effective interaction between colloids often have no counterpart in atomic physics. Yet, if we ignore all these subtleties, there are similarities. Both atoms and colloids have an effective hard-core diameter: σ_a for atoms and σ_c for colloids. Typically, $\sigma_c/\sigma_a = \mathcal{O}(10^3)$. The characteristic interaction energies between colloids ε_c are of the order of the thermal energy $k_B T$. For atomic solids, the interaction energy ε_a depends on the nature of the inter-atomic interaction: it may vary from a value comparable to $k_B T$ for van der Waals crystals, to a value of the order of electron-volts for covalently bonded materials (e.g., diamond). Knowing the characteristic sizes and interaction energies of the particles is enough to give an order-of-magnitude estimate of various physical properties (basically, this is simply an over extension of van der Waals's Law of Corresponding States). For instance, the elastic constants of a solid have the dimensions [energy/volume]. This means that the elastic constants of a dense colloidal suspension are of the order kT/σ_c^3. For an atomic van der Waals solid, the elastic constants are of the order kT/σ_a^3. In other words, the force needed to deform a colloidal crystal is a factor $\sigma_c^3/\sigma_a^3 \approx 10^9$ smaller than for an atomic crystal held together by dispersion forces (and these are the softest atomic crystals). Clearly, colloidal matter is very easily deformable: it is indeed 'soft matter'.

1.1.8 Polydispersity

All atoms of a given type are identical. They have the same size, weight and interaction strength. This is usually not true for colloids. In fact, all synthetic colloids are to some degree polydisperse, i.e., they do not all have the same size (or mass, shape or refractive index). This polydispersity is usually a complicating factor: it makes it more difficult to interpret experimental data (e.g., x-ray or neutron scattering, or dynamic light scattering). In addition, it may broaden phase transitions and, in some cases, even completely wipe out certain phases. However, polydispersity is not all bad; it also leads to interesting new physics. For instance, polydispersity may sometimes induce a new phase that is not stable in the monodisperse limit (Bates and Frenkel 1998). In general, the effect of polydispersity on the stability of phases is most pronounced in the high-density limit. In that limit, polydispersity may lead to a frustration of the local packing.

2 Colloidal phase behaviour

In Section 1, I explained that the interactions between colloids can often be *tuned*. It is possible to make (uncharged, refractive-index matched, sterically stabilised) colloids that have a steep repulsive interaction and no attraction. These colloids behave like the hard-core models that have been studied extensively in computer simulation. But it is also possible to make (charged) colloids with smooth, long-ranged repulsion. And, using, for instance, added polymer to induce a depletion interaction, colloids can be made with variable ranged attractions. Finally, colloids need not be spherical. It is possible to make colloidal rods and disks. In the following, I briefly discuss some of the interesting consequences that this freedom to 'design' the colloid-colloid interaction has for the phase behaviour.

2.1   Entropic Phase transitions

The second law of thermodynamics tells us that any spontaneous change in a closed system results in an increase of the entropy, S. In this sense, all spontaneous transformations of one phase into another are entropy driven. However, this is not what the term 'entropic phase transitions' is meant to describe. It is more common to consider the behaviour of a system that is not isolated, but can exchange energy with its surroundings. In that case, the second law of thermodynamics implies that the system will tend to minimise its Helmholtz free energy $F = E - TS$, where E is the internal energy of the system and T the temperature. Clearly, a system at constant temperature can lower its free energy in two ways: either by *increasing* the entropy S or by *decreasing* the internal energy E. To gain a better understanding of the factors that influence phase transitions, we must look at the statistical mechanical expressions for entropy. The simplest starting point is to use Boltzmann's expression for the entropy of an isolated system of N particles in volume V at an energy E,

$$S = k_B \ln \Omega, \tag{27}$$

where k_B, the Boltzmann constant, is a constant of proportionality. Ω is the total number of (quantum) states that is accessible to the system. In the remainder of this chapter, I choose my units such that $k_B = 1$. The usual interpretation of Equation 27 is that Ω, the number of accessible states of a system, is a measure for the 'disorder' in that system. The larger the disorder, the larger the entropy. This interpretation of entropy suggests that a phase transition from a disordered to a more ordered phase can only take place if the loss in entropy is compensated by the decrease in internal energy. This statement is completely correct, provided that we use Equation 27 to *define* the amount of disorder in a system. However, we also have an *intuitive* idea of order and disorder: we consider crystalline solids 'ordered' and isotropic liquids 'disordered'. This intuitive picture suggests that a spontaneous phase transition from the fluid to the crystalline state can only take place if the freezing lowers the internal energy of the system sufficiently to outweigh the loss in entropy; i.e., the ordering transition is 'energy driven'. In many cases, this is precisely what happens. It would, however, be a mistake to assume that our intuitive definition of order always coincides with the one based on Equation 27. In fact, the aim of this paper is to show that many 'ordering' transitions that are usually considered to be energy-driven may, in fact, be entropy driven. I stress that the idea of entropy-driven phase transitions is an old one. However, it has only become clear during the past few years that such phase transformations may not be interesting exceptions, but the rule!

In order to observe 'pure' entropic phase transitions, we should consider systems for which the internal energy is a function of the temperature, but not of the density. Using elementary statistical mechanics, it is easy to show that this condition is satisfied for classical hard-core systems. Whenever these systems order at a fixed density and temperature, they can only do so by increasing their entropy (because, at constant temperature, their internal energy is fixed). Such systems are conveniently studied in computer simulations. But increasingly, experimentalists — in particular, colloid scientists — have succeeded in making real systems that behave very nearly as ideal hard-core systems. Hence, the phase transitions discussed here can, and in many cases do, occur in nature. In the following, I list examples of entropic ordering in hard-core systems. But I stress that the list is far from complete.

2.1.1 Computer simulation of (liquid) crystals

The earliest example of an entropy-driven ordering transition is described in a classic paper by Onsager (1949), on the isotropic-nematic transition in a (three-dimensional) system of thin hard rods. Onsager showed that on compression, a fluid of thin hard rods of length L and diameter D *must* undergo a transition from the isotropic fluid phase, where the molecules are translationally and orientationally disordered, to the nematic phase. In the latter phase, the molecules are translationally disordered, but their orientations are, on average, aligned. This transition takes place at a density such that $(N/V)L^2D = \mathcal{O}(1)$. Onsager considered the limit $L/D \to \infty$. In this case, the phase transition of the hard-rod model can be found exactly (see e.g., Kayser and Raveche 1978). At first sight, it may seem strange that the hard rod system can *increase* its entropy by going from a disordered fluid phase to an orientationally ordered phase. Indeed, due to the orientational ordering of the system, the orientational entropy of the system decreases. However, this loss in entropy is more than offset by the increase in translational entropy of the system: the available space for any one rod increases as the rods become more aligned. In fact, we shall see this mechanism returning time-and-again in ordering transitions of hard-core systems: the entropy *decreases* because the density is no longer uniform in orientation or position, but the entropy *increases* because the free-volume per particle is larger in the ordered than in the disordered phase.

The most famous, and for a long time controversial, example of an entropy-driven ordering transition is the freezing transition in a system of hard spheres. This transition had been predicted by Kirkwood in the early fifties (Kirkwood 1951) on the basis of an approximate theoretical description of the hard-sphere model. As this prediction was quite counterintuitive and not based on any rigorous theoretical results, it met with wide-spread scepticism until Alder and Wainwright (Alder and Wainwright 1957) and Wood and Jacobson (Wood and Jacobson 1957) performed numerical simulations of the hard-sphere system that showed direct evidence for this freezing transition. Even then, the acceptance of the idea that freezing could be an *entropy* driven transition, came only slowly (Percus 1963). However, by now, the idea that hard spheres undergo a first-order freezing transition is generally accepted. Interestingly, although the hard-sphere model was originally devised as an idealised and highly unrealistic model of an atomic fluid, it is now realised that this model provides a good description of certain classes of colloidal systems (Pusey 1991). At this stage, we know a great deal about the phase behaviour of hard spheres. Since the work of Hoover and Ree (Hoover and Ree 1968), we know the location of the thermodynamic freezing transition, and we now also know that the face-centred cubic phase is more stable than the hexagonal close-packed phase (Bolhuis *et al.* 1997), be it by only $10^{-3}kT$ per particle. To understand how little this is, consider the following: if we would use calorimetric techniques to determine the relative stability of the *fcc* and *hcp* phases, we would find that the free-energy difference amounts to some 10^{-11} cal/cm^3! Moreover, computer simulations allow us to estimate the equilibrium

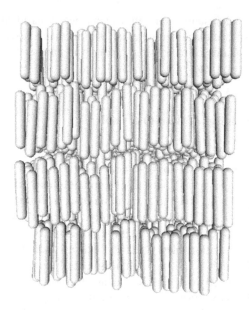

Figure 4. *Snapshot of a hard-core smectic liquid crystal.*

concentration of point defects (in particular, vacancies) in hard-sphere crystals (Bennett and Alder 1970). At melting, this concentration is small, but not very small (of the order of one vacancy per 4000 particles).

The next surprise in the history of ordering due to entropy came in the mid-eighties when computer simulations (Frenkel *et al.* 1988) showed that hard-core interactions alone could also explain the formation of more complex liquid crystals. In particular, it was found that a system of hard spherocylinders (i.e., cylinders with hemispherical caps) can form a smectic liquid crystal, in addition to the isotropic liquid, the nematic phase and the crystalline solid (Bolhuis and Frenkel 1997). In the smectic (A) phase, the molecules are orientationally ordered but, in addition, the translational symmetry is broken: the system exhibits a one-dimensional density modulation. Subsequently, it was found that some hard-core models could also exhibit columnar ordering (Veerman and Frenkel 1992). In the latter case, the molecules assemble in liquid-like stacks, but these stacks order to form a two-dimensional crystal. In summary, hard-core interaction can induce orientational ordering and one-, two- and three-dimensional positional ordering. This is rather surprising because, in particular for the smectic and the columnar phase, it was generally believed that their formation required specific energetic interactions.

2.1.2   To boil or not to boil...

Why do liquids exist? We are so used to the occurrence of phenomena such as boiling and freezing that we rarely pause to ask ourselves if things could have been different. Yet the fact that liquids must exist is not obvious *a priori*. This point is eloquently made in an essay by V. F. Weisskopf (Weisskopf 1977):

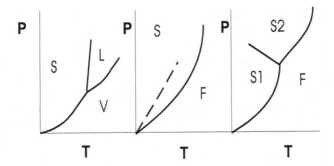

Figure 5. *Phase diagram of a system of spherical particles with a variable ranged attraction. As the range of attraction decreases (left to right), the liquid-vapour curve moves into the metastable regime (middle, dashed line). For very short-ranged attraction (less than 5% of the hard-core diameter), a first-order iso-structural solid-solid transition (right; S1-S2) appears in the solid phase (Bolhuis and Frenkel 1994, Bolhuis et al. 1994). Phase diagrams of the middle type are common for colloidal systems, but rare for simple molecular systems. A possible exception is C_{60} (Hagen et al. 1993).*

...The existence and general properties of solids and gases are relatively easy to understand once it is realized that atoms or molecules have certain typical properties and interactions that follow from quantum mechanics. Liquids are harder to understand. Assume that a group of intelligent theoretical physicists had lived in closed buildings from birth such that they never had occasion to see any natural structures. Let us forget that it may be impossible to prevent them to see their own bodies and their inputs and outputs. What would they be able to predict from a fundamental knowledge of quantum mechanics? They probably would predict the existence of atoms, of molecules, of solid crystals, both metals and insulators, of gases, but most likely not the existence of liquids.

Weisskopf's statement may seem a bit bold. Surely, the liquid-vapour transition could have been predicted *a priori*. This is a hypothetical question that can never be answered. But, as I shall discuss here, in colloidal systems there may exist an analogous phase transition that has not yet been observed experimentally and that was found in simulation before it had been predicted. To set the stage, let us first consider the question of the liquid-vapour transition. In his 1873 thesis, van der Waals gave the correct explanation for a well-known, yet puzzling feature of liquids and gases, namely that there is no essential distinction between the two: above a critical temperature T_c, a vapour can be compressed continuously all the way to the freezing point. Yet, below T_c, a first-order phase transition separates the dilute fluid (vapour) from the dense fluid (liquid) (Rowlinson 1988). It is due to a the competition between short-ranged repulsion and longer-ranged attraction. From the work of Longuet-Higgins and Widom (Longuet-Higgins and Widom 1964), we now know that the van der Waals model (molecules are described as hard spheres with an infinitely weak, infinitely long-ranged attraction (Hemmer and Lebowitz 1976)) is even richer than originally expected; it exhibits not only the liquid-vapour transition, but also crystallisation.

The liquid-vapour transition is possible between the critical point and the triple point, and the temperature of the critical point in the van der Waals model is about a factor two large than that of the triple point. There is, however, no fundamental reason why this transition should occur in every atomic or molecular substance, nor is there any rule that forbids the existence of more than one fluid-fluid transition. Whether a given compound will have a liquid phase depends sensitively on the range of the intermolecular potential; as this range is decreased, the critical temperature approaches the triple-point temperature, and when T_c drops below the latter, only a single stable fluid phase remains. In mixtures of spherical colloidal particles and non-adsorbing polymer, the range of the attractive part of the effective colloid-colloid interaction can be varied by changing the size of the polymers (see Section 1.1.6). Experiment, theory and simulation all suggest that when the width of the attractive well becomes less than approximately one third of the diameter of the colloidal spheres, the colloidal 'liquid' phase disappears (Figure 5, middle).

Finally, for *very* short-range attractions, a first-order iso-structural solid-solid transiton appears (Figure 5, right). Phase diagrams of this type have, thus far, not been observed in colloidal systems. Nor had they been predicted before the simulations appeared (suggesting that Weisskopf was right).

3   Colloid dynamics

For the computer simulator, the study of colloid dynamics is a challenge. The reason is that colloid dynamics span a wide range of timescales. No single simulation can cover all timescales simultaneously. In the following, I shall discuss two aspects of colloid dynamics that clearly illustrate the timescale problem. The first is colloidal *hydrodynamics*. The second is homogeneous nucleation of a new phase from a metastable phase.

3.1   Hydrodynamic effects in colloidal suspensions

Colloid dynamics is a field in its own right (see Dhont 1996). Clearly, I cannot cover this field in a few pages. I therefore wish to focus on a few simple concepts that are useful when thinking about the dynamics of colloidal particles. The analogy between colloids and atoms that was useful when discussing the static properties of colloidal matter breaks down completely when discussing the dynamics. The reason is that atoms in a dilute gas phase move *ballistically* and colloids in a dilute suspension move *diffusively*. In order to understand the motion of colloids, we have to consider the hydrodynamic properties of the surrounding solvent. Just imagine what would happen if kinetic gas theory applied to the motion of colloids: then the frictional force acting on a spherical colloid would be caused by independent collisions with the solvent molecules, and we would find that the frictional force is proportional to the velocity of the colloid, v (which is correct) and the effective area of the colloid (πa^2) (which is wrong). In fact, the true frictional force on a colloid moving at a constant velocity v is given by the Stokes expression

$$F_{frict} = -6\pi\eta a v, \tag{28}$$

where η is the viscosity of the solvent and a the radius of the colloid. The Stokes relation can be derived from hydrodynamics; however, this derivation does not make it intuitively obvious why the friction is proportional to a rather than to a^2. Here, I shall give a

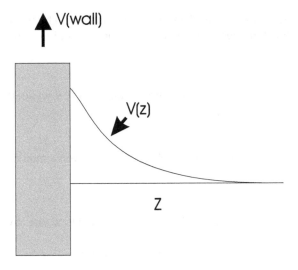

Figure 6. *When a wall is suddenly given a tangential velocity v_{Wall}, the transverse velocity field penetrates diffusively into the bulk fluid.*

handwaving derivation that is more intuitively appealing (be it that the answer is not quite right). We start with the assumption that the time evolution of any flowfield $\mathbf{u}(r,t)$ in the solvent obeys the Navier-Stokes equation for an incompressible fluid,

$$d_s \left(\frac{\partial \mathbf{u}(r,t)}{\partial t} + \mathbf{u}(r,t).\nabla \mathbf{u}(r,t) \right) = \eta \nabla^2 \mathbf{u}(r,t) - \nabla p(r,t),$$

where $\mathbf{u}(r,t)$ is the flow velocity at point r and time t, d_s is the mass density of the solvent and $p(r,t)$ is the hydrostatic pressure. I shall consider the case that $\mathbf{u}(r,t)$ is 'small' (low Reynolds-number regime (see Dhont 1996)). Then we can neglect the $\mathbf{u}(r,t).\nabla \mathbf{u}(r,t)$ term. Let us now consider the situation that the solvent is in contact with a flat surface (see Figure 6).

Initially, both the fluid and the wall are at rest. At time $t = 0$, the wall is given a tangential velocity v_{Wall}. We assume that this velocity is parallel to the y-direction. The normal to the surface defines the z-direction. In this geometry, the equation of motion for the flow field reduces to

$$d_s \frac{\partial u_y(z,t)}{\partial t} = \eta \nabla_z^2 u_y(z,t).$$

But this is effectively a *diffusion* equation for the transverse velocity. The 'diffusion coefficient' is equal to $(\eta/d_s) \equiv \nu$. This 'diffusion coefficient' for transverse momentum is called the *kinematic viscosity*. The larger ν, the faster the transverse momentum diffuses away from its source. Diffusion equations typically show up when we consider the transport of a quantity that is conserved, such as mass, energy or, in this case, momentum.

Let us now use this concept of diffusing momentum to estimate the frictional drag on a sphere. To simplify matters, I shall pretend that the transverse momentum is a *scalar* rather than a *vector*. Clearly, this is wrong, but it will not affect the qualitative answer.

A moving sphere acts as a source of transverse momentum. The transverse momentum flux j_T is related to the gradient in the transverse velocity field (v_T) by

$$j_T = -\eta \nabla v_T.$$

In steady state, $\nabla^2 v_T(r) = 0$. If the transverse velocity were a scalar, the solution to this equation would be

$$v_T(r) = v_0 \frac{a}{r}, \tag{29}$$

where v_0 is the velocity of the colloidal sphere. The transverse momentum current density is then

$$j_T = \eta v_0 \frac{a}{r^2}.$$

The frictional force on the sphere equals the negative of the total rate at which momentum flows into the fluid

$$F_{frict} = -4\pi r^2 j_T = -4\pi \eta a v_0, \tag{30}$$

which is almost Stokes law (the factor 4π instead of 6π is due to our cavalier treatment of the vectorial character of the velocity).

This trivial example shows that the conservation of momentum is absolutely crucial for the understanding of colloid dynamics. A second result that follows almost immediately from Equation 29 is that the flow velocity at a distance r from a moving colloid decays as $1/r$. Through this velocity field, one colloid can exert a drag force on another colloid. This is the so-called *hydrodynamic interaction* — this interaction is very long-ranged, as it decays as $1/r$. Again, for a correct derivation, I refer the reader to Dhont's book (Dhont 1996).

Having established a simple language to talk about colloid dynamics, we can make estimates of the relevant timescales that govern the time evolution of a colloidal system. The shortest timescale, τ_s, is usually not even considered. It is the time-scale on which the solvent behaves as a *compressible* fluid. If we set a colloid in motion, this will set up a density disturbance. This density modulation will be propagated away as a sound wave (carrying with it one third of the momentum of the colloid). This sound wave will have moved away after a time $\tau_s = a/c_s$ (where c_s is the velocity of sound). Typically, $\tau_s = \mathcal{O}(10^{-10})$s. The next timescale is the one associated with the propagation of hydrodynamic interactions: τ_H. It is of the order of the time it takes the transverse momentum to diffuse a typical interparticle distance: $\tau_H = \mathcal{O}(\rho^{-2/3}/\nu)$, where ρ is the number density of the colloids. In dense suspensions, the typical inter-particle distance is comparable to the diameter of the colloids, and then $\tau_H = \mathcal{O}(a^2/\nu)$. Usually, this timescale is of the order of 10^{-8}s. Next, we get the timescale for the decay of the initial velocity of a colloid. If we assume (somewhat inconsistently, as it will turn out) that this decay is determined by Stokes law, we find that the decay of the velocity of a colloid occurs on a timescale $\tau_v = \mathcal{O}(M_c/\eta a)$, where M_c is the mass of a colloid. If we write $M_c = (4\pi a^3 d/3)$, where d_c is the mass density of the colloid, then we can write $\tau_v = \mathcal{O}(d_c a^2/\eta)$. In a dense suspension, $\tau_v = (d_c/d_s)\tau_H$. This means that, for a neutrally buoyant colloid, there is no separation in timescales between τ_v and τ_H. The final timescale in colloid dynamics is the one associated with appreciable displacements of the colloids. As the colloids move diffusively, and as the diffusion constant is related to the Stokes friction constant by $D = kT/(6\pi\eta a)$, the time it takes a colloid to diffuse over a distance comparable to its own radius is $\tau_R = \mathcal{O}(a^2/D) \sim \mathcal{O}(\eta a^3)$. τ_R is of the order of milliseconds to seconds. Clearly, there is a wide timescale separation between τ_R and the other times. For times that are much longer than τ_v and τ_H, we can pretend that the colloids perform uncorrelated Brownian motion. However, this is not quite correct: even though the hydrodynamic interactions have long decayed, they mean that the effective diffusion constant of every colloid depends on the instantaneous configuration of its neighbours. This is one of the reasons why the theory of colloid dynamics is not simple (Dhont 1996).

One aspect of diffusion is of particular interest in the context of biomolecular processes. It is the fact that the probability of two molecules meeting each other within a given time depends very strongly on the dimensionality of the space within which the molecules move. In particular, the probability of meeting each other in a given time interval is much higher if two molecules move on a two-dimensional surface (e.g., in a membrane) than when they diffuse freely through the cell. To see this, consider the volume swept out by a molecule of diameter a in a d-dimensional space. The time it takes the molecule to diffuse over its own diameter is of the order $\tau_D \approx a^2/D$, where D is the coefficient of self-diffusion of the molecule. The volume swept out by the molecule in N diffusion times is $\sim Na^d$. Suppose that the molecules diffuse in a (d-dimensional) volume L^d. The fraction of the volume swept out by the molecule after N diffusion times is (approximately)

$$f = 1 - \exp[-Na^d/(L^d)]$$

Typically, the number of diffusion times it takes for two (identical) molecules to meet is such that

$$Na^d/L^d = 1$$

The time that this takes is

$$t_m = N\tau_D = (L/a)^d \times a^2/D$$

Let us compare the time it takes two molecules to meet in $2D$ and in $3D$:

$$\frac{t_m^{3D}}{t_m^{2D}} = L/a$$

(assuming that the diffusion coefficients are the same in $2D$ and $3D$). The preceding equation shows that it takes a factor L/a longer for molecules to meet in $3D$ than in $2D$. For a typical protein in a typical cell, this factor is easily of the order of 10^3. That is why membranes are such good meeting places for biomolecules.

3.2   Homogeneous nucleation in colloidal suspensions

It is well known that liquids can be supercooled before they freeze and vapours can be supersaturated before they condense. A homogeneous phase can be supercooled because the only route to the more stable state is via the formation of small *nuclei*. The free energy of such nuclei is determined not only by the difference in chemical potential between vapour and liquid, which drives the nucleation process, but also by the surface free energy. The surface free energy term is always positive because of the work that must be done to create an interface. Moreover, initially this term dominates, and hence the free energy of a nucleus increases with size. Only when the droplet has reached a certain 'critical' size, the volume term takes over, and the free energy decreases. It is only from here on that the nucleus grows spontaneously into a bulk liquid. In classical nucleation theory (CNT) (Volmer and Weber 1926, Becker and Döring 1935), it is assumed that the nuclei are compact, spherical objects that behave like small droplets of bulk phase. The free energy of a spherical liquid droplet of radius R in a vapour is then given by

$$\Delta G = 4\pi R^2 \gamma + \frac{4}{3}\pi R^3 \rho \Delta\mu, \tag{31}$$

where γ is the surface free energy, ρ is the density of the bulk liquid and $\Delta\mu$ is the difference in chemical potential between bulk liquid and bulk vapour. Clearly, the first term on the right-hand side of Equation 31 is the surface term, which is positive, and the second term is the volume term, which is negative; the difference in chemical potential is the driving force for the nucleation process. The height of the nucleation barrier can easily be obtained from the preceding expression, yielding

$$\Delta G^* = \frac{16\pi\gamma^3}{3\rho^2 \Delta\mu^2}. \tag{32}$$

This equation shows that the barrier height depends not only on the surface free energy γ (and the density ρ), but also on the difference in chemical potential $\Delta\mu$. The difference in chemical potential is related to the supersaturation. Hence, the height of the free-energy barrier that separates the stable from the metastable phase depends on the degree of supersaturation. At coexistence, the difference in chemical potential is zero, and the height of the barrier is infinite. Although the system is equally likely in the liquid and vapour phases, once the system is one state or the other, it will remain in this state; it simply cannot transform into the other state.

Macroscopic thermodynamics dictates that the phase that is formed in a supersaturated system is the one that has the lowest free energy. However, nucleation is an essentially dynamic process, and therefore one cannot expect *a priori* that, on supersaturating the system, the thermodynamically most stable phase will be formed. In 1897, Ostwald (Ostwald 1897) formulated his step rule, stating that the crystal phase that is nucleated from the melt need not be the one that is thermodynamically most stable, but the one that is closest in free energy to the fluid phase. Stranski and Totomanow (Stranski and Totomanow 1933) reexamined this rule and argued that the nucleated phase is the phase that has the lowest free-energy barrier of formation, rather than the phase that is globally stable under the conditions prevailing. The simulation results discussed in the following suggest that, even on a *microscopic* scale, something similar to Ostwald's step rule seems to hold.

3.2.1   Coil-globule transition in the condensation of dipolar colloids?

The formation of a droplet of water from the vapour is probably the best-known example of homogeneous nucleation of a polar fluid. However, the nucleation behaviour of polar fluids is still poorly understood. In fact, while classical nucleation theory gives a reasonable prediction of the nucleation rate of nonpolar substances, it seriously overestimates the rate of nucleation of highly polar compounds, such as acetonitrile, benzonitrile

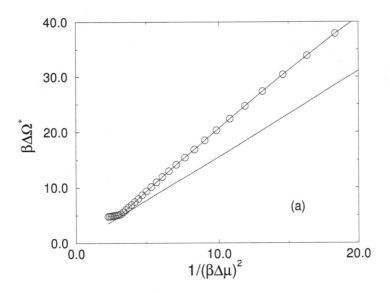

Figure 7. *Comparison of the barrier height between the simulation results (open circles) and classical nucleation theory (straight solid line) for a Stockmayer fluid with reduced dipole moment $\mu^* = \mu/\sqrt{\epsilon\sigma^3} = 4$ and reduced temperature $T^* = k_B T/\epsilon = 3.5$. The chemical potential difference $\Delta\mu$ is the difference between the chemical potential of the liquid and the vapour.*

and nitrobenzene (Wright *et al.* 1993, Wright and El-Shall 1993). In order to explain the discrepancy between theory and experiment, several nucleation theories have been proposed. It has been suggested that in the critical nuclei, the dipoles are arranged in an anti-parallel head-to-tail configuration (Wright *et al.* 1993, Wright and El-Shall 1993), giving the clusters a non-spherical, prolate shape, which increases the surface-to-volume ratio and thereby the height of the nucleation barrier. In the oriented dipole model introduced by Abraham (Abraham 1970), it is assumed that the dipoles are perpendicular to the interface, yielding a size-dependent surface tension due to the effect of curvature of the surface on the dipole-dipole interaction. However, in a density-functional study of a weakly polar Stockmayer fluid (see the definition in Equation 33), it was found that on the liquid (core) side of the interface of critical nuclei, the dipoles are not oriented perpendicular to the surface, but parallel (Talanquer and Oxtoby 1993).

We have studied the structure and free energy of critical nuclei, as well as pre-and postcritical nuclei, of a highly polar Stockmayer fluid (ten Wolde *et al.* 1998). In the Stockmayer system, the particles interact via a Lennard-Jones pair potential plus a dipole-dipole interaction potential

$$v(\mathbf{r}_{ij}, \mu_i, \mu_j) = 4\epsilon \left[\left(\frac{\sigma}{r_{ij}} \right)^{12} - \left(\frac{\sigma}{r_{ij}} \right)^6 \right]$$
$$- 3(\mu_i \cdot \mathbf{r}_{ij})(\mu_j \cdot \mathbf{r}_{ij})/r_{ij}^5 + \mu_i \cdot \mu_j/r_{ij}^3. \tag{33}$$

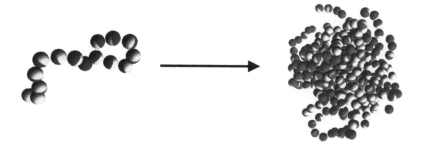

Figure 8. *Left: Sub-critical nucleus in a supercooled vapour of dipolar spheres. The dipolar particles align head-to-tail. Right: Critical nucleus. The chain has collapsed to form a more-or-less compact, globular cluster.*

Here, ϵ is the Lennard-Jones well depth, σ is the Lennard-Jones diameter, μ_i denotes the dipole moment of particle i and $\mathbf{r}_{ij}$ is the vector joining particle i and j. We have studied the nucleation behaviour for $\mu^* = \mu/\sqrt{\epsilon\sigma^3} = 4$, which is close to the value of water.[4]

We have computed (ten Wolde *et al.* 1998) the excess free energy $\Delta\Omega$ of a cluster of size n in a volume V, at chemical potential μ and at temperature T, from the probability distribution function $P(n)$

$$\beta\Delta\Omega(n, \mu, V, T) \equiv -\ln[P(n)] = -\ln[N_n/N]. \tag{34}$$

Here, N_n is the average number of clusters of size n and N is the average total number of particles. As the density of clusters in the vapour is low, the interactions between them can be neglected. Thus, we can obtain the free-energy barrier at any desired chemical potential μ' from the nucleation barrier measured at a given chemical potential μ via

$$\beta\Delta\Omega(n, \mu', V, T) = \beta\Delta\Omega(n, \mu, V, T) - \beta(\mu' - \mu)n$$
$$+ \ln\left[\rho(\mu')/\rho(\mu)\right], \tag{35}$$

where $\rho = N/V$ is the total number density in the system.

Figure 7 shows the comparison between the simulation results and CNT for the height of the barrier. Clearly, the theory underestimates the height of the nucleation barrier. As the nucleation rate is dominated by the height of the barrier, our results are in qualitative agreement with the experiments on strongly polar fluids (Wright *et al.* 1993, Wright and El-Shall 1993), in which it was found that CNT overestimates the nucleation rate. But, unlike the experiments, the simulations allow us to investigate the microscopic origins of the breakdown of classical nucleation theory.

In classical nucleation theory, it is assumed that the smallest clusters are already compact, more or less spherical objects. In a previous simulation study on a typical nonpolar fluid, the Lennard-Jones fluid, we found that this is a reasonable assumption (ten Wolde and Frenkel 1998), even for nuclei as small as ten particles. However, the interaction potential of the Lennard-Jones system is isotropic, whereas the dipolar interaction potential is anisotropic. On the other hand, the bulk liquid of this polar fluid is isotropic.

---

[4]The context should make clear when μ refers to dipole moment and when it means the chemical potential.

We find that the smallest clusters that initiate the nucleation process are not compact spherical objects, but chains in which the dipoles align head-to-tail (Figure 8). In fact, we find a whole variety of differently shaped clusters in dynamical equilibrium: linear chains, branched-chains and 'ring-polymers'. Initially, when the cluster size is increased, the chains become longer. But, beyond a certain size, the clusters collapse to form a compact globule. The Stockmayer fluid is a simple model system for polar fluids, and the mechanism that we describe here might not be applicable for all fluids that have a strong dipole moment. However, it is probably not a bad model for colloids with an embedded electrical or magnetic dipole. The simulations show that the presence of a sufficiently strong permanent dipole may drastically change the pathway for condensation.

3.2.2 Crystallisation near a critical point

Proteins are notoriously difficult to crystallise. The experiments indicate that proteins only crystallise under very specific conditions (McPherson 1982, Durbin and Feher 1996, Rosenberger 1996). Moreover, the conditions are often not known beforehand. As a result, growing good protein crystals is a time-consuming business. Interestingly, there seems to exist a similarity between the phase diagram of globular proteins and of colloids with a short-ranged attractive interactions (Rosenbaum *et al.* 1996). In fact, a series of studies (Berland *et al.* 1992, Asherie *et al.* 1996, Broide *et al.* 1996, Muschol and Rosenberger 1997) show that the phase diagram of a wide variety of proteins is of the kind shown in Figure 5 (middle). Rosenbaum and Zukoski (Rosenbaum *et al.* 1996, Rosenbaum and Zukoski 1996) observed that the conditions under which a large number of globular proteins can be made to crystallise map onto a narrow temperature range of the computed fluid-solid coexistence curve of colloids with short-ranged attraction (Hagen and Frenkel 1994). If the temperature is too high, crystallisation is hardly observed at all, whereas if the temperature is too low, amorphous precipitation, rather than crystallisation, occurs. Only in a narrow window around the metastable critical point, high-quality crystals can be formed. In order to grow high-quality protein crystals, the quench should be relatively shallow, and the system should not be close to a glass transition. Under these conditions, the rate-limiting step in crystal nucleation is the crossing of the free-energy barrier. Using simulation, it is possible to study the nucleation barrier, and the structure of the critical nucleus in the vicinity of this metastable critical point (ten Wolde and Frenkel 1997).

We performed simulations on a model system for particles with a short-ranged attraction for a number of state points near the metastable critical point. These state-points were chosen such that on the basis of classical nucleation theory, the same height of the barrier could be expected. In order to compute the free-energy barrier, we have computed the free energy of a nucleus as a function of its size. However, we first have to define what we mean by a 'nucleus'. As we are interested in crystallisation, it might seem natural to use a crystallinity criterion. However, as mentioned, we expect that crystallisation near the critical point is influenced by critical density fluctuations. We therefore used not only a crystallinity criterion, but also a density criterion. We define the size of a high-density cluster (be it solid- or liquidlike) as the number of connected particles, N_ρ, that have a significantly higher local density than the particles in the remainder of the system. The number of these particles that is also in a crystalline environment is denoted by N_{crys}. In

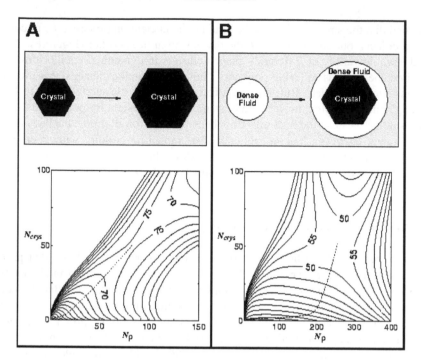

Figure 9. *Contour plots of the free-energy landscape along the path from the metastable fluid to the critical crystal nucleus for our system of spherical particles with short-ranged attraction. The curves of constant free energy are drawn as a function of N_ρ and N_{crys} (see text) and are separated by $5k_BT$. If a liquidlike droplet forms in the system, we expect N_ρ to become large, whereas N_{crys} remains essentially zero. In contrast, for a normal crystallite, we expect that N_ρ is proportional to N_{crys}. Figure A shows the free-energy landscape well below the critical temperature ($T/T_c = 0.89$). The lowest free-energy path to the critical nucleus is indicated by a dashed curve. Note that this curve corresponds to the formation and growth of a highly crystalline cluster.*
Figure B: Idem, but now for $T = T_c$. In this case, the free-energy valley (dashed curve) first runs parallel to the N_ρ axis (formation of a liquidlike droplet), and moves towards a structure with a higher crystallinity (crystallite embedded in a liquidlike droplet). The free-energy barrier for this route is much lower than the one shown in Figure A.

our simulations, we have computed the free-energy 'landscape' of a nucleus as a function of the two coordinates N_ρ and N_{crys}.

Figure 9 shows the free-energy landscape for $T = 0.89\,T_c$ and $T = T_c$. The free-energy landscapes for the other two points are qualitatively similar to the one for $T = 0.89\,T_c$ and will not be shown here. We find that away from T_c (both above and below), the path of lowest free energy is one where the increase in N_ρ is proportional to the increase in N_{crys} (Figure 9A). Such behaviour is expected if the incipient nucleus is simply a small crystallite. However, around T_c, critical density fluctuations lead to a striking change in the free-energy landscape (Figure 9B). First, the route to the critical nucleus leads through a region where N_ρ increases while N_{crys} is still essentially zero.

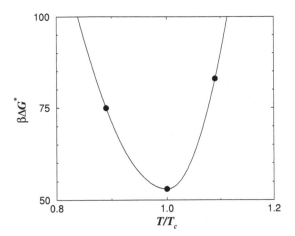

Figure 10. *Variation of the free-energy barrier for homogeneous crystal nucleation, as a function of T/T_c, in the vicinity of the critical temperature. The solid curve is a guide to the eye. The nucleation barrier at $T = 2.23T_c$ is 128 k_BT and is not shown in this figure. The simulations show that the nucleation barrier goes through a minimum around the metastable critical point (see text).*

In other words, the first step towards the critical nucleus is the formation of a liquidlike droplet. Then, beyond a certain critical size, the increase in N_ρ is proportional to N_{crys}, that is, a crystalline nucleus forms inside the liquidlike droplet. Clearly, the presence of large density fluctuations close to a fluid-fluid critical point has a pronounced effect on the route to crystal nucleation. But, more importantly, the nucleation barrier close to T_c is much lower than at either higher or lower temperatures (Figure 10). The observed reduction in ΔG^* near T_c by some 30 k_BT corresponds to an increase in nucleation rate by a factor 10^{13}.

Finally, let us consider the implications of this reduction of the crystal nucleation barrier near T_c. An alternative way to lower the crystal nucleation barrier would be to quench the solution deeper into the metastable region below the solid-liquid coexistence curve. However, such deep quenches often result in the formation of amorphous aggregates (George and Wilson 1994, Rosenbaum *et al.* 1996, Rosenbaum and Zukoski 1996, Ilett *et al.* 1995, Poon *et al.* 1995, Poon 1997, Muschol and Rosenberger 1997). Moreover, in a deep quench, the thermodynamic driving force for crystallisation ($\mu_{liq} - \mu_{cryst}$) is also enhanced. As a consequence, the crystallites that nucleate will grow rapidly and far from perfectly (Durbin and Feher 1996). Thus, the nice feature of crystal nucleation in the vicinity of the metastable critical point is that crystals can be formed at a relatively small degree of undercooling. It should be stressed that nucleation will also be enhanced in the vicinity of the fluid-fluid *spinodal*. Hence, there is more freedom in choosing the optimal crystallisation conditions. Finally, I note that in colloidal (as opposed to protein) systems, the system tends to form a gel before the metastable fluid-fluid branch is reached. A possible explanation for the difference in behaviour of proteins and colloids has been proposed (Noro *et al.* 1999).

3.2.3 'Microscopic' step rule

Ostwald formulated his step rule more than a century ago (Ostwald 1897) on the basis of macroscopic studies of phase transitions. The simulations suggest that also on a microscopic level, a 'step rule' may apply and that metastable phases may play an important role in nucleation. We find that the structure of the precritical nuclei is that of a metastable phase (chains/liquid). As the nuclei grow, the structure in the core transforms into that of the stable phase (liquid/fcc-crystal). Interestingly, in the interface of the larger nuclei, traces of the structure of the smaller nuclei are retained.

Acknowledgements

The work of the FOM Institute is part of the research program of FOM and is made possible by financial support from the Netherlands Organisation for Scientific Research (NWO).

References

Abraham F F, 1970, *Science* **168** 833.
Alder B J and Wainwright T E, 1957, *J Chem Phys* **27** 1208.
Asakura S and Oosawa F, 1954, *J Chem Phys* **22** 1255.
Asakura S and Oosawa F, 1958, *J Polym Sci* **33** 183.
Asherie N, Lomakin A and Benedek G B, 1996, *Phys Rev Lett* **77** 4832.
Bates M A and Frenkel D, 1998, *J Chem Phys* **109** 6193.
Becker R and Döring W, 1935, *Ann Phys* **24** 719.
Bennett C H and Alder B J, 1970, *J Chem Phys* **54** 4796.
Berland C R, Thurston G M, Kondo M, Broide M L, Pande J, Ogun O O and Benedek G B, 1992, *Proc Natl Acad Sci USA* **89** 1214.
Bolhuis P G and Frenkel D, 1994, *Phys Rev Lett* **72,** 221.
Bolhuis P G, Hagen M H J and Frenkel D, 1994, *Phys Rev E* **50** 4880.
Bolhuis P G, Frenkel D, Mau S-C and Huse D A, 1997, *Nature* **388** 235.
Bolhuis P and Frenkel D, 1997, *J Chem Phys* **106** 666.
Broide M L, Tominc T M and Saxowsky M D, 1996, *Phys Rev E* **53** 6325.
Dhont J K G, 1996, *An Introduction to Dynamics of Colloids* (Elsevier, Amsterdam).
Durbin S D and Feher G, 1996, *Ann Rev Phys Chem* **47** 171.
Frenkel D, Lekkerkerker H N W and Stroobants A, 1988, *Nature* **332** 822.
Frenkel D and Louis A A, 1992, *Phys Rev Lett* **68** 3363. In fact, in a different context, this model had been introduced by Widom B, 1967, *J Chem Phys* **46** 3324.
Gast A P, Hall C K and Russel W B, 1983, *J Colloid Interface Sci* **96** 251.
George A and Wilson W W, 1994, *Acta Crytallogr D* **50** 361.
Hagen M H J, Meijer E J, Mooij G C A M, Frenkel D and Lekkerkerker H N W, 1993, *Nature*, **365** 425.
Hagen M H J and Frenkel D, 1994, *J Chem Phys* **101** 4093.
Hemmer P C and Lebowitz J L, 1976, in *Critical Phenomena and Phase Transitions 5b*, eds Domb C and Green M (Academic Press, New York).
Hoover W G and Ree F H, 1968, *J Chem Phys* **49** 3609.
Ilett S M, Orrock A, Poon W C K, and Pusey P N, 1995, *Phys Rev E* **51** 1344.

Israelachvili J, 1992, *Intermolecular and Surface Forces*, second edition (Academic Press, London).

Kayser R F and Raveche H J, 1978, *Phys Rev A* **17** 2067.

Kirkwood J E, 1951, in *Phase Transformations in Solids* 67, eds Smoluchowski R, Mayer J E and Weyl W A (Wiley, New York).

Kusaka I, Wang Z-G, and Seinfeld J H, 1998, *J Chem Phys* **108** 3446.

Lekkerkerker H N W, Poon W C K, Pusey P N, Stroobants A and Warren P B, 1992, *Europhysics Lett* **20** 559.

McPherson A, 1982, *Preparation and analysis of protein crystals* (Krieger Publishing, Malabar).

Longuet-Higgins H C and Widom B, 1964, *Mol Phys* **8** 549.

Meijer E J and Frenkel D, 1991, *Phys Rev Lett* **67** 1110.

Meijer E J and Frenkel D, 1994, *J Chem Phys* **100** 6873.

Meijer E J and Frenkel D, 1995, *Physica A* **213** 130-137.

Muschol M and Rosenberger F, 1997, *J Chem Phys* **107** 1953.

Noro M, Kern N and Frenkel D, 1999 *Europhys Lett* **48** 332.

Onsager L, 1949, *Ann NY Acad Sci* **51** 627.

Onsager L, 1944, *Phys Rev* **65** 117.

Ostwald W, 1897, *Z Phys Chem* **22** 289.

Percus J K, ed, 1963, *The Many-Body Problem* (Interscience, New York).

Poon W C K, Pirie A D and Pusey P N, 1995, *Faraday Discuss* **101** 65.

Poon W C K, 1997, *Phys Rev E* **55** 3762.

Pusey P N, 1991, in *Liquids, freezing and glass transition*, eds Hansen J P, Levesque D and Zinn-Justin J, p 763 (North-Holland, Amsterdam).

Rosenberger F, 1996, *J Crystal Growth* **166** 40.

Rosenbaum D, Zamora P C and Zukoski C F, 1996, *Phys Rev Lett* **76** 150.

Rosenbaum D and Zukoski C F, 1996, *J Crystal Growth* **169** 752.

Rowlinson J S, 1988, *Studies in Statistical Mechanics*, Vol. XIV, ed Lebowitz J L, (North Holland, Amsterdam).

Stranski I N and Totomanow D, 1933 *Z Phys Chem* **163** 399.

Talanquer V and Oxtoby D W, 1993, *J Chem Phys* **99** 4670.

ten Wolde P R and Frenkel D, 1997, *Science* **277** 1975.

ten Wolde P R and Frenkel D, 1998, *J Chem Phys* **109** 9901.

ten Wolde P R, Oxtoby D W and Frenkel D, 1998, *Phys Rev Lett* **81** 3695.

Veerman J A C and Frenkel D, 1992, *Phys. Rev. A* **45** 5633.

Verwey E J and Overbeek J Th G, 1948, *Theory of the Stability of Lyophobic Colloids* (Elsevier, Amsterdam).

Volmer M and Weber A, 1926, *Z Phys Chem* **119** 227.

Vrij A, 1976, *Pure Appl Chem* **48** 471.

Weisskopf V F, 1977, *Trans NY Acad Sci II* **38** 202.

Wood W W and Jacobson J D, 1957, *J Chem Phys* **27** 1207.

Wright D, Caldwell R, Moxeley C and El-Shall M S, 1993, *J Chem Phys* **98** 3356.

Wright D and El-Shall M S, 1993, *J Chem Phys* **98** 3369.

The physics of floppy polymers

Patrick B Warren

Unilever R&D Port Sunlight
Wirral CH63 3JW, UK

patrick.warren@unilever.com

1 Introduction

The title of this chapter obviously deserves some explanation. The development of classical polymer physics can be traced perhaps to the ready availability of synthetic polymers in the early part of the 20th century, such as polyethylene, polystyrene, nylon, etc. Pioneers such as Staudinger, Kuhn, and especially Flory, led the way in proposing theories and models to understand the experimentally measured properties of these polymers. These synthetic polymers are essentially floppy objects. Their properties are largely determined by the fact that they have a huge number of possible conformations. Thus, rubber elasticity, rheology (flow behaviour) and so on are determined by *entropy*, in other words, changes in the number of conformations accessible to the chains.

Globular proteins—far and away the most important class of biopolymers—are usually not thought of as being floppy objects though. Their structures are prescribed (α-helix, β-sheet, etc.), and they have definite shapes in order to perform their functions. Nevertheless, globular proteins do change shape (allostery!), and it is now becoming appreciated that floppiness may be an important characteristic for a wider class of proteins than is apparent from a perusal of the crystallographic databases (the bias in perception is due to the fact that some rigidity is a pre-requisite for solving protein structures by crystallographic methods).[1] Of course, there are biopolymers that are clearly floppy, such as the nucleic acids (DNA and RNA), polysaccharides like cellulose and starch and the oft-neglected fibrous proteins such as silk and gelatin.

This chapter will therefore introduce some of the basic concepts of floppy polymer physics, starting with the equilibrium properties of single chains, moving on to the equilibrium properties of many-chain systems and finally briefly discussing polymer dynam-

---

[1] For more on globular proteins, see Poon's chapter in this volume, while Smith in his chapter discusses the importance of dynamics to protein function.

Patrick B Warren

Figure 1. *A random walk (left) and a polymer under an applied force (right).*

ics. The aim is to introduce some of the basic language of polymer physics without necessarily going into all the details. No discussion of polymer crystallisation or the glass transition has been included. These aspects are clearly important in polymer processing in the plastics industry, for example, but perhaps not so relevant for biopolymers.

2   Statistical physics of single chains

The fundamental model of a floppy polymer molecule is a random walk. Let us consider a random walk of N steps, illustrated in Figure 1 (left). The end-end vector $\mathbf{R}$ is a useful characteristic of the size. For any particular conformation, it is given by $\mathbf{R} = \sum_{i=1}^{N} \mathbf{l}_i$ where the $\mathbf{l}_i$ are the individual steps. The mean and variance of the end-end vector distribution follow easily from this:

$$\langle \mathbf{R} \rangle = \sum_{i=1}^{N} \langle \mathbf{l}_i \rangle = 0, \quad \langle \mathbf{R}^2 \rangle = \sum_{i,j=1}^{N} \langle \mathbf{l}_i \cdot \mathbf{l}_j \rangle = \sum_{i=1}^{N} \langle \mathbf{l}_i^2 \rangle = Nl^2. \tag{1}$$

In writing this, we assume that there is no preferred direction, thus $\langle \mathbf{l}_i \rangle = 0$, and that each step is uncorrelated with the previous steps, thus $\langle \mathbf{l}_i \cdot \mathbf{l}_j \rangle = \langle \mathbf{l}_i \rangle \cdot \langle \mathbf{l}_j \rangle = 0$. The root-mean-square (rms) step length $l \equiv \langle \mathbf{l}_i^2 \rangle^{1/2}$ has been introduced.

The key result is that the size of the chain, as characterised by the rms end-end vector $\langle \mathbf{R}^2 \rangle^{1/2}$ for instance, grows like $N^\nu l$ where $\nu = 1/2$. Later we will see that non-trivial values of ν are possible. For this case though, $\nu = 1/2$ is known as the ideal random walk exponent, and a polymer which obeys this law is said to behave as an ideal chain. One can show that all sensible measures of the chain size obey the same law, for example the radius of gyration as measured in a scattering experiment is $R_G = N^{1/2}l/\sqrt{6}$.

Equation 1 gives the first and second moments of the end-end vector distribution. What about the distribution $P(\mathbf{R})$ itself? The answer is that $P(\mathbf{R})$ is a Gaussian, and the reason lies in a mathematical result known as the central limit theorem. This theorem states that the probability distribution for the sum of a large number of random variables tends to a Gaussian, as the number of variables increases (under some mild restrictions). Since our random walk is precisely such a sum, and a Gaussian is completely determined by the first and second moments, we can immediately write down the probability distribution:

$$P(\mathbf{R}) \sim \exp\left(-\frac{3\mathbf{R}^2}{2Nl^2}\right). \tag{2}$$

The '3' in this comes from the number of space dimensions. This result is the basis for a large number of formal developments. One simple application is to show that polymers behave as 'entropic springs'. Let us consider applying a force $\pm\mathbf{f}$ to the ends of the polymer coil as in Figure 1 (right). A conformation with an end-end vector $\mathbf{R}$ now acquires a Boltzmann weight $\exp(\mathbf{f}\cdot\mathbf{R}/k_\mathrm{B}T)$ where k_B is the Boltzmann constant, and T the temperature. From this, we see that the end-end vector distribution is modified to become

$$P'(\mathbf{R}) \sim \exp\left(-\frac{3\mathbf{R}^2}{2Nl^2} + \frac{\mathbf{f}\cdot\mathbf{R}}{k_\mathrm{B}T}\right). \tag{3}$$

A simple way to proceed is to complete the square,

$$\frac{3\mathbf{R}^2}{2Nl^2} - \frac{\mathbf{f}\cdot\mathbf{R}}{k_\mathrm{B}T} = \frac{3}{2Nl^2}\left(\mathbf{R} - \frac{Nl^2}{3k_\mathrm{B}T}\mathbf{f}\right)^2 - \frac{\mathbf{f}^2Nl^2}{6(k_\mathrm{B}T)^2}. \tag{4}$$

The last term can be thrown away as it will be cancelled in the normalisation of $P'(\mathbf{R})$. From this, we immediately see that $\langle\mathbf{R}\rangle = (Nl^2/3k_\mathrm{B}T)\mathbf{f}$. We can rewrite this as $\mathbf{f} = (3k_\mathrm{B}T/Nl^2)\langle\mathbf{R}\rangle$, showing that the force is proportional to the (mean) extension. Thus, the polymer follows Hooke's law. Moreover the Hooke's law constant is proportional to temperature, which signals the entirely entropic origin of the effect; in other words, the polymer behaves as an entropic spring. One application of this is to rubber elasticity, where the prediction is that the elastic modulus $G \sim nk_\mathrm{B}T$ where n is the number density of chains between crosslinks.

Experimentally, single-molecule chain stretching experiments can be performed and Hooke's law holds for small extensions, but obviously deviations occur when the extension becomes comparable to the backbone length of the polymer. The deviations from Hooke's law can be analysed to obtain valuable information about the microscopic chain properties.

Our random walk model shows that the mean square end-end distance $\langle\mathbf{R}^2\rangle = Nl^2$, but what is N and l for a real polymer? To answer this, let the backbone length be L. Clearly, $N \propto L$, so $\langle\mathbf{R}^2\rangle \propto L$ also. This allows us to define a length $l_\mathrm{K} \equiv \langle\mathbf{R}^2\rangle/L$, so that $\langle\mathbf{R}^2\rangle = Ll_\mathrm{K}$. If we also define $N = L/l_\mathrm{K}$, we see that $\langle\mathbf{R}^2\rangle = Nl_\mathrm{K}^2$. The length l_K is known as the Kuhn length, and is a characteristic property of the polymer chain. In these terms, a real polymer is equivalent to a random walk with step length given by the Kuhn length l_K, and number of steps or 'Kuhn segments' equal to L/l_K.

We can illustrate this with a worked example. The Kuhn length for polyethylene glycol (PEG, also known as PEO, 'O' for 'oxide') is 1.8 nm. Let us estimate the size of PEG20k where the '20k' is the molecular weight. To solve this problem, we need the backbone length of PEG20k. The PEG monomer is $-CH_2CH_2O-$, with a molecular weight 44, and contributing 0.44 nm to the backbone length (from one C-C bond of length 1.5 Å and two C-O bonds of length 1.45 Å each). PEG20k, therefore, has 20k/44 $\approx$ 450 monomers, and the backbone length is $L = 450 \times 0.44 \approx 200$ nm. It follows that there are $N = 200/1.8 \approx 110$ Kuhn segments, and an estimate of the size of PEG20k is $\sqrt{110} \times 1.8 \approx 20$ nm.

Various detailed models of chains exist, such as the freely-jointed chain, or the fixed valance angle model, etc. Of these, the worm-like chain model (also known as the Kratky-Porod model) is particularly interesting. This model is appropriate for stiff polymers such as DNA. In this model, the polymer is treated as a filament that follows some path in space, $\mathbf{r}(s)$, where s measures distance along the backbone. Introduce the tangent vector $\mathbf{t}(s) = \partial\mathbf{r}/\partial s$. For a given conformation, we assign a Boltzmann weight $e^{-U/k_\mathrm{B}T}$

d	ν (Flory)	ν (exact)	exact method
1	1	1	(fully extended chains)
2	3/4	3/4	(conformal field theory)
3	3/5	0.588 ± 0.001	(diagrammatic resummation)
≥ 4	1/2	1/2	(ideal chains)

Table 1. *Exponent in $R \sim N^\nu$ for the polymer excluded volume problem in various space dimensions. Results are shown from an argument due to Flory given in the main text, and from more exact methods (Vanderzande 1998).*

where $U = (\kappa/2) \int_0^L ds\,(\partial t/\partial s)^2$ is the bending energy, $|\partial t/\partial s|$ is the curvature and κ an elastic modulus. Many properties of the worm-like chain model can be solved exactly, for instance, one can prove the tangent correlation function decays exponentially, $\langle t(0) \cdot t(s) \rangle = \exp(-s/l_P)$, with $l_P = \kappa/k_B T$ being the 'persistence length', i.e., the distance along the backbone over which the chain loses the memory of its orientation. From $R = \int_0^L t(s)\,ds$, one can derive $\langle R^2 \rangle = 2l_P^2(\exp(-L/l_P) - 1 + L/l_P)$. This has two limiting behaviours. For $L \ll l_P$, one has $\langle R^2 \rangle = L^2$ (short worm-like chains behave as rigid rods). For $L \gg l_P$, one has $\langle R^2 \rangle = 2Ll_P$. Thus, the Kuhn length for a worm-like chain is $l_K = 2l_P$. (For more details on the worm-like chain, see MacKintosh's chapter in this volume.)

As an example, DNA has a persistence length of about 50 nm. Each base pair (bp) contributes 3.4 Å to the backbone length. The chromosome in *E. coli* is 4.64 Mbp in length. Let us estimate the native size of the chromosome. The backbone length is $L = 0.34 \times 4.64 \times 10^6 \approx 1.6 \times 10^6$ nm (i.e., about 1.6 mm!). The Kuhn length is $2 \times 50 = 100$ nm, so the number of Kuhn segments is $1.6 \times 10^6/100 \approx 1600$. Hence, an estimate of the native size is $\sqrt{1600} \times 100 \approx 4000$ nm $= 4\,\mu$m (we ignore the fact that the chromosome is a circular loop of DNA). This exceeds the size of *E. coli* itself, which explains why *E. coli* takes active measures to reduce the genome size, such as using gyrase enzymes to twist the DNA into a 'plectonemic' state.

Now we turn to an important aspect of the polymer problem, and one which changes the exponent ν. This is the effect of 'excluded volume', in other words, the consequence of the fact that polymers occupy space. The simplest way to think about this problem is in terms of a self-avoiding walk. A more sophisticated approach though is to impose an energetic penalty for self-intersection. Universality ensures that both approaches are ultimately the same. Table 1 shows how the exponent changes due to excluded volume for polymers in varying space dimensions. It may seem peculiar to talk about d-dimensional polymers, etc., but thinking about the excluded volume problem in this way brings out the close connection with critical phenomena in statistical physics. Indeed one can argue that the modern era of polymer physics starts with the landmark paper by de Gennes (1972), which showed how the properties of polymers were related to a certain class of spin models in statistical physics.

To set up the excluded volume problem in energetic terms, we first estimate the concentration of segments in the polymer coil to be $c \sim N/R^3$ (here and later we use scaling arguments, so R is any typical measure of the spatial extent of the randomly coiled chain). We then suppose that the energy density due to self-interactions is $vk_BTc^2/2$, where v, known as the Edwards' excluded volume parameter, is the second virial coefficient between polymer segments in units of k_BT. The units of v are a microscopic volume, e.g., $v \sim l^3$ where $l \approx l_K$.

Let us estimate the self-interaction energy E for a polymer,

$$\frac{E}{k_BT} \sim R^3 \times \frac{vc^2}{2} \sim v\frac{N^2}{R^3}. \tag{5}$$

The factor R^3 is used as an estimate of the chain volume. If we additionally suppose the chain obeys ideal random walk statistics, $R \sim N^{1/2}l$, we find that $E \sim (vk_BT/l^3)N^{1/2}$. This grows indefinitely with N, therefore excluded volume effects are *always* important for sufficiently long chains.

If we repeat the analysis in d dimensions, we find $E/k_BT \sim (v/l^d)N^{(4-d)/2}$, which shows that excluded volume effects are unimportant for $d > 4$ and marginal at $d = 4$. Thus, $d_c = 4$ is the 'upper critical dimension' for the excluded volume problem and chains in $d \geq 4$ are expected to be ideal, as in Table 1.

A beautifully simple argument due to Flory gives an indication of how the exponent should change. We use the Gaussian coil result for $P(\mathbf{R})$ to estimate the loss in the number of chain conformations due to chain swelling. In entropy terms, this corresponds to $S = k_B \log P \sim -k_B R^2/Nl^2$. This is added to self-interaction energy from Equation 5 to obtain an estimate for the total free energy, $F = E - TS \approx vk_BTN^2/R^3 + k_BTR^2/Nl^2$. We minimise this, $\partial F/\partial R = 0$, to find that $R \sim (v/l^3)^{1/5}N^{3/5}l$. Thus, Flory predicts the exponent is $\nu = 3/5$. This result can be generalised to d dimensions to find $\nu = 3/(d+2)$ (an exercise for the reader). As is shown in Table 1, Flory's prediction coincides with the exact results in $d = 1, 2$ and 4 dimensions, and is close to the exact result in $d = 3$. However, it goes wrong (badly) if we try to calculate the free energy of swollen polymer for instance.

Before concluding this section, we consider what happens for different values of the Edwards' excluded volume parameter v, which can be varied by changing the temperature, for instance. Three cases are usually distinguished: $v \sim l^3$ is the 'good solvent' case where coils are fully swollen; $v \approx 0$ is the so-called θ-solvent condition where chains are ideal; and $v < 0$ is the poor solvent condition where chains are collapsed. If $v < 0$, inspection of the Flory free-energy estimate suggests there is nothing to stop the polymer coil collapsing to a point $R = 0$. In reality, many-body interactions stop the collapse (this can be captured in a minimal theory by including the third virial coefficient). It turns out that the 'coil-globule' transition for $v < 0$ can be a sudden jump, like a vapour-liquid transition. The globule state looks like a model for globular proteins, however we shall see in a moment that such collapsed polymer coils must be very dilute. Real globular proteins are stabilised by other effects such as the presence of hydrophilic or charged surface groups.

3   Statistical physics of many chains

In the previous section, we discussed the properties of an isolated chain, using the random walk model as a basis. In this section, we discuss the properties of many chains together; in other words, the properties of polymer solutions and polymer melts. Again, many of the fundamental ideas are due to Flory and his coworkers.

We start with Flory-Huggins theory, which is a model for the free energy of a polymer solution or melt. It is traditionally based on a lattice model, where the lattice spacing l is taken to be of the order of the Kuhn length. Let us suppose that the lattice volume is V, so the number of lattice sites is V/l^3. The polymers have N Kuhn segments, and let the segment concentration be c. The number of polymers per unit volume is then $\rho = c/N$ (the total number of polymers is Vc/N), and the fraction of sites occupied by polymers (the polymer volume fraction) is $\phi = l^3 c$.

With these definitions, various equivalent methods can be used to estimate the number of ways Vc/N polymers can be inscribed on the lattice and obtain the configurational entropy S. Without going into the details (see de Gennes (1979), for instance), the result is

$$-\frac{l^3 S}{V k_B} = \frac{\phi}{N} \log \phi + (1 - \phi) \log(1 - \phi) + A\phi + B. \tag{6}$$

In this, A and B are unimportant constants. We add to this a mean field estimate of the energy E. The approach is very similar to Bragg-Williams theory for alloys, and a host of other mean field models in statistical physics. We write

$$E = \frac{V}{2l^3} [\epsilon_{pp} \phi^2 + \epsilon_{ss}(1 - \phi)^2 + 2\epsilon_{sp} \phi(1 - \phi)]. \tag{7}$$

In the first term, for example, $V\phi^2/2l^3$ is an estimate of the number of polymer-polymer contacts and ϵ_{pp} is the energy per contact. We combine this with the entropy estimate to arrive at the Flory-Huggins free energy:

$$\frac{l^3 F}{V k_B T} = \frac{\phi}{N} \log \phi + (1 - \phi) \log(1 - \phi) + \chi \phi(1 - \phi). \tag{8}$$

We have defined

$$\chi k_B T = \epsilon_{sp} - (\epsilon_{ss} + \epsilon_{pp})/2. \tag{9}$$

The free energy in Equation 8 is a free energy of mixing, with the constants A and B in Equation 6 chosen such that $F \to 0$ at $\phi \to 0$ and $\phi \to 1$. With this choice, the energy term only depends on the so-called Flory χ-parameter, defined in Equation 9. The χ-parameter is seen to be a measure of the chemical dissimilarity between the solvent and the polymer.

We can make a connection to the excluded volume problem as follows. Expand Equation 8 about small ϕ to obtain

$$\frac{l^3 F}{V k_B T} = \frac{\phi}{N} \log \phi + A\phi + \frac{(1 - 2\chi)\phi^2}{2} + \cdots \tag{10}$$

where A is an unimportant constant. In the excluded volume approach, we can write a similar virial expansion of the free energy,

$$\frac{F}{V k_B T} = \rho \log \rho + \frac{vc^2}{2} + \cdots \tag{11}$$

The first term in this is for an ideal gas of polymers at a number density $\rho = c/N$, and the second term accounts for the second virial coefficient between polymer segments. Comparing this with the Flory-Huggins expansion, we see that

$$v = (1 - 2\chi)l^3. \tag{12}$$

This is a most important result. It shows, for example, that good solvent conditions correspond to $\chi \approx 0$, θ-solvent conditions to $\chi = 1/2$, and poor solvent conditions to $\chi > 1/2$. From Equation 9, we expect that $\chi \sim 1/T$; thus, χ should increase (the solvent quality gets poorer) with decreasing temperature. It turns out this is true for many polymers in organic solvents, but is often untrue for aqueous systems (e.g., PEO in water).

If the polymer and solvent are chemically identical, all the energies in Equation 9 are the same and $\chi = 0$ (the so-called 'athermal' solvent case). Thus, a polymer dissolved in a solvent of its own monomers is expected to be fully swollen, which is a somewhat counterintuitive result.

A more counterintuitive result is the so-called 'Flory theorem', which states that a polymer dissolved in a solvent of equal polymers (in other words a polymer in a melt) is ideal, and not swollen at all. Various proofs of this can be constructed: via a spin-model mapping, via Edwards calculation of screening of the excluded volume interaction, or via a simple extension to Flory-Huggins theory. This last approach, although not rigorous, is quite interesting. We generalise Flory-Huggins theory to consider polymers of length N and volume fraction ϕ, in a solvent of other polymers of length M and volume fraction $1 - \phi$. The required generalisation is quite straightforward and is

$$\frac{l^3 F}{V k_{\mathrm{B}} T} = \frac{\phi}{N} \log \phi + \frac{(1 - \phi)}{M} \log(1 - \phi) + \chi \phi (1 - \phi). \tag{13}$$

We now make the expansion about small ϕ to get

$$\frac{l^3 F}{V k_{\mathrm{B}} T} = \frac{\phi}{N} \log \phi + A\phi + \frac{1}{2}\left(\frac{1}{M} - 2\chi\right)\phi^2 + \dots \tag{14}$$

Comparing with Equation 11, we see that in this case $v = (1/M - 2\chi)l^3$. Thus, v is reduced if the solvent is polymerised. If all the polymers are chemically the same, $\chi = 0$ and $v = l^3/M$. To prove the Flory theorem from this, we use Equation 5 from the previous section. This shows that the excluded volume interaction is only a small perturbation to ideal chain statistics ($E \ll k_{\mathrm{B}} T$) if $v/l^3 \ll N^{-1/2}$. Applying this to the present mixture, excluded volume is unimportant if $1/M \ll N^{-1/2}$, in other words, if $N \ll M^2$. If the polymers are all of the same length as well as chemically identical, then $N = M$ and the condition $N \ll M^2$ is trivially satisfied. Thus, polymers in a melt are ideal.

We now turn to another aspect of Flory-Huggins theory. This is the prediction that is made for the phase behaviour of polymer solutions. Whilst the entropic term in Equation 8 always favours mixing, we see that the energetic term favours demixing if $\chi > 0$. In fact, the free energy develops a double minimum if χ becomes large enough, shown in Figure 2, and Flory-Huggins theory predicts liquid-liquid demixing, shown in Figure 3. Let us write $f = l^3 F/V k_{\mathrm{B}} T$ as a dimensionless free-energy density. Then a

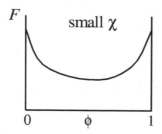

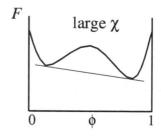

Figure 2. *Flory-Huggins free energy for small (left) and large (right) χ-parameters. On the right, the 'double tangent construction' is illustrated, the points of common tangency giving the densities of coexisting phases.*

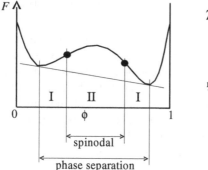

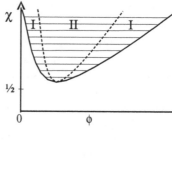

Figure 3. *Phase behaviour of a polymer solution predicted by Flory-Huggins theory.*

condition that the free energy has a double minimum is that there is a region where the second derivative $d^2f/d\phi^2 < 0$. The boundary of this 'spinodal' region is the spinodal line $d^2f/d\phi^2 = 0$ as indicated in Figure 3. The minimum value of χ on the spinodal curve is the point where additionally $d^3f/d\phi^3 = 0$. This point is a fluid-fluid demixing critical point (Ising universality class). Solving $d^2f/d\phi^2 = d^3f/d\phi^3 = 0$ for the Flory-Huggins free energy shows that the demixing critical point occurs at $\phi_c \approx N^{-1/2}$ and $\chi_c \approx 1/2 + N^{-1/2}$, where $N \gg 1$ is assumed. Thus, we see that increasing N shifts the critical point to small volume fractions and closer to θ-solvent conditions. The corresponding critical Edwards excluded volume parameter is $v_c = -l^3 N^{-1/2}$. Thus, a small negative virial coefficient between segments will result in phase separation (i.e., collapsed chains in a solution are dilute).

Flory-Huggins theory correctly indicates that increasing χ, i.e., decreasing temperature since $\chi \sim 1/T$ usually, or increasing N favours demixing, but being a mean-field theory there are obviously some things it does not get right, for example, the shape of the coexistence curve in the vicinity of the critical point. Less obviously, the prediction for the solubility in bad solvent conditions (i.e., the dilute-solution coexistence curve) is poor. This is because in bad solvent conditions, the mean-field estimate of the energy is inappropriate for polymers, which are essentially dense collapsed coils.

As another example, we can use the extended Flory-Huggins theory to predict miscibility in polymer blends. We leave as an exercise for the reader the proof that in the symmetric case ($N = M$), Equation 13 has a demixing critical point at $\phi_c = 1/2$ and $\chi_c = 2/N$. Thus, a very small positive χ-parameter (a small chemical incompatibility) results in phase separation in a polymer blend. In fact, even deuterated and non-deuterated polymers may phase separate!

The third aspect of Flory-Huggins theory that we shall consider is the prediction that is made for the osmotic pressure of a polymer solution. Recall that the osmotic pressure is the pressure difference required to maintain equilibrium across a semi-permeable membrane that seperates the solution from a reservoir of solvent. In these practically incompressible systems, it can be shown that the osmotic pressure Π is (minus) the volume derivative of the free energy, $\Pi = -\partial F/\partial V$. Osmotic pressure is here defined to be a thermodynamic quantity, but don't get hung up on this! It is capable of exerting real mechanical forces, as will be testified by any bacterium that has burst after being placed in distilled water.

Flory-Huggins theory, Equation 8, predicts that $l^3 \Pi / k_{\mathrm{B}} T = \phi/N - \log(1-\phi) - \phi - \chi \phi^2$, or on expanding for small ϕ,

$$\frac{l^3 \Pi}{k_{\mathrm{B}} T} = \frac{\phi}{N} + \left(\frac{1}{2} - \chi\right)\phi^2 + \ldots \tag{15}$$

Let us focus on the good solvent case ($\chi = 0$). We notice that there is a crossover at $\phi^* \sim 1/N$. If $\phi < \phi^*$, we have $\Pi = \phi k_{\mathrm{B}} T / N l^3 = \rho k_{\mathrm{B}} T$. In this regime, the van 't Hoff law is obeyed (osmotic pressure equals $k_{\mathrm{B}} T$ times the number density of objects in solution). The van 't Hoff law can be used to determine the molecular weight of the polymer, but the fact that it only obtains for very low polymer volume fractions was a stumbling block in the early days of polymer science. For $\phi > \phi^*$, Flory-Huggins predicts a regime where $\Pi \sim \phi^2$ is independent of N.

It turns out that these predictions of Flory-Huggins theory are nearly, but not quite, right. The modern approach to these problems originated with a mapping of polymer solution statistics onto another spin model by des Cloizeaux (1975). In this approach, ϕ^* is the overlap volume fraction for the coils, and the regime where $\phi^* \ll \phi \ll 1$ is known as the semi-dilute solution regime. Briefly, ϕ^* can be computed as the point where $R^3 \times (\phi^*/N l^3) \sim 1$. This results in $\phi^* \sim N^{1-3\nu}$, or $\phi^* \sim N^{-4/5}$ using the Flory value for the swelling exponent. In the semi-dilute regime, it is still true that the osmotic pressure does not depend on N (the chain ends do not count). If we suppose that power law scaling holds in the semi-dilute regime, $\Pi \sim \phi^\alpha$, and demand continuity of the osmotic pressure with the van 't Hoff law as one approaches ϕ^*, we can derive $\alpha = 3\nu/(3\nu-1)$. With the Flory value for ν, this predicts $\Pi \sim \phi^{9/4}$. Other aspects of the theory of semi-dilute solutions indicate how the chain shrinks from being fully swollen at $\phi \le \phi^*$, to being ideal as $\phi \to 1$ (the Flory theorem). If we assume $R \sim N^\nu (\phi/\phi^*)^{-\beta}$, where β is an exponent characterising the shrinkage, then an interesting exercise is to prove that $\beta = (2\nu - 1)/(6\nu - 2)$ (*i.e.* $\beta = 1/8$ if $\nu = 3/5$).

4   Polymer dynamics

The previous two sections focussed on the equilibrium properties of floppy polymers, either as individual chains, or as many chains together in a solution. In this section, we visit some of the concepts that are used to understand the dynamics of floppy polymers. Let us start by recapitulating some aspects of Brownian motion. A colloid particle jiggles about in a fluid due to the fluctuating pressure field. The mean square displacement grows linearly with time, $\langle r^2 \rangle = 6Dt$ where D is the classical diffusion coefficient that describes the diffusive spreading of a dilute collection of such particles in a suspension. By considering the fluxes in a sedimentation equilibrium, Einstein showed that $D = k_B T/\xi$ where ξ is the drag coefficient on the particle (the drag force required to move the particle at unit velocity through the fluid). For a spherical particle for instance, Stokes' law says that $\xi = 6\pi\eta a$ where η is the fluid viscosity and a the radius. (See Frenkel's chapter in this volume for an explanation of the physics behind Stoke's law.) The combined result, $D = k_B T/6\pi\eta a$, is known as the Stokes-Einstein relation.

One of the ways to describe Brownian motion is to add a random force into the equation of motion of the particle. The resulting 'Langevian equation' generates stochastic (random) trajectories. The statistics of the random force are prescribed by a fluctuation-dissipation theorem such that the Einstein relation is satisfied. By way of an aside, be aware that the trajectory of a Brownian particle in an external force field does not have the same statistics as the path of a polymer in the same external field. Rather, the polymer path is closely analogous to the trajectory of a quantum particle in the external field.

Now we turn to polymer dynamics. We start with an isolated polymer chain. Under the influence of Brownian motion, it wriggles about in solution, continually changing its shape. The most common approach is to introduce a fluctuating bead and spring model, known in its most basic form as the Rouse model. In the Rouse model, we write down a Langevin equation for the bead positions $\mathbf{r}_i$ ($i = 1 \ldots N$):

$$\xi \frac{d\mathbf{r}_i}{dt} = \mathbf{f}_i^{(R)} - k(\mathbf{r}_i - \mathbf{r}_{i-1}) + k(\mathbf{r}_{i+1} - \mathbf{r}_i). \tag{16}$$

In this, each bead has a drag coefficient ξ, and is subjected to a random force $\mathbf{f}_i^{(R)}$. There is a spring potential $k(\mathbf{r}_{i+1} - \mathbf{r}_i)^2/2$ between adjacent beads (obviously Equation 16 needs correcting for the first and last beads in the chain). It turns out that it is legitimate to ignore bead inertia. We have also ignored hydrodynamic interactions between beads, which is known as the free-draining approximation and can be a severe limitation on the applicability of the model. The model is constructed so that the equilibrium distribution for the beads is an ideal chain, so we have also ignored excluded volume effects.

Equation 16 can be solved exactly in terms of normal modes. In this context, the normal modes are known as Rouse modes. The normal modes look like

$$\mathbf{r}_i = \mathbf{A}_p \cos\left(\frac{(p-1)\pi i}{N}\right) \exp(-t/\tau_p) \tag{17}$$

where $p = 1 \ldots N$ labels the modes, and τ_p is the mode relaxation time. Collectively, the set of τ_p are described as the 'relaxation spectrum'.

The first mode ($p = 1$) corresponds to all the beads moving collectively. Because each bead contributes a drag ξ, the total drag is $N\xi$. Therefore, one can define a diffusion

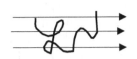

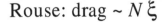

Rouse: drag $\sim N\xi$ Zimm: drag $\sim 6\pi\eta R \sim N^\nu \xi$

Figure 4. *In the Rouse model (left), a free-draining approximation is made and the friction is proportional to the number of beads. In the Zimm model (right), the streamlines are screened from the interior of the polymer so the friction scales with the polymer size.*

coefficient $D = k_B T/N\xi$ and identify τ_1 with the structural relaxation time $R^2/D \sim N^2 l^2 \xi/k_B T$ (we have used $R \sim N^{1/2}l$ for an ideal chain). This mode corresponds to centre-of-mass diffusion for the chain.

For $p > 1$, the Rouse modes correspond to N/p beads moving in concert. The resulting drag coefficient is $N\xi/p$, and the relaxation time is $\tau_p \sim R_p^2/D_p$. The length scale in this is $R_p \sim (N/p)^{1/2}l$, the typical size of an ideal chain of N/p beads. Working through the algebra, the relaxation spectrum is $\tau_p = \tau_1/p^2$. One power of p^{-1} comes from the drag reduction, and another comes from R_p^2. This result applies to the case of an ideal chain. An approximate way to take into account excluded volume is to suppose that $R_p \sim (N/p)^\nu$, in which case the spectrum becomes $\tau_p = \tau_1/p^{1+2\nu}$.

Unfortunately, as we have already alluded to, the Rouse model is not applicable to isolated polymers in solution because of the neglect of hydrodynamic interactions. A Langevin equation including hydrodynamic interactions can be written down, but cannot be solved without approximations. One idea is that the Rouse modes survive and can be used as the basis of an approximate theory, provided one adjusts the drag coefficient to take into account fluid entrainment. The resulting model is known as the *Zimm model*.

In the Zimm model, the drag on a collection of beads is reduced because fluid is entrained, i.e., fluid streamlines are screened from the interior of the cluster, as shown in Figure 4. The drag on a chain of beads is approximated by Stokes' law. Write $R \sim N^\nu$ to allow for the effects of excluded volume (for ideal chains, set $\nu = 1/2$) and eliminate the liquid viscosity by writing $\xi \sim k_B T/\eta l$ (i.e., the beads obey Stokes-Einstein with a radius of order l). Then the Stokes' drag $6\pi\eta R \sim N^\nu \xi$. Let us apply this first to the chain diffusion coefficient. We now expect this to go as $D \sim N^{-\nu}$ (Zimm) compared to $D \sim N^{-1}$ (Rouse). This reduced dependency of the chain self diffusion coefficient on the polymer molecular weight has been confirmed experimentally. The relaxation spectrum for the Zimm model can also be calculated. From $\tau_p \sim R_p^2/D_p$, it is found to be $\tau_p = \tau_1/p^{3\nu}$. One power of $p^{-\nu}$ comes from the drag, and the remaining two powers from R_p^2.

As an application of all this, we consider how the mean square displacement of a monomer grows with time. At short times, the bead does not realise that it is part of a larger structure; thus, the diffusion is quite rapid. As time progresses, more and more of the polymer become dynamically connected to the bead, and a sub-diffusive law is obtained, $\langle r^2 \rangle \sim t^\gamma$ with $\gamma < 1$. At long times, $\langle r^2 \rangle \sim t$ is recovered because eventually the bead diffuses with the diffusion coefficient of the whole chain. We can use the relaxation spectrum to determine γ as follows. At a given time t, modes with $\tau_p < t$ will have relaxed, and modes with $\tau_p > t$ will be effectively frozen (unrelaxed). Since $\tau_p = \tau_1/p^{3\nu}$ (Zimm model), this means that modes with $p < (t/\tau_1)^{-1/3\nu}$ will be relaxed. We expect that the mean square displacement of the bead $\langle r^2 \rangle \sim R_p^2$, where R_p corresponds to the relaxed mode with the largest spatial size; this is the mode with $p \approx (t/\tau_1)^{-1/3\nu}$. Since $R_p^2 \sim (N/p)^{2\nu}$, this means that $\langle r^2 \rangle \sim (t/\tau_1)^{2/3}$. Thus, the Zimm model predicts $\gamma = 2/3$, independent of ν. This prediction has been confirmed for DNA, for instance. It is left as an exercise for the reader to show that the Rouse model predicts $\gamma = 2\nu/(2\nu + 1)$, which only coincides with the Zimm model for $\nu = 1$ (rods).

The Rouse model may be useless for isolated polymers in solution, but it turns out that it is applicable to polymers in melts and in semi-dilute solutions. This is because hydrodynamic interactions are screened out by the presence of the other polymers and, moreover, polymers in a melt are ideal (Flory theorem). In a melt or solution though, if the polymers become long enough, the effects of entanglements start to become important. These slow the dynamics down in a way that we now describe rather briefly. It is important to note that there is an entanglement threshold, for example, entanglements in a melt only start to be important for polymers above the 'entanglement molecular weight', which is basically an experimentally determined characteristic.

The basic idea, illustrated in Figure 5, is that an entangled chain escapes by sliding back and forth along its own backbone. This dynamical mode is known as *reptation*, and one can approximate the confining entanglements by a tube. The polymer slides back and forth along the tube, and the ends of the polymer create a new tube. The reptation time τ_r is the time it takes for the tube to be completely renewed by this process. The dependence on the polymer molecular weight can be estimated as follows. The length of the tube L will be proportional to the backbone length of the polymer; thus, $L \sim N$. The reptation time is $\tau_r \sim L^2/D_r$ where the diffusion coefficient for slithering backwards and forwards is $D_r \sim 1/N$ (i.e., hydrodynamic interactions are screened, and the friction is expected to be proportional to the backbone length). Extracting the N-dependence, therefore, the reptation time is expected to scale as $\tau_r \sim N^3$.

We can use this to estimate the chain diffusion coefficient for entangled polymers in a melt. Chain diffusion is slowed because chains are confined to move in the tubes. After every time period τ_r though, the tube is completely renewed and the polymer will have moved a distance comparable to its own size. Thus, over a long time t, the polymer will take t/τ_r random walk steps of size $R \sim N^{1/2}l$ (polymers in a melt are ideal!). The mean square displacement follows as $\langle r^2 \rangle \sim (t/\tau_r)R^2$. We thus identify the self diffusion coefficient $D \sim R^2/\tau_r \sim 1/N^2$. Thus, the self diffusion coefficient in a melt is predicted to cross over from $D \sim 1/N$ (Rouse) to $D \sim 1/N^2$ (tube model) as the entanglement threshold is passed with increasing N. This crossover has been seen experimentally.

Another famous application of the tube model is to the rheology of polymer melts. If we imagine suddenly shearing a polymer melt in a step-strain experiment, the initial response will be rubber-like, as the entanglements act as temporary cross links. The stress that is set up will decay away though, as the polymers reptate out of their deformed tubes into new equilibrium tubes. This 'visco-elastic' response is very characteristic of polymer melts and solutions, and is one of the many varied rheological phenomena that can be observed.

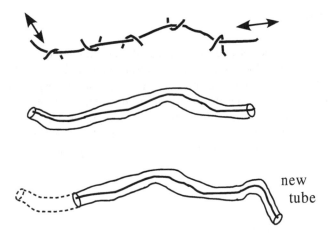

Figure 5. *Above the entanglement threshold, a chain escapes by sliding back and forth between the entanglements (top). The entanglements can be modelled by a tube (middle), which is gradually refreshed by diffusion (bottom).*

If instead of a step-strain experiment, we make a steady strain experiment, then a steady state will be reached in which the rate at which stress grows due to the deformation is balanced by the rate at which stress decays due to the relaxation processes in the system. For the stress build up,

$$\text{stress} = G \times \text{strain} \quad \Rightarrow \quad \frac{d(\text{stress})}{dt} = G \times \text{strain rate} \tag{18}$$

where G is the modulus for the initial elastic response. Similarly, for the stress decay,

$$\frac{d(\text{stress})}{dt} = -\frac{\text{stress}}{\tau} \tag{19}$$

where τ is the relaxation time. In steady state, these balance so that

$$G \times \text{strain rate} = \frac{\text{stress}}{\tau} \quad \Rightarrow \quad \text{stress} = G\tau \times \text{strain rate.} \tag{20}$$

We recognise this as the definition of viscosity; thus,

$$\eta \approx G\tau. \tag{21}$$

This is an extremely useful relation, which often allows us to estimate the viscosity of a complex fluid, given some basic knowledge of the dynamics (the simple approach we have used to derive Equation 21 is known as the *Maxwell model*).

For a polymer melt, the initial elastic response is rubber-like, and dependent only on the entanglement density (see discussion after Equation 4). Thus, G is not expected to depend significantly on the polymer molecular weight. On the other hand, the stress relaxation is determined by the tube renewal time, $\tau = \tau_r \sim N^3$. Putting these together, we expect to see the viscosity $\eta \sim N^3$. This rapid increase in viscosity with molecular weight is a classical prediction of the tube model, and is quite close to the experimentally observed $\eta \sim N^{3.4}$ law. The discrepancy can be attributed to neglected effects in the basic version of the tube model, such as constraint release (disappearance of entanglements as other polymers move out of the way) and tube contour length fluctuations.

Bibliography

This discussion on polymer physics has necessarily been rather concise. For additional details, the excellent introductory text books available are by Doi (1995), by Grosberg and Khokhlov (1997) and most recently by Rubinstein and Colby (2003). For more advanced material, one can consult the classic monographs by Flory (1953, 1969), de Gennes (1979), and Doi and Edwards (1986). A comprehensive account of polymer solutions is given by des Cloizeaux and Jannink (1991). Very readable lecture notes on polymer statistical physics from a modern point of view is by Vanderzande (1998).

References

de Gennes P G, 1972, *Phys Lett* **38A** 339.
de Gennes P G, 1979, *Scaling concepts in polymer physics* (Cornell, Ithaca, New York).
des Cloizeaux J, 1975, *J Physique* **36** 281.
des Cloizeaux J and Jannink G, 1991, *Polymers in solution* (OUP, Oxford).
Doi M, 1995, *Introduction to polymer physics* (OUP, Oxford).
Doi M and Edwards S F, 1986, *The theory of polymer dynamics* (OUP, Oxford).
Flory P J, 1953, *Principles of polymer chemistry* (Cornell, Ithaca, New York).
Flory P J, 1969, *Statistical mechanics of chain molecules* (Interscience, New York).
Grosberg A Yu and Khokhlov A R, 1997, *Giant molecules* (Academic, San Diego).
Rubinstein M and Colby R H, 2003, *Polymer physics* (OUP, Oxford).
Vanderzande C, 1998, *Lattice models of polymers* (CUP, Cambridge).

Self-assembly and properties of lipid membranes

Peter D Olmsted

School of Physics & Astronomy, University of Leeds
Leeds LS2 9JT, UK

p.d.olmsted@leeds.ac.uk

1   Introduction

Lipid bilayers are among the most important 'construction materials' for the cell (Alberts *et al.* 2002). A bilayer membrane constitutes a flexible barrier that separates the interior and exterior of a cell, encapsulates the nucleus and can perform a number of roles such as acting as a functional host for protein production. Membranes appear in flat (plasma membrane), spherical (vesicular transport), tubular (transport) or tortuous and possibly bicontinuous (Endoplasmic Reticulum, ER and Golgi apparatus) forms in the cell, depending on function and composition. Lipid membranes comprise a vast variety of lipids and other constituents, of which cholesterol is a prime player, whose composition determines the large-scale properties necessary for the function of the particular membrane. Such properties include flexibility, stiffness with respect to bending and stretching, viscosity and fluidity, overall shape and degree of internal order. Most biological membranes contain a large area fraction, up to 30% or more in some cases, of membrane proteins, which themselves influence membrane properties and whose specific functions act in concert with the local lipid compositions to direct their biological role. The subjects of membrane elasticity and dynamics, membrane-protein interactions and membrane-filament interactions are vast and of vital importance to a concrete biophysical understanding of cell processes.

The biology and physics (Safran *et al.* 1994) communities have pursued overlapping studies of membranes over the past few decades, with the natural segmentation into more specific (biological) and general (physics) studies. The physicists are often entranced by the variety of topological and phase behaviours possible in membranes. However, with new experimental techniques that include better visualisation of dynamical processes in cells, there are increasingly more opportunities for interdisciplinary cross-fertilisation. In this chapter, I will outline a small corner of our understanding of the physics of biological membranes, with an eye towards an eventual understanding of their material properties and functions.

2   The constituents of lipid bilayer membranes

Bilayers comprise a variety of components, including many species of two-tailed phospholipids, cholesterol and other sterols and many proteins, depending on the particular membrane and its function (Alberts *et al.* 2002). The building blocks are as follows:

- **Phospholipids** have a polar head group and two hydrophobic hydrocarbon tails. This amphiphilic molecule provides the primary impetus for forming the bilayer to avoid hydrophobic contacts. Typical tails lengths are from 14–20 carbons, and regulate thickness and membrane stability. The different sizes of head groups lead to different overall lipid shapes, which we will see leads to different possibilities for regulation of membrane shape and function. The most common phospholipids include *phosphatidylcholine* (PC), *phosphatidylethanolamine* (PE) and *phosphatidylserine* (PS).

  One of the tails of lipids such as PE, PC and PS is typically fully saturated (e.g., 18:0), while the other typically has one or sometimes more double bonds (denoted, e.g., 18:1 or 18:2). Fully saturated chains are crystalline (gel) at higher temperatures and are less fluid than their unsaturated counterparts. Hence, the melting (or *main* transition) temperature may be regulated by a suitable blend of lipids.

- **Sphingomyelin** (SM) is another phospholipid, which deserves special mention because the linkage between head and tail contains an OH group, which can act both as hydrogen-bond donor and acceptor, in contrast to the lipids above. Moreover, sphingomyelin and related *sphingolipids* generally have two fully saturated chains and, thus, a relatively high melting point and a less fluid phase.

- **Cholesterol** has a very small polar head, 4 planar rings and a relatively short hydrocarbon tail. The role and mechanism by which cholesterol influences membrane behaviour is still not clear, although certain points deserve mention. The planar structure encourages packing of less saturated tails, which leads to a 'preference' for SM. Moreover, it is thought by some that the special nature of the SM headgroup further encourages specific SM/cholesterol interactions. This remains a controversial area (McConnell and Vrljic 2003, Radhakrishnan and McConnell 1999, Holopainen *et al.* 2004, Pandit *et al.* 2004, de Almeida *et al.* 2003, Slotte 1999).

In addition to these basic building blocks, membranes, of course, are full of proteins of all sorts, including transmembrane proteins that span both bilayer leaflets; glycosylphosphatidylinositol (GPI) anchored proteins that participate in one leaflet; ion channels, ATPases, and many others. An important current question is how the different proteins

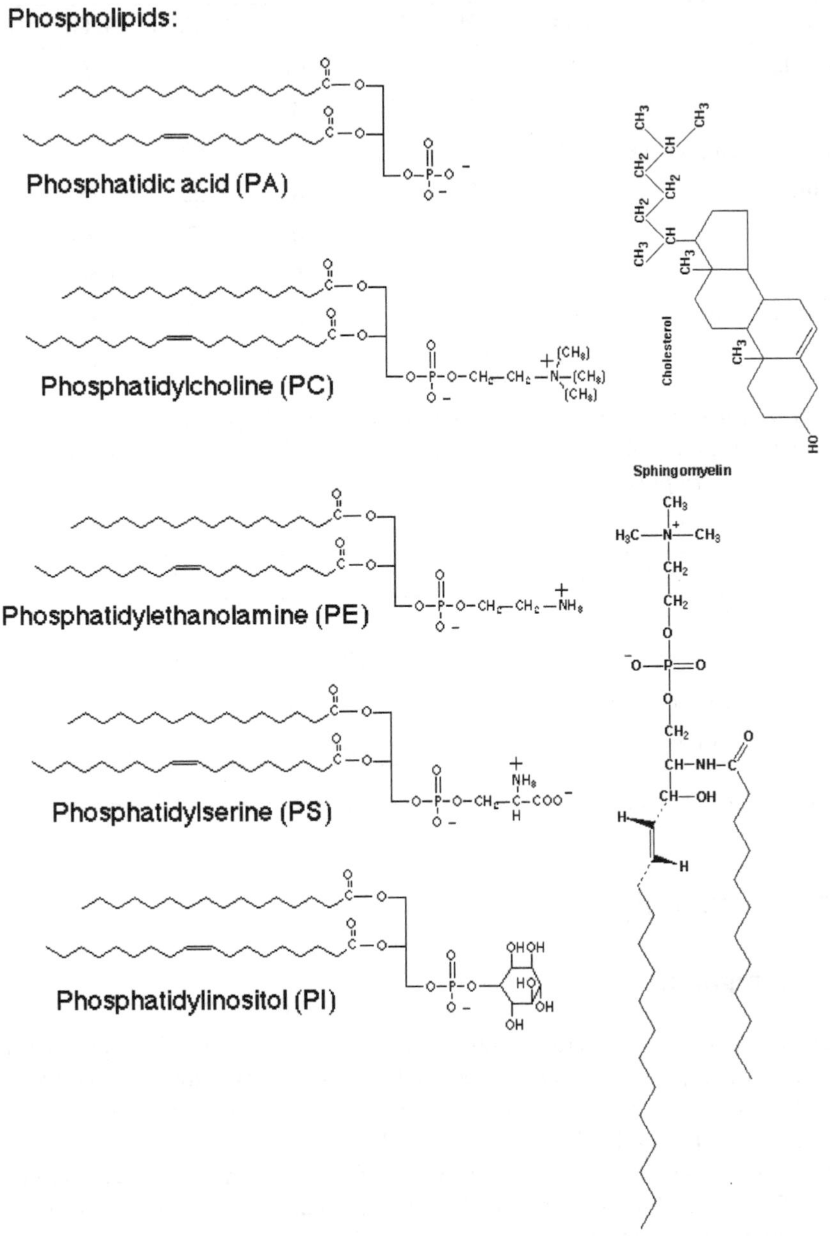

Phospholipids:

Figure 1. *Some common phospholipids and cholesterol.*

interact with and are influenced by the local membrane composition (Mayor and Rao 2004, Sharma *et al.* 2004, Varma and Mayor 1998). Membrane compositions vary greatly (see Table 2) with, broadly, more cholesterol towards the plasma membrane (rather than nucleus) portion of the cell.

Membrane	% PC	% PE	%PS	% SM	% Chol	% GL	% other
Liver plasma	24	7	4	19	17	7	24
Liver nuclear	51	12	3	3	7		24
Red blood cell plasma	17	18	7	18	23	3	13
Myelin	10	15	9	8	22	28	8
Mitochondrian[†]	39	25	2	0	3	trace	21
ER	40	17	5	5	6	trace	27
E. Coli bacterium	0	70	trace	0	0	0	30
Golgi	45	18	6	7	9	0	15

Table 1. *Approximate compositions (as fraction of total lipid/sterol content) of some membranes. Sources: Alberts et al. 2002, Maeda 2004.* [†] *Both inner and outer membranes. GL: Glycolipids.*

Number of carbons/chain	14	15	16	17	18	20
Outer mitochondrial	<1	27	25	14	14	16
Inner mitochondrial	<1	27	22	16	16	19
Plasma membrane	1	37	31	6	13	11
Golgi	1	35	23	9	18	15

Table 2. *Percent by weight of carbon numbers of lipids in some membranes (Boal 2002).*

3 Self assembly

3.1 Aggregation

Bilayers are only one of several self-assembled structures formed from phospholipids. Self assembled objects are usually either rod-like (one dimensional), plate-like (2D), spherical (3D) or more complex (3D periodic or random structures such as those formed in diblock copolymers (Hamley 1998) or possibly the Golgi). Self assembly is ubiquitous in biology: examples of one-dimensional self assembly include G actin into F actin (Howard 2001), tubulin into microtubules and misfolded proteins into amyloid fibrils. Bilayers are a 2D example, while micelles are a 3D example. We will see that spherical micelles are not typically formed by two-tailed phospholipids.

Self assembly is regulated by the chemical potential of exchange between aggregates of different sizes. It can be conveniently studied by minimising the total free energy F given by

$$F = \sum_N \frac{X_N}{N} \left\{ k_B T \left(\ln \frac{X_N}{N} - 1 \right) + E_N \right\}, \tag{1}$$

where X_N is the volume fraction of aggregates of number N and E_N is the energy of

	Name	$N : n_u$	Example
Unsaturated	palmitoleic	16:1	DPPC
	oleic	18:1	DOPC
Saturated	lauric	12:0	DLPC
	myristic	14:0	DMPC
	palmitic	16:0	
	stearic	18:0	DSPC
	arachidic	20:0	DAPC
Mixed	palmitoyl-oleoyl	16:0-18:1	POPC

Table 3. *Some common synthetic phospholipids. DPPC implies dipalmitoyl phosphatidylcholine. N is the carbon number per chain and n_u is the number of unsaturated bonds/chain. PO** refers to two different tails.*

the aggregate. The total amphiphile volume fraction is $\phi = \sum_{N=0}^{\infty} X_N$. By minimising over F subject to the constraint of fixed volume fraction, one finds the following relation between monomers and N-mers:

$$X_N = N \left[X_1 e^{\left(\frac{\varepsilon_1 - \varepsilon_N}{k_B T} \right)} \right]^N, \tag{2}$$

where $\varepsilon_N \equiv E_N / N$. Aggregation occurs for $\varepsilon_N < \varepsilon_1$. The entire distribution is obtained by requiring $\phi = \sum_{N=0}^{\infty} X_N$, and the resulting chemical potential satisfies $e^{\mu N / k_B T} = \frac{X_N}{N} e^{N \varepsilon_N k_B T}$. In either case, the critical micelle concentration ϕ_{CMC}, above which an appreciable fraction of amphiphile incorporates into micelles, is given roughly when the term in brackets in Equation 2 is of order one, for $\phi = X_1$:

$$\phi_{\text{CMC}} = e^{-\delta\varepsilon / k_B T}, \qquad\qquad \delta\varepsilon = \varepsilon_1 - \varepsilon_N. \tag{3}$$

For $\phi > \phi_{\text{CMC}}$, the micelles act as a reservoir to accept excess amphiphiles, as well as to regulate the monomer chemical potential in solution to be roughly constant,

$$\mu|_{\phi > \phi_{\text{CMC}}} \simeq \mu_0(T) - k_B T \ln \phi_{\text{CMC}}, \tag{4}$$

where $\mu_0(T)$ is the reference chemical potential. The regulation or buffering of μ by a reservoir of self-assembled materials could be of practical use in the cell.

The form of the distribution depends critically on how the additional gain per monomer ε_N depends on aggregation number N (Israelachvili 1998). This, in turn, is closely tied to the geometry of the aggregate, which is itself determined by the shapes and energies of the individual monomers. We will consider three geometries: spherical (3D) and linear (1D) micelles, and bilayers (2D).

Spherical micelles (3D) — For spherical micelles, packing constraints dictate that there be a preferred aggregation number M, so that $\varepsilon_N \simeq \varepsilon_M + \Lambda (N - M)^2 + \dots$. In this case, the distribution X_M is sharply peaked about M for any reasonably large

M, of order 10, say. This follows from Equation 2. Typical aggregation numbers for single-tailed surfactants such as SDS are of order 10–100.

Cylindrical micelles (1D) — For cylindrical micelles, the energy to add additional monomers is only the end cap energy. In the limit of an infinite rod, the additional cost to add another monomer is zero, leading to $\varepsilon_N = \varepsilon_\infty + \alpha k_B T/N$, where α is proportional to the energy of having two endcaps rather than the infinite rod. In this case, the length distribution is highly polydisperse, and well described by

$$X_N = A\,N e^{-N/M},\qquad\qquad(5)$$

where the characteristic rod length M is proportional to the mean $\langle N \rangle$.

Bilayer aggregates (2D) — In this case, the energy of an N-aggregate differs from that of the infinite aggregate by the edge energy, which scales as the aggregate length, or $\sqrt{N}$. Hence, the energy per particle in an aggregate is $\varepsilon_N = \varepsilon_\infty + \alpha k_B T/\sqrt{N}$, where α is related to the edge energy. The aggregate distribution, Equation 2, is thus given by

$$X_N = N\,[X_1 e^\alpha]^N\,e^{-\alpha\sqrt{N}}.\qquad\qquad(6)$$

For $\phi > \phi_{CMC}$, the concentration of monomers remains very close to ϕ_{CMC}, so that $X_N \simeq N e^{-\alpha N}$. This distribution is vanishing for typical values of $\alpha \gtrsim 1$, so that there are very few micelles, which are small! Israelachvili has shown that this implies a 'condensation' of monomers into a single very large bilayer aggregate that consumes all excess monomers above the CMC (Israelachvili 1998). Hence, bilayer-forming amphiphiles quickly organise themselves into large bilayer structures at very small concentrations. Note that, depending on the bending energies, growing bilayer sheets may roll up into vesicles to avoid the costly edge energy at the expense of bending energy.

3.2   Molecular considerations and packing

Amphiphiles have different molecular shapes: a single-tailed surfactant with a large head group resembles a cone, while a double-tailed phospholipid can easily be cylindrical or, if there are enough unsaturated bonds and perhaps a small head, have an inverted cone shape. Cones are more apt to pack into spheres, while cylinders will pack most easily into bilayers. The critical shape factors can be easily estimated.

Consider amphiphiles with head area a, volume v and maximum length d. The number of amphiphiles on the surface of a sphere is $N = 4\pi R^2/a$, which, in a spherical micelle, must equal the number contained within the resulting volume, assuming space filling, $N = 4\pi R^3/(3v)$. This leads to $R = 3v/a$, which cannot exceed d. Hence, the condition for forming a spherical micelles is

$$\text{Packing parameter }\;\frac{v}{ad}\equiv p<\frac{1}{3}\qquad\qquad\text{spheres.}\qquad(7a)$$

Similar arguments for packing amphiphiles into other geometries yield

$$\frac{1}{3} < p < \frac{1}{2} \qquad\qquad \text{cylinders} \qquad\qquad (7\text{b})$$

$$\frac{1}{2} < p < 1 \qquad\qquad \text{bilayers} \qquad\qquad (7\text{c})$$

$$1 < p \qquad\qquad \text{inverted phases.} \qquad (7\text{d})$$

Most two-tailed phospholipids fall into the bilayer class, while surfactants such as SDS and Triton form spherical micelles. SDS also forms cylindrical micelles at higher salt (which provides screening to reduce the effective area/head group). PC, PS and SM all have bilayer-forming shapes with $p < 1$, while PE has a packing parameter closer to $p \simeq 1$ (Israelachvili 1998). Increasing either the tail length or the degree of unsaturation for fixed head group size (e.g., from DMPC to DOPC) leads to larger p.

Given a packing parameter in the correct range, lipids will preferentially self-assemble into a bilayer. Each monolayer itself has a spontaneous curvature, typically away from the water (for $p < 1$). Assembly into a bilayer flattens each monolayer and places them under an effective stress. This and other issues are discussed in more detail in the chapter by Kozlov in this volume.

The different lipids are believed to play a role in the organisation of membrane proteins into the bilayer. Different shape factors p, for example, can facilitate the insertion of proteins with different shapes. Lipid tail lengths can be used to accommodate proteins of different hydrophobic lengths; this can be used, for example, to counteract the tendency of long proteins to tilt in an otherwise thin membrane with relatively short tails. Topological changes in membranes, such as budding, tubulation, and invagination, can be enhanced (or discouraged) by using lipids with specific shapes ideal for the task at hand.

4 Bilayer membrane phases

A pure bilayer is fluid at high temperatures, and solidifies at lower temperatures at the so-called *main* transition temperature T_m. In between, there are several possible phases: see Koynova and Caffrey (1998) for a comprehensive survey of the PC family. There is some confusion in the literature about the different possible lipid phases obtainable upon cooling. However, the various possibilities include:

1. L_α: Fluid, or *liquid crystalline*, phase. At high temperature, the tails are liquid and disordered. There is an appreciable orientational order parameter S (which could be measured from the C-C bond orientations relative to the layer normal) due to the stretching necessary to relieve the hydrophobic free-energy cost of the tails. The transition into the fluid phase from below is often denoted the main transition.

2. L_β gel phase (Tristram-Nagle *et al.* 2002). Below T_m, the tails would attempt to be crystalline, which is likely to be frustrated by the head groups.

3. L'_β or *tilted* phase. The gel phase can tilt relative to the layer normal if the head group is too large for efficient packing.

4. P'_β phase. The tilted phase can develop an asymmetric ripple at higher temperatures, with a wavelength of the order of 100 Å (Sun *et al.* 1996), at temperatures slightly above the L'_β phase. One explanation for this is a coupling between molecular tilt and layer bend, together with chirality (Lubensky and MacKintosh 1993). The transition into this phase from the L'_β phase has also been called the 'pre-main' temperature.

The main transition temperature increases with increasing chain length, because the chains are less inhibited by the aqueous surface. Note that the main transition temperature for PCs, as least, has been extrapolated to infinite chain length, to reach $T_{m\infty} \simeq 430$ K, close to the melting temperature of bulk polyethylene (Koynova and Caffrey 1998). Increasing the degree of unsaturation also frustrates chain packing, inhibiting the gel/crystal phase and decreasing the main transition temperature. The resulting membrane is generally more fluid and has lower elastic moduli. Mixtures of lipids can thus lead to very complex phase diagrams, analogous to those found in metallurgy. In the absence of cholesterol, phase coexistence generally occurs between fluid and ordered (gel/crystal) phases. Liquid-liquid phase separation is generally quite difficult to obtain in these materials because the tails are typically chemically very similar, and have many (entropic) degrees of freedom. Hence, head group incompatibilities generally do not induce phase separation into separate liquids.

However, cholesterol can induce liquid-liquid phase separation. A typical model system contains equal moles of cholesterol, DOPC and a saturated lipid such as SM (de Almeida *et al.* 2003, Veatch and Keller 2003, Dietrich *et al.* 2001, Bagatolli and Gratton 2000). The two liquid phases have different degrees of orientational order, with the more ordered phase thought to be richer in cholesterol and saturated lipids. The liquid ordered phase was first suggested in NMR by Vist and Davis (Davis and Vist 1990) on cholesterol-DMPC mixtures. Cholesterol can have several possible effects on the lipid phases: (1) The planar ring structure encourages saturated chains to pack close by, leading to a segregation of species and enhanced tail order; (2) cholesterol will act as an impurity on any gel/crystal phase, both because it is a different molecule and because it is generally shorter than most lipids, so flexible lipids can more easily fill in excess space inside the cholesterol molecules. This should disrupt gel/crystal phases and increase T_m; (3) the small OH polar head may have specific interactions with certain lipids (such as SM), which may further induce segregation.

There have been few complete studies of ternary phase diagrams of membranes (Veatch and Keller 2003, de Almeida *et al.* 2003, Dietrich *et al.* 2001, Bacia *et al.* 2004, Kahya *et al.* 2004), but the general findings are that cholesterol stabilises a liquid-ordered phase rich in saturated lipids and cholesterol, and allows for two-phase regions of gel-fluid and fluid-fluid, as well as possibly a three-phase region. The nature of the liquid-order phase is still controversial; it has been suggested to be an ordinary liquid, to consist of 'clusters' of, for example, associated lipid/cholesterol molecules (Radhakrishnan and McConnell 1999) or to consist of a 'superlattice' of cholesterol and lipids at a particular stoichiometric ratio (Somerharju *et al.* 1999). Models for this behaviour include fairly detailed phenomenological models that incorporate the acyl chain degrees of freedom (Pink *et al.* 1980, Nielsen *et al.* 1999, Ipsen *et al.* 1987), as well as more coarse-grained Landau-type approaches, based on an orientational or structural order parameter (Priest 1980, Komura *et al.* 2004).

5 Membrane energies

Fluid membranes have attracted much interest from physicists as an ideal model two-dimensional system with a rich variety of possible energies, dynamics and topologies. Almost embarrassingly, cells provide a fertile playground within which to explore these possible types of behaviour. The starting point for any quantitative description is the relevant energies. Because membranes retain their integrity under deformation and flows, they can be described by a minimal number of parameters in much the same way that a cohesive solid may be described in terms of its shear and compression moduli. Hence, we are led to describe the energy of a membrane in terms of its area A, and its shape, parametrised by the local curvature. The fundamental energetic quantities are outlined next, and are discussed in more detail in Kozlov's chapter in this volume.

5.1 Surface tension, frame tension

The surface or frame tension γ penalises overall increases in free energy G to linear order in the area:

$$dG_{\text{surf}} = \gamma \, dA, \tag{8}$$

where A is the surface area and dG is the Gibbs free energy. If a membrane equilibrates by self assembly to reach a local minimum in the free energy as a function of membrane area, then the surface tension γ vanishes. This condition, thus, holds for equilibrium membrane structures such as a free lamellar phase.

A vesicle generally does not have vanishing surface tension. It will contain a certain number of molecules in the bilayer, as well as a volume inside. Changes in membrane shape give rise to a change in area, but must be performed in such a way as to constrain the total number of molecules in the layer, together with the enclosed volume. These constraints lead to a tension γ conjugate to area changes, as well as a pressure difference. Hence, a vesicle shape is given by examining the fluctuations in the following energy,

$$G = G_0 - \tilde{\gamma}A - pV, \tag{9}$$

where $\tilde{\gamma}$ and p, respectively, enforce total membrane area (assuming no stretching) and enclosed volume. This tension, which is due to constraints placed on the membrane shape, is often called a 'frame tension', and plays the same role mathematically as a surface tension. One way to tune $\tilde{\gamma}$ is by osmotically changing a vesicle's volume.

A Gibbs-Duhem relation applies to membranes, similar to the one for bulk fluids, which relates γ to the chemical potential:

$$d\gamma + \sum_i c_i \, d\mu_i = 0, \tag{10}$$

where c_i is the concentration of species on the surface. Hence, the surface tension is intimately related to the monomer chemical potential. The surface tension γ regulates the addition of more lipid to the membrane at the expense of other forms of lipid, such as free monomers or micelles, while the frame tension $\tilde{\gamma}$ parametrises the work done in increasing the area against a specific form of external constraint. In practice, the two forms are mathematically similar, both physically lead to the same effects and are denoted by the same symbol (here, γ) in most situations.

The plasma membrane can thus be expected to be under some tension, depending on the local conditions of the cell, as well as because it is generally not in equilibrium, and is subject to external influences such as osmotic pressure differences, an elastic attached (and fluctuating) cytoskeleton, external hydrodynamic drag (rolling red blood cell), etc. Other membranes under tension include tubules formed from the Golgi during dynamic processes.

5.2 Stretching elasticity

For strong in-plane deformations, a stretching modulus resists changes to the area per head group. This is primarily due to the large hydrophobic cost of exposing the acyl tail groups. The energy is given by

$$\frac{G_{str}}{A_0} = \frac{1}{2}k_s \left(\frac{A - A_0}{A_0}\right)^2, \tag{11}$$

where A_0 is the original area and A the area after deformation. (k_s is referred to as E in Kozlov's chapter in this volume). For inhomogeneous stretching, this is given by

$$G_{str} = \frac{1}{2}k_s \int d^2A_0 \left(\frac{a - a_0}{a_0}\right)^2, \tag{12}$$

where a_0 and a are the local areas per head group before and after deformation. A typical value for k_s for DOPC is $k_s = 0.2$ J/m$^2 \simeq 400k_BT/$ nm^2 at room temperature; this is of the order of twice the hydrophobic energy at the water/hydrocarbon interface.

5.3 Bending elasticity

The bending elasticity penalises shape changes due to internal deformations (changes in chain stretching, inter-head group distances, etc.). It was first written on symmetry grounds by Helfrich (1973), and is given by

$$\frac{G_{\text{bend}}}{\text{Area}} = \frac{1}{2}\kappa \left(\frac{1}{R_1} + \frac{1}{R_2} - C_0\right)^2 + \bar{\kappa}\frac{1}{R_1 R_2}, \tag{13}$$

where R_1 and R_2 are the two (local) principal curvatures. The phenomenological constants here are the *mean curvature modulus* κ, the *Gaussian curvature modulus* $\bar{\kappa}$ and the *spontaneous curvature* C_0. The curvature moduli have units of energy, and the mean curvature modulus has values $\kappa \simeq 30k_BT$ for lipid bilayers.

The Gaussian curvature term has the remarkable property that it is a topological invariant for a given surface: the Gauss-Bonnet theorem states that

$$\int \frac{1}{R_1 R_2} dA = 4\pi (n_s - g), \tag{14}$$

where n_s is the number of distinct disconnected membranes and g is the total number of handles (a torus and a coffee cup have one handle each). Hence, for processes that do not change topology, the Gaussian curvature term can be ignored. However, it does play a role for topological changes. For example, the energy of forming a spherical bud is

$$G_{bud} = 4\pi \left(2\kappa + \bar{\kappa}\right). \tag{15}$$

Remarkably, the scaling of the bending moduli (both scale with energy) implies that the total bending energy of a structure is independent of its size. For a single sheet with an edge, the Gauss-Bonet theorem is

$$\int \frac{1}{R_1 R_2} \, dA = 2\pi - \oint_{\mathbf{R}(s)} ds \, k_g, \tag{16}$$

where

$$k_g = \frac{d^2\mathbf{R}}{ds^2} \cdot \left(\widehat{n} \times \frac{d\mathbf{R}}{ds}\right) \tag{17}$$

is the geodesic curvature. Here, $\mathbf{R}(s)$ is the curve that bounds the membrane patch and $\widehat{n}$ is the normal to the membrane.

The Helfrich free energy is the minimal energy necessary for describing deformations: for situations (such as necking or budding) that require very sharp bends, this energy should be extended to higher-order powers of the total $(R_1^{-1} + R_2^{-1})$ and Gaussian $((R_1 R_2)^{-1})$ curvatures. In addition to these fundamental energies, membranes that possess degrees of internal order such as lipid tilt will have anisotropic bending moduli and elasticity associated with the deformation of the tilt direction (Fournier 1999).

The spontaneous curvature of a symmetric bilayer vanishes by symmetry. However, most bilayers in a biological context have an asymmetry: the inside/outside of the plasma membrane, the interior and exterior of the Golgi apparatus and ER. These asymmetries may include different chemical environments, ionic strengths, species and concentrations of proteins and dynamic interactions. Hence, it is likely that membrane compositions are asymmetric across the two leaflets of the bilayer in most cases, which would then result in a spontaneous curvature. The modification of C_0 (e.g., due to helper proteins) may aid topological changes such as budding and tubulation.

6   Fluctuations

Fluctuations around a flat state are most easily calculated within the so-called *Monge gauge*, in terms of a single height field $h(x, y)$. In this representation, the local area is given by $dA = 1 + \frac{1}{2}(\nabla h)^2 + \ldots$ and the curvature by $\nabla^2 h$. So, ignoring the Gaussian curvature, the energy relative to the flat state is given by

$$G = \int dx \, dy \left[\frac{1}{2}\gamma \left(\nabla h\right)^2 + \frac{1}{2}\kappa \left(\nabla^2 h - C_0\right)^2\right]. \tag{18}$$

Excess area — For small bending moduli, membranes under weak tension have appreciable thermal fluctuations. The *excess area* relative to the flat state can be calculated from Equation 18 and the equipartition theorem to yield

$$\delta A \equiv A - A_0 = \frac{1}{2} \int dx dy \, (\nabla h)^2 = \frac{1}{2} L^2 \sum_q \langle |h(q)|^2 \rangle \tag{19a}$$

$$= \frac{k_B T}{8 \pi \kappa} \ln \left[\frac{\left(\frac{\pi}{a}\right)^2 + \gamma/\kappa}{\left(\frac{\pi}{L}\right)^2 + \gamma/\kappa} \right], \tag{19b}$$

where $h(r) = \sum_q h(q) e^{i\mathbf{q} \cdot \mathbf{r}}$, a is a microscopic cutoff of order the headgroup spacing, and L the lateral size of the membrane. In this case, the excess area is entirely due to thermal fluctuations, and the tension γ 'irons out' these fluctuations. At extremely high tensions the layer begins to stretch, and the area is further limited by the stretching modulus k_s:

$$\delta A = \frac{k_B T}{8 \pi \kappa} \ln \left[\frac{\left(\frac{\pi}{a}\right)^2 + \gamma/\kappa}{\left(\frac{\pi}{L}\right)^2 + \gamma/\kappa} \right] + \frac{\gamma}{k_s}. \tag{20}$$

Evans and co-workers have exploited this relation to measure the bending and stretching moduli using vesicles under known tension applied by a micropipette (Evans and Needham 1987).

The persistence of a tensionless membrane can be easily calculated from Equation 18. The local normal vector is $\hat{\boldsymbol{n}} = -\nabla h$, and the correlation $g_n(\mathbf{r})$ of normal vectors separated by a distance $\mathbf{r}$ is given by

$$g_n(\mathbf{r}) = \langle |\hat{\boldsymbol{n}}(\mathbf{r}) - \hat{\boldsymbol{n}}(\mathbf{0})|^2 \rangle \tag{21a}$$

$$= 2 \sum_q q^2 \langle |h(q)|^2 \rangle (1 - \cos \mathbf{q} \cdot \mathbf{r}) \simeq \frac{k_B T}{\kappa} \ln \left(\frac{r}{a} \right), \tag{21b}$$

from which we can identify the de Gennes-Taupin persistence length (de Gennes and Taupin 1982),

$$\xi \simeq a e^{\frac{4 \pi \kappa}{3 k_B T}}. \tag{22}$$

For most biological membranes with $\kappa \simeq 30 k_B T$ this length is very large, so the membranes can be treated as essentially flat, with very small fluctuations.

7 Domains, shapes and other current issues

As mentioned earlier, membranes take many shapes, including tubules (Golgi, ER), vesicles and flat membranes. Moreover, the many components in membranes can lead to various forms of phase separation, whose domains then influence shapes through their mechanical properties (elasticity, spontaneous curvature, degree of gel or liquid crystalline order). There is an enormous amount of work on domain and shape formation, and only a small amount will be mentioned here. Many workers have studied aspects of membrane shape transitions using the fundamental bending, stretching and tension energies noted earlier: only a few are noted here, for reasons of space (Lipowsky and

Dimova 2003, Jülicher and Lipoesky 1996, Jülicher *et al.* 1993, Kumar *et al.* 2001). This includes budding on vesicles and flat sheets, kinetics of bud formation and the interplay with other internal degrees of freedom. Despite work going back a decade or so, it is only with the advent of confocal fluorescent imaging of labelled giant vesicles undergoing phase separation that concrete experimental data is now available (Bagatolli and Gratton 2000, Dietrich *et al.* 2001, Veatch and Keller 2003, 2002, Kahya *et al.* 2004, Baumgart *et al.* 2003). In addition to myriad studies on model mixtures of synthetic lipids, recent experiments on native lipid mixtures extracted from pulmonary membranes have shown phase separation into two fluid phases (de la Serna *et al.* 2004). Another area of considerable interest now includes tubule formation and dynamics under tension (Roux *et al.* 2002, Upadhyaya and Sheetz 2004, Allain *et al.* 2004), both of homogeneous and mixed lipid membranes.

There are many other areas of active interest and unresolved questions. The liquid ordered phase stabilised by cholesterol has been posited as a candidate phase for *lipid rafts* (Brown and London 1998). Lipid rafts are certain detergent-resistant membrane compositions that are thought by some to play a vital role in signalling and other processes. It is not yet clear how large these structures are, should they exist, or whether they are a stirred macrophase-separated state, a microphase-separated state or only due to proteins that may organise them. This is a large motivation for many studies of model mixed membranes, with the standard mixture being roughly equimolar amounts of cholesterol and a saturated and an unsaturated lipid.

The broad picture of elasticity given here is only the beginning of the story. It remains to relate quantitatively the macroscopic moduli to the degrees of freedom of specific lipids (chain length, degree of saturation, headgroup size, cholesterol content, ...), as well as address the influence of protein inclusions. There has already been much work in both of these directions (Zemel *et al.* 2004, May 2000, Sens and Safran 2000, Dan *et al.* 1994).

References

Alberts B, Johnson A, Lewis J, Raff M, Roberts K and Walter P, 2002, *Molecular Biology of the Cell*, fourth edition (Garland Science, New York).

Allain J M, Storm C, Roux A, Amar M B and Joanny J F, 2004, Fission of a multiphase membrane tube, *Phys Rev Lett* **93** 158104.

Bacia K, Scherfeld D, Kahya N and Schwille P, 2004, fluorescence correlation spectroscopy relates rafts in model and native membranes, *Biophys J* **87** 1034–1043.

Bagatolli L A and Gratton E, 2000, Two photon fluorescence microscopy of coexisting lipid domains in giant unilamellar vesicles of binary phospholipid mixtures, *Biophys J* **78** 290–305.

Baumgart T, Hess S T and Webb W W, 2003, Imaging coexisting fluid domains in biomembrane models coupling curvature and line tension, *Nature* **425** 821–824.

Boal D, 2002, *Mechanics of the Cell* (Cambridge University Press, Cambridge).

Brown D A and London E, 1998, Functions of lipid rafts in biological membranes, *Annu Rev Cell Dev Biol* **14** 111–136.

Dan N, Berman A, Pincus P and Safran S A, 1994, Membrane-induced interactions between inclusions, *J Phys II (France)* **4**(10) 1713–1725.

Davis J H and Vist M R, 1990, Phase equilibria of cholesterol/dipalmitoylphosphatidylcholine mixtures: 2h nuclear magnetic resonance and differential scanning calorimetry, *Biochem* **29** 451–464.

de Almeida R F M, Fedorov A and Prieto M, 2003, Sphingomyelin/phosphatidylcholine/cholesterol phase diagram: Boundaries and composition of lipid rafts, *Biophys J* **85** 2406–2416.

de Gennes P G and Taupin C, 1982, Microemulsions and the flexibility of oil/water interfaces, *J Phys Chem* **86** 2294–2304.

de la Serna J B, Perez-Gil J, Simonsen A C and Bagatolli L A, 2004, Cholesterol rules - direct observation of the coexistence of two fluid phases in native pulmonary surfactant membranes at physiological temperatures, *J Biol Chem* **279** 40715–40722.

Dietrich C, Bagatolli L A, Volovyk Z N, Thompson N L, Levi M, Jacobson K and Gratton E, 2001, Lipid rafts reconstituted in model membranes, *Biophys J* **80** 1417–1428.

Evans E and Needham D, 1987, Physical properties of surfactant bilayer membranes: thermal transitions, elasticity, rigidity, cohesion and colloidal interactions, *J Phys Chem.* **91** 4219.

Fournier J B, 1999, Microscopic membrane elasticity and interactions among membrane inclusions: interplay between the shape, dilation, tilt and tilt-difference modes, *Eur Phys J B* **11** 261–272.

Hamley I W, 1998, *The Physics of Block Copolymers* (Oxford University Press, Oxford).

Helfrich W H, 1973, Elastic properties of lipid bilayers: theory and possible experiments, *Z Naturforsch* **28c** 693–703.

Holopainen J M, Metso A J, Mattila J P, Jutila A and Kinnunen P K J, 2004, Evidence for the lack of a specific interaction between cholesterol and sphingomyelin, *Biophys J* **86** 1510–1520.

Howard J, 2001, *Mechanics of Motor Proteins and the Cytoskeleton* (Sinauer Associates, Sunderland MA).

Ipsen J H, Karlstrom G, Mouritsen O G, Wennerstrom H and Zuckermann M J, 1987, Phase-equilibria in the phosphatidylcholine-cholesterol system, *Biochim Biophys Acta* **905** 162–172.

Israelachvili J N, 1998, *Intermolecular and Surface Forces* (Academic Press, London).

Jülicher F and Lipowsky R, 1996, Shape transformations of vesicles with intramembrane domains, *Phys Rev E* **53** (3) 2670–2683.

Jülicher F, Seifert U and Lipowsky R, 1993, Phase-diagrams and shape transformations of toroidal vesicles, *J Phys II (France)* **3** (11) 1681–1705.

Kahya N, Scherfeld D, Bacia K and Schwille P, 2004, Lipid domain formation and dynamics in giant unilamellar vesicles explored by fluorescence correlation spectroscopy, *J Struct Biol* **147** 77–89.

Komura S, Shirotori H, Olmsted P D and Andelman D, 2004, Lateral phase separation in mixtures of lipids and cholesterol, *Europhys Lett* **67** 321–327.

Koynova R and Caffrey M, 1998, Phases and phase transitions of the phosphatadycholines, *Biochem Biophys Acta* **1376** 91–145.

Kumar P B S, Gompper G and Lipowsky R, 2001, Budding dynamics of multicomponent membranes, *Phys Rev Lett* **86** 3911–3914.

Lipowsky R and Dimova R, 2003, Domains in membranes and vesicles, *J Phys Condes Matter* **15** S31–S45.

Lubensky T C and MacKintosh F C, 1993, Theory of ripple phases of lipid bilayers, *Phys Rev Lett* **71** (10) 1565–1568.

Maeda M, 2004, Membranes and transport, in *Medical Biochemistry* chapter 7, eds Baynes J and Dominiczak M H (Mosby, Basildon).

May S, 2000, Theories on structural perturbations of lipid bilayers, *Curr Opin Colloid Interface Sci* **5** 244–249.

Mayor S and Rao M, 2004, Rafts: Scale-dependent, active lipid organization at the cell surface, *Traffic* **5** 231–240.

McConnell H M and Vrljic M, 2003, Liquid-liquid immiscibility in membranes, *Annu Rev Biophys Biomolec Struct* **32** 469–492.

Nielsen M, Miao L, Ipsen J H, Zuckermann M J and Mouritsen O G, 1999, Off-lattice model for the phase behavior of lipid-cholesterol bilayers, *Phys Rev E* **59** 5790–5803.

Pandit S A, Vasudevan S, Chiu S W, Mashl R J, Jakobsson E and Scott H L, 2004, Sphingomyelin-cholesterol domains in phospholipid membranes: Atomistic simulation, *Biophys J* **87** 1092–1100.

Pink D A, Green T J and Chapman D, 1980, Raman scattering in bilayers of saturated phosphatidylcholines, *Biochemistry* **19** 349–356.

Priest R G, 1980, Landau phenomenological theory of one and two component phospholipid bilayers, *Mol Cryst Liq Cryst* **60** 167–184.

Radhakrishnan A and McConnell H M, 1999, Condensed complexes of cholesterol and phospholipids, *Biophys J* **77** 1507–1517.

Roux A, Cappello G, Cartaud J, Prost J, Goud B and Bassereau P, 2002, A minimal system allowing tubulation with molecular motors pulling on giant liposomes, *Proc Natl Acad Sci USA* **99** 5394–5399.

Safran S A, 1994, *Statistical Thermodynamics of Surfaces, Interfaces and Membranes* (Addison-Wesley, Reading MA).

Sens P and Safran S A, 2000, Inclusions induced phase separation in mixed lipid film, *Eur Phys J E* **1** 237–248.

Sharma P, Varma R, Sarasij R C, Ira, Gousset K, Krishnamoorthy G, Rao M and Mayor S, 2004, Nanoscale organization of multiple gpi-anchored proteins in living cell membranes, *Cell* **116** 577–589.

Slotte J P, 1999, Sphingomyelin-cholesterol interactions in biological and model membranes, *Chem Phys Lipids* **102** 13–27.

Somerharju P, Virtanen J A and Cheng K H, 1999, Lateral organisation of membrane lipids - the superlattice view, *BBA-Mol Cell Biol Lipids* **1440** 32–48.

Sun W J, Tristram-Nagle S, Suter R M and Nagle J F, 1996, Structure of the ripple phase in lechthin bilayers, *Proc Nat Acad Sci* **93** 7008–7012.

Tristram-Nagle S, Liu Y F, Legleiter J and Nagle J F, 2002, Structure of gel phase dmpc determined by x-ray diffraction, *Biophys J* **83** 3324–3335.

Upadhyaya A and Sheetz M P, 2004, Tension in tubulovesicular networks of golgi and endoplasmic reticulum membranes, *Biophys J* **86** 2923–2928.

Varma R and Mayor S, 1998, GPI-anchored proteins are organized in submicron domains at the cell surface, *Nature* **394** 798–801.

Veatch S L and Keller S L, 2002, Organization in lipid membranes containing cholesterol, *Phys Rev Lett* **89** 268101.

Veatch S L and Keller S L, 2003, Separation of liquid phases in giant vesicles of ternary mixtures of phospholipids and cholesterol, *Biophys J* **85** 3074–3083.

Zemel A, Ben-Shaul A and May S, 2004, Membrane perturbation induced by interfacially adsorbed peptides, *Biophys J* **86** 3607–3619.

Some aspects of membrane elasticity

Michael M Kozlov

Department of Physiology and Pharmacology, Tel Aviv University
Ramat Aviv 69978, Tel Aviv, Israel

michk@post.tau.ac.il

1 Introduction

Phospholipids referred in the following as lipids serve as building blocks of membranes of cells and cell organels. Stimulated by their biological relevance, the membranes consisting of lipid molecular layers are the subject of extensive experimental and theoretical studies. Recent successful applications of advanced experimental techniques revealed a broad spectrum of membranes shapes formed spontaneously by various classes of lipids in the process of their self-assembly, and made it possible to study quantitatively the elastic properties of the membranes of different configurations. Attempts to understand the physics of these systems require development of sophisticated theoretical models treating the elastic behaviour of lipid membranes characterised by high and inhomogeneous curvatures. Consideration of some of these theories is the goal of this overview. In the following, I give a short phenomenological description of structures formed by the lipid molecules and survey the main theoretical ideas involved in the analysis of these structures.

1.1 Physical origin and spontaneous shapes of lipid monolayers

Phospholipids belong to the class of substances referred to as the amphiphiles. A lipid molecule, as the molecules of other amphiphiles, is composed of a hydrophilic polar head and one or two hydrophobic hydrocarbon chains. The combined hydrophilic-hydrophobic nature of amphiphiles determines their propensity to self-organise in molecular monolayers, where all the molecules are oriented in the same direction, the polar heads towards

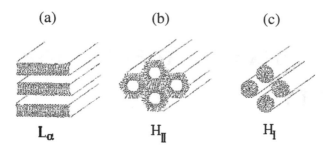

Figure 1. *Illustration of mesophases of surfactants. (a) Lamellar phase. (b) Inverted hexagonal (H$_{II}$) phase. (c) Normal hexagonal (H$_I$) phase.*

one side and the hydrophobic chains towards the other side of the monolayer. An effective driving force of self-organisation of amphiphiles is called the hydrophobic effect (Tanford 1980, Cevc and Marsh 1987). In this section I briefly review the shapes of the spontaneously formed monolayers, which are not subjected to any external force.

The geometry of an amphiphile monolayer can be qualitatively described by its thickness δ, area A, and characteristic radius r of its curvature. In the following I will characterise the monolayer curvature in detail. The thickness of a monolayer is approximately equal to the characteristic length of an amphiphile molecule and is of the order of 1 nm. The area depends on the conditions of monolayer formation. In the present work I will consider extended monolayers where linear dimensions (square root of the area) are much larger than the thickness. Such monolayers can be treated as surfaces.

The curvature depends mainly on the kind of amphiphile building the monolayer and, as I will show, varies in a very broad range. A natural limit of possible curvatures of a monolayer is given by its thickness δ. Indeed, the radius of curvature should not be smaller than the thickness, $r \geq \delta$. Therefore, I will use δ as a measure of the monolayer bending. The monolayer will be regarded as strongly bent if its radius is close to the limit, $r/\delta \simeq 1$, and as weakly bent in the case $r/\delta \gg 1$.

As curvature plays an important role in the present study, I consider in more detail the related phenomenology.

1.2   Amphiphile monolayers in water and oil system

In a system of water and a hydrophobic liquid (oil), the amphiphile monolayers cover the interfaces between the two immiscible fluids so that the hydrocarbon chains contact the hydrophobic phase while the polar heads are exposed to water. The amphiphile monolayer gives rise to a drastic decrease of the surface tension and a related growth of the area of the interface between water and oil. This results in formation of such systems as emulsions and microemulsions. The latter proved to be promising for industrial applications and, therefore, have been extensively investigated during the recent period (for a review, see Safran 1994). The amphiphile monolayers in microemulsions are stable thermodynamically and in general strongly curved. The bending can be in either direction and is usually controlled by the temperature of the system. At high temperatures the monolayers are bent towards the hydrophobic phase so that droplets of water in oil are formed (Winsor I microemulsion). At low temperatures the bending is directed towards water, and the resulting phase consists of droplets of oil in water (Winsor II microemulsion). In the intermediate range of temperatures the monolayer adopts a sponge-like configuration (bicontinuous microemulsion), where the local shape of the monolayer is saddle-like. It is important to stress that at all temperatures, independent of the direction of the bending, the absolute characteristic value of the radius of curvature of the monolayer surface is of an order of 10 nm, i.e., the amphiphile monolayers in microemulsions may be viewed as strongly bent.

1.3   Amphiphile monolayers in pure water

If the amphiphile molecules are added to pure water (in the absence of hydrophobic liquid), they self-assemble in such a way that the hydrophobic tails are protected by the polar heads from contact with water. The resulting structures, called mesophases, are characterised by particular shapes of the constituting monolayers and depend of the amphiphile type. The physics of these systems was studied by using various amphiphiles known in physical chemistry and the whole range of lipids constituting the cell membranes (for a review, see Gelbart *et al.* 1993, Lipowsky and Sackmann 1995).

If the amount of water is limited (or in special stabilising conditions), the amphiphile mesophases form lyotropic liquid crystals (for a review, see Luzzati 1968, Gruner 1989, Seddon and Templer 1993, Rand and Fuller 1994, Koynova and Caffrey 1994), while in excess water the mesophases can be seen as isotropic solutions of amphiphile aggregates. As a large number of mesophases has been described, I will mention here just some of them characterised by the simplest and most common shapes of the monolayers. The amphiphiles can be classified according to the type of mesophases that they form (Israelachvili 1985).

The class of so-called 'bilayer' amphiphiles includes various common lipids such as lecithins and many other amphiphilic compounds. The molecules of this kind tend to self-organise in weakly curved monolayers. Coupling of such monolayers by the hydrophobic effect results in the formation of flat bilayer membranes. In a limited amount of water the flat bilayers build a lamellar phase consisting of a stack of alternating membranes and water layers, the latter of a thickness of few nanometres, see Figure 1(a) (Rand, Web site). If the water content of the system strongly exceeds the amount of amphiphile, the resulting phase is a suspension of vesicles, i.e., closed 'bags' formed by a lipid bilayer (Lasic 1995). Since an extensive literature exists on vesicles (Lasic 1995), I will just mention that in most cases the vesicles are large so that their bilayers can be regarded as practically flat. However, it is possible to produce small vesicles whose membranes are considerably bent (radius of curvature of 20 – 30 nm). Although the global thermodynamic stability of vesicles is still a matter of debate, in practice, they often last for long periods of time.

The amphiphiles attributed to the next class can be called 'hexagonal'. A typical example of a lipid belonging to this group is dioleoyl-phosphatidyl-ethanolamine (DOPE). A monolayer formed spontaneously by a 'hexagonal' amphiphile is strongly curved and has the shape of thin cylinder having the radius of the order of a few nanometres. The polar heads of the amphiphile molecules are oriented towards the internal space of the cylinder, which is filled with water. By convention, this direction of monolayer bending is referred to as the inverted direction. The long inverted cylinders are packed parallel to each other so that their cross sections form a two-dimensional hexagonal lattice. The resulting mesophase is called the inverted hexagonal (H_{II}) phase (Figure 1(b)).

To the third class belong the amphiphiles that can be called 'micellar'. They comprise lysolipids, a particular type of lipids having only one hydrocarbon chain per molecule. In addition, a wide range of well-known surfactants (detergents), whose molecules have relatively small hydrophobic parts and large polar heads, can be classified as micellar substances. The monolayers of these compounds are highly curved in the direction opposite to that of the 'hexagonal' amphiphiles and tend to form the cylinders or spheres with internal volumes filled by the hydrocarbon chains. This direction of bending is conventionally defined as normal. The radii of the resulting monolayers are prescribed by the monolayer thickness ($\simeq$ 1nm). In a limited amount of water such compounds can form the normal hexagonal (H_I) phases consisting of hexagonally packed cylinders (Figure 1(c)). In excess water they form suspensions of micelles. While the problem of possible shapes of the micelles is not yet completely solved, the existence of long cylindrical micelles has been proved by different experimental methods (for a review, see Lichtenberg *et al.* 2000).

Finally, there exists a broad class of amphiphiles forming bicontinuous cubic phases (for a review, see Seddon and Templer 1995), which in some cases are stable in excess water. To a good approximation, the membranes of these cubic phases form minimal surfaces (Nitsche 1989). The characteristic radii of curvature of cubic-phase monolayers are of the order of several nanometres, which means they are strongly curved.

These short phenomenological considerations aim to illustrate the wide spectrum of shapes adopted spontaneously by amphiphile monolayers. In particular, they show that the degree of monolayer spontaneous bending varies from low values in the case of the flat bilayers of lamellar phases to high values in the hexagonal, micellar and cubic phases, where the radii of curvature approach the monolayer thickness.

1.4   Experimental studies of elasticities of strongly curved membranes

The spontaneous shapes considered above are assumed by the amphiphile monolayers in the process of self-assembly and may be regarded as shapes of minimal elastic energy. Deviations of the monolayer shape from the spontaneous one will, therefore, increase the free energy of the system. Common reasons for such deformations are thermal fluctuations. Another possibility is the deformation of the membrane by external forces. The free energy of deformations is controlled by the elastic coefficients and plays a basic role in analysis of statistical and mechanical behaviour of the membranes.

While various experimental methods have been applied to measure the elasticities of the weakly bent membranes of lipid vesicles and biological cells (for reviews, see Evans and Skalak 1979, Helfrich 1990), the elastic properties of the strongly curved monolayers of hexagonal and cubic mesophases have not been accessible to measurements for a long time. This has changed with the development of methods allowing the study of deformations as a function of external forces (for a review, see Parsegian *et al.* 1986).

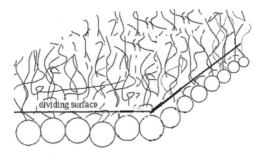

Figure 2. *Dividing surface of an element of interface.*

The main idea of these methods is to equilibrate the water inside the lipid phase with a second phase of known water activity. In one realisation of this idea, called osmotic stress method, the lipid sample consisting of the inverted hexagonal or cubic phase is exposed to a solution of large polymers, which cannot penetrate the space among the membranes. Equilibration of the chemical potential of water molecules between the lipid phase and the polymer solution results in an osmotic pressure compressing the lipid phase. The value of the applied osmotic pressure is controlled by the concentration of the polymer solution, while the resulting deformation of the lipid phase is measured with very high accuracy by x-ray diffraction (for a review, see Rand and Fuller 1994).

Another procedure to control the water content, called the gravimetric method, consists in equilibrating the lipid phase with a vapour of known relative humidity (for a review, see Parsegian *et al.* 1986).

The two methods provide the relationships between the lipid-phase deformations and the applied pressure. However, the interpretation of the results in terms of the elastic coefficients of a strongly curved monolayer requires a detailed theoretical analysis.

2   Gibbs' description

2.1   Dividing surface

The problem of dealing with the elastic properties of a layer of small but finite thickness is not specific to amphiphile monolayers. It originated from attempts to treat all kinds of transition regions between immiscible liquid phases. Such interfaces are usually few molecules (at least several angstroms) thick and have an internal structure and physical properties different from those of bulk phases that they separate. Therefore, I will discuss here the elasticity of interfaces in general before turning to the amphiphile monolayer as a particular interface with a thickness equal to the length of the amphiphile.

The theory of interfaces has been addressed first by Gibbs and later by many others (Gibbs 1876, 1878, for a review, see Murphy 1966). Two ways to approach this problem have been suggested. The first consists in considering the interface as a thin layer of a special volume phase separated from the bulk phases by two boundary surfaces. The

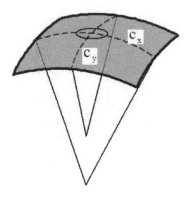

Figure 3. *The principle curvatures, C_x and C_y, of a curved surface.*

second way, which I will mainly use in this overview, has been proposed by Gibbs and may be called the surface excess approach.

In Gibbs' approach one selects inside the transition region a surface characterising the orientation and location of the interface, referred to as the dividing surface (Figure 2). The interface is treated as a geometric surface coinciding with the dividing surface and is characterised by physical quantities to be defined below. In the case of an amphiphile monolayer, the dividing surface may be placed inside the monolayers with its normal vector pointing from the hydrocarbon tails towards the polar heads.

To define the effective thermodynamical properties of the interface such as its entropy S^s, internal energy U^s and masses of the components m_i^s, Gibbs defined a reference system by extrapolating the bulk phases with their unchanged properties up to the dividing surface. The excess thermodynamical values S^s, U^s and m^s, i.e., the differences between the thermodynamical values of the real system and those of the reference system, are assigned to the dividing surface.

Note that in this scheme only the orientation of the dividing surface is fixed, while the exact position of the dividing surface is arbitrary. Therefore, one can choose from an infinite number of dividing surfaces that locally are parallel to each other. Gibbs' method can be applied to any dividing surface. However, as discussed in the following, there are physical reasons for selection of particular dividing surfaces, which simplify the whole description.

2.2   Gibbs' energy of interface

Consider an element of the dividing surface of area A. The shape of the element can be characterised by two principal curvatures c_x and c_y, which are supposed to be homogeneous over the element surface (Figure 3). The three variables, A, c_x and c_y, completely determine the local geometric state of the surface.

A thermodynamic equation relating the changes of the internal energy of the interface to variations of its entropy, the masses of the components and the geometrical characteristics has been introduced by Gibbs in the form

$$dU^s = T \cdot dS^s + \sum_i \mu_i \cdot dm_i^s + \gamma \cdot dA + C_x \cdot dc_x + C_y \cdot dc_y. \qquad (1)$$

The contributions to Equation 1 specific to the thermodynamics of interfaces are related to the changes of the geometrical characteristics of the dividing surface and are given by the last three terms. The parameters associated in Equation 1 with the geometric variables are the tension γ (called surface tension or lateral tension, respectively, depending on whether or not the components exchange between the interface and the bulk phases), and C_x and C_y, which have meanings of bending moments. The tension and the bending moments represent the interfacial stresses. Our notations differ slightly from Gibbs'.

All the variables entering Equation 1 depend on the position of the dividing surface. For the geometric variables, one can see this directly by comparing the area and curvatures of the dividing surfaces positioned at different levels across the interface. It is important to emphasise that Equation 1 is valid for an arbitrary dividing surface.

Gibbs showed that Equation 1 can be simplified by a suitable choice of the dividing surface. He rewrote it in the form

$$U^s = T \cdot dS^s + \sum_i \mu_i \cdot dm_i^s + \gamma \cdot dA + \frac{1}{2}(C_x + C_y) \cdot d(c_x + c_y) + \frac{1}{2}(C_x - C_y) \cdot d(c_x - c_y),$$
$$(2)$$

and considered the case of slight deviations of the surface from the initially flat shape with $c_x = 0, c_y = 0$. Considering the transition to spherical shapes, $c_x = c_y$, Gibbs demonstrated the existence of a particular dividing surface for which the sum of the bending moments vanishes, $C_x + C_y = 0$. The last term in Equation 2 has been shown to be negligibly small if the radius of curvature strongly exceeds the thickness of the transition region. Treating the interface in terms of this dividing surface called the 'Gibbs' surface of tension', one has to consider only the effect of tension, while the contributions of the bending vanish.

In modern studies on interfacial phenomena, Gibbs' theory has been reformulated in terms of variables that are more convenient for applications than the principal curvatures. The energy has been expressed as function of the total curvature $J = c_x + c_y$ and the Gaussian curvature $K = c_x \cdot c_y$. An advantage of these variables becomes evident if one considers the complicated geometries of the surface different from spherical or cylindrical ones. J and K are independent invariants, i.e., scalars that can be derived from the tensor of curvature. On the other hand, the description in terms of the total and Gaussian curvatures is completely equivalent to that using the principal curvatures as variables. It proved to be convenient to present the bending moments corresponding to J and K as intensive thermodynamical variables so that Gibbs' equation expressed through the new variables is

$$dU^s = T \cdot dS^s + \sum_i \mu_i \cdot dm_i^s + \gamma \cdot dA + C_1 \cdot A \cdot dJ + C_2 \cdot A \cdot dK. \qquad (3)$$

The stresses C_1 and C_2 will be called in this overview the first and the second bending moments, respectively. They are related to Gibbs' bending moments by the relationships $C_x = A\left[C_1 + \frac{1}{2}C_2(J - \sqrt{J^2 - 4K})\right]$ and $C_y = A\left[C_1 - \frac{1}{2}C_2(J + \sqrt{J^2 - 4K})\right]$.

Obviously, the interfacial stresses can be defined as the partial derivatives of the interfacial energy,

$$\gamma = \frac{\partial U^s}{\partial A}|_{S^s,m_i^s,J,K} \qquad (4)$$

$$C_1 = \frac{1}{A}\frac{\partial U^s}{\partial J}|_{S^s,m_i^s,A,K} \qquad (5)$$

$$C_2 = \frac{1}{A}\frac{\partial U^s}{\partial K}|_{S^s,m_i^s,A,J}. \qquad (6)$$

Analogously to the Gibbs' surface of tension, a particular dividing surface can be found where one of the bending moments, C_1 or C_2, or a particular combination of them vanishes and Equation 3 is simplified.

Gibbs' interfacial theory (Equations 1 – 3) is based on the tension and bending moments is general, and does not involve any assumptions limiting its applications. However, the application of this theory to the analysis of particular problems requires further development of the Gibbs approach often at the expense of generality.

One of the main directions of such development is related to description of energetics and equilibrium shapes of interfaces characterised by small or even vanishing lateral tension and high curvatures, such as the amphiphile monolayers forming the vesicles, the hexagonal and cubic phases or the micelles. With regard to these systems, Gibbs' theory in its initial form is insufficient for the following reasons.

First, it accounts for changes of the energy only in the first order in the variations of area and curvatures (called in the following the deformations of the interface). In the cases of vanishing stresses of the amphiphile monolayers, the contributions to the energy are of higher-order in the deformations and are related to the monolayer elastic moduli. Description of the higher-order contributions to the energy goes beyond Gibbs' consideration.

Another reason is that Gibbs' approach in its generality does not use explicitly any information about the structure of the interface. In particular, it does not give any insight into possible dependence of the tension and the bending moments on the distribution of microscopic stresses inside the interface. On the other hand, the experiments provide in many cases information on the changes of interactions between the polar heads or on modifications of packing of the hydrocarbon chains. Examples for that are the alteration of electric charge of polar heads of amphiphile molecules by the change of the ionic strength or pH of the electrolyte solution facing the monolayer, and modifications of the hydrocarbon chains by introducing the unsaturated bonds. In the framework of the original Gibbs' approach, one cannot explicitly account for the changes of energy related to specific modifications of the interfacial structure.

The last and more formal reason is that Gibbs' surface of tension falls within the interface only if the tension is sufficiently large. In the cases of small tensions this surface is placed far away from the physical location of the interface, and in the limit of vanishing tension it is infinitely removed. Moreover, the negligibility of the second terms in Equation 2 requires that the radius of interfacial curvature be small comparing to the thickness of the interface; in the opposite case this contribution has to be taken into account. As a result, it is in most cases inconvenient to use Gibbs' surface of tension for treatment of the amphiphile monolayers, and Equations 1 – 3 cannot be simplified.

Another direction of the development of the original Gibbs' approach is related to analysis of thermodynamical equilibrium of the interfaces. Equations of equilibrium with respect to the deformations are called equations of shape as their solutions predict the equilibrium forms of the interfaces. These predictions can be checked experimentally and, therefore, the equations of shape play a primary role in the theory on interfaces.

2.3 Elastic moduli of interface

The interfacial tension γ and the bending moments C_1 and C_2 entering the general thermodynamic equation (Equation 3) are functions of all thermodynamic variables, namely, of the interfacial entropy S^s, masses of the components m_i^s, the area A, the total curvature J and the Gaussian curvature K. I will assume S^s and m_i^s to be constant and concentrate on the functions on the geometrical variables,

$$\gamma = \gamma(A, J, K), \quad C_1 = C_1(A, J, K), \quad C_2 = C_2(A, J, K). \tag{7}$$

I will consider these functions in the elastic approximation. This means that I will assume the deformations to be small and account only for the changes of the interfacial stresses (Equation 7) to first order in the deformations only. The relationships between the stresses and the deformations are determined in this approximation by the elastic moduli of the interface. The related contributions to the energy are of the second order in the deformations and are called the elastic contributions. The elastic moduli are the material parameters of the system determined solely by its internal structure.

The elastic moduli related to the changes of the area A and the total curvature J, called the stretching modulus and the bending modulus, respectively, have been defined in the literature, and their values have been determined experimentally.

2.3.1 Stretching modulus

Gibbs was the first to address the question of elasticity of a liquid film formed by adsorption of at least two components on an interface (Gibbs 1876/78). He defined the stretching modulus E of such an interface as a quantity determining the variation of the tension γ with a change of the area A in the form

$$E = 2A \cdot \frac{\partial \gamma}{\partial A}. \tag{8}$$

The derivative in Equation 8 is calculated at constant numbers of adsorbed molecules of all components.

In considerations of the flat lipid membranes (for reviews, see Evans and Skalak 1979, Helfrich 1990) the stretching modulus, E, has been defined by introducing the elastic energy F_A of change of the membrane area in the form

$$F_A = \frac{1}{2} \cdot E \cdot \frac{(A - A_0)^2}{A_0}, \tag{9}$$

where A_0 is the initial area of the unstressed membrane. This definition is practically the same as that of Gibbs, as can be seen by determining the tension γ from Equation 9 according to Equation 4 and comparing it with Equation 8. The only difference is that the factor two is dropped in the latter definition so that the stretching modulus is

$$E = A \cdot \frac{\partial \gamma}{\partial A}. \tag{10}$$

Experimentally, the stretching modulus has been studied for flat lipid bilayers of various compositions and its values proved to be of the order of 200 dyn/cm.

2.3.2 Bending modulus

The elastic modulus of bending denoted as κ was first introduced by Helfrich (1973) for weakly bent lipid membranes. In the Helfrich model the initial state of the membrane is that of zero curvatures, $J = 0$ and $K = 0$, and the deformation results in a shape with principal curvatures small compared to the inverse thickness of the membrane, $|c_1|, |c_2| \ll \delta^{-1}$. The model accounts for the contributions to the energy up to the second order in the small parameter $c \cdot \delta$ so that the energy per unit area is

$$f_b = \frac{1}{2}\kappa(J - J_s)^2 + \bar{\kappa} \cdot K, \tag{11}$$

where besides the bending modulus κ, two other membrane characteristics are introduced and called the spontaneous curvature J_s and the modulus of the Gaussian curvature $\bar{\kappa}$.

Interpretation of the Helfrich model in terms of Gibbs' approach shows that the spontaneous curvature J_s and the modulus of the Gaussian curvature $\bar{\kappa}$ determine the first and the second bending moments in the initial flat state of the membrane. Indeed, the first bending moment resulting from Equation 11, according to Equation 5, is

$$C_1 = \kappa \cdot (J - J_s), \tag{12}$$

so that its value in the initial flat state is proportional to the spontaneous curvature, $C_1(J = 0) = -\kappa \cdot J_s$.

The second bending moment (Equation 6) determined from the Helfrich energy (Equation 11) is simply equal to the modulus of the Gaussian curvature

$$C_2 = \bar{\kappa}, \tag{13}$$

and is assumed to be independent on the geometrical variables of the surface.

The Helfrich bending modulus κ expressed in terms of the first bending moment (Equation 12) determines the change of C_1 with the change of the total curvature J,

$$\kappa = \frac{\partial C_1}{\partial J}. \tag{14}$$

Summarising, the Helfrich model accounts for the stresses in the initial flat state of the membrane expressed by the non-vanishing first, C_1, and second, C_2, bending moments. The bending modulus κ determines the variations of the first bending moment with the total curvature. However, the validity of the Helfrich model for the bending energy is restricted to small curvatures. Therefore, the dependence of the membrane stresses on the Gaussian curvature is not introduced. Moreover, the dependence of the tension γ on the total curvature J as well as the variations of the bending moments C_1 and C_2 with the area A are not considered. Helfrich's elastic model of weakly bent membrane has been extended by Mitov (1978), who considered the energy including the terms of higher orders in principal curvatures.

A definition of the bending modulus for the strongly curved lipid monolayers forming the cylinders of the inverted hexagonal (H_{II}) phases has been given by Gruner (1985) in a way formally similar to that of Helfrich but dealing with a different physical situation. In contrast to the case of flat stressed membrane considered by Helfrich, the initial state of the cylindrical monolayer is strongly curved and is supposed to be completely relaxed so that the first bending moment C_1 vanishes. The total curvature J_0 in this state of zero bending energy has been originally called the intrinsic curvature, but in later studies it is often referred to as the spontaneous curvature in spite of the obvious difference of these two notions.

The Gruner bending modulus κ of strongly curved cylindrical monolayer determines the deviation of the bending moment C_1 from zero with variation of the total curvature J,

$$C_1 = \kappa \cdot (J - J_0). \tag{15}$$

The energy of the monolayer deformation is given by

$$f_b = \frac{1}{2}\kappa(J - J_0)^2.$$

The Gruner model is valid for high curvatures comparable to the inverse thickness of the monolayer $J \simeq \delta^{-1}$. The quantity assumed to be small is the relative change of the curvature, $\left|\frac{(J-J_0)}{J_0}\right| \ll 1$. This model is, however, limited by consideration of cylindrical shapes of a monolayer, and the only deformation taken into account is the change of the total curvature J. Therefore, it does not describe any contributions to the energy from the changes of the Gaussian curvature and the area.

2.4 Further developments

In spite of the progress achieved in the description of interface elasticities, important questions remained open in particular in connection with strongly curved membranes.

First, the moduli of bending and stretching considered earlier do not give the complete description of the elastic properties of the interfaces. One still needs to account for the elastic coefficients related to the changes of the Gaussian curvature K and for the elasticities describing the effects of deformations involving the simultaneous changes of two different geometrical variables.

Second, the values of the elastic moduli depend on the position of the dividing surface chosen to describe the interface. Indeed, the stretching and bending elastic moduli have been shown to be directly related to the tension and the first bending moment, whose values, as noted above, depend on the choice of the dividing surface. Therefore, one can expect that a particular dividing surface exists for which the description of the elastic properties of the interface has a highly simple form.

A series of works (Kozlov *et al.* 1989, Kozlov and Markin 1990, Kozlov and Winterhalter 1991a,1991b, Kozlov *et al.* 1992, Kozlov *et al.* 1994, Andelman *et al.* 1994, Leikin *et al.* 1996) concentrated on the specific features of strongly curved amphiphile monolayers, where a radius of curvature can be close to the monolayer thickness. The spontaneous state of the amphiphile monolayer has been defined as the state of vanishing stresses, whose geometry is characterised by the spontaneous values of the total curvature, J_s, Gaussian curvature, K_s, and the area, A_s. Deformation of the monolayer with respect to the spontaneous state results in stresses whose values are related to the variations of A, J and K by the elastic moduli of the monolayer. By introducing the set of six independent elastic moduli, the elastic energy of the membrane has been determined up to the contributions of the second order in deformations. This description is related to a choice of the Gibbs dividing surface inside a monolayer, all elastic moduli depending strongly on the position of this surface. A neutral surface has been defined for which the deformations of bending and stretching are energetically decoupled. The relationships between the different elastic moduli have been illustrated by using a simplest model of distribution of microscopic rigidities over the monolayer thickness.

The developed approach has been applied to the study of the elastic properties of the cylindrically curved monolayers constituting the inverted hexagonal (H_{II}). The analysis is based on the experimental data obtained by osmotic stress and gravimetric methods (Rand and Fuller 1994).

By determining the relationships between the sets of the elastic moduli for different dividing surfaces, a way has been found to locate the position of the neutral surface. Analysis of experimental results on dioleoyl-phosphatidyl-ethanolamine (DOPE) has permitted the determination of the position of the neutral surface in the strongly curved cylindrical monolayers of this lipid and the moduli of bending and stretching. Treatment of the experimental data on the mixtures of DOPE and an electrically charged lipid dioleoyl-phosphatidyl-serine (DOPS) has been performed to analyse the effects of electric charge of the molecules constituting the cylindrical monolayer on the elastic moduli and the position of the neutral surface.

Application of this approach to analysis of elastic behaviour of the DOPE mesophases subjected to osmotic pressure accounted for the unusual structural change in which DOPE undergoes a reentrant hexagonal-lamellar-hexagonal transition sequence induced by the continuously growing osmotic pressure. A model of this effect has been suggested based on a delicate balance between elastic and hydration energies of the H_{II} and lamellar phases. The results show not just qualitative but excellent quantitative agreement with the experimental data, thus justifying the approach used to describe the system.

Another phenomenon analysed within the framework of the developed elastic theory is the transition between flat bilayers and cylindrical micelles driven by spontaneous curvature of the amphiphile monolayer. Consideration of a monolayer consisting of a mixture of a 'bilayer' lipid and a 'micellar' amphiphile has shown that, in accordance with the experimental results, such system undergoes a first-order phase transition between the flat and the strongly curved shapes. The phase state of the system has been demonstrated to be controlled by the composition of the system, and the resulting phase diagram was in qualitative agreement with the experimental one.

3 Description in terms of microscopic properties

To account for the properties of the interface structure, model assumptions should be made that, on the one hand, reduce the generality of the theory but, on the other, enable the expression of the tension γ, the bending moments C_1 and C_2 and the elastic moduli in terms of microscopic interactions inside the interface.

This approach, called local thermodynamics, was first developed to describe the surface tension and bending moments of interfaces between fluid phases (for a review, see Murphy 1966). The interface is considered as a transition zone, which can be characterised in each point by a set of thermodynamical variables. As the properties of the transition zone change across its thickness, the thermodynamical variables depend on the position inside the interface. Local stresses are effectively characterised by a pressure tensor P^n_m, which is analogous to the usual thermodynamical pressure P, but accounts for a possible anisotropy of the interface properties. The indices of the P^n_m can adopt three values corresponding to the normal and two tangential directions to the plane of the interface. To complete the description of the transition zone, it is suggested that it has a definite thickness and the pressures in the bulk phases separated by the interface are equal to P_o and P_i.

For the cases of interfaces between fluid phases (for example, between water and oil) the pressure tensor is assumed to have a diagonal form

$$P^n_m = \begin{pmatrix} P_n & 0 & 0 \\ 0 & P_T & 0 \\ 0 & 0 & P_T \end{pmatrix}, \tag{16}$$

where P_n is the component normal to the plane of the interface, while P_T are the components in the tangential directions. Owing to fluidity of the phases composing the interface, the non-diagonal components of the pressure tensor are supposed to vanish independently on the shape of the interface, and the two components in the tangential directions are assumed to be equal.

The interfacial tension and the bending moments are determined with respect to a fixed dividing surface and are expressed in terms of the pressure tensor (Equation 16) and the pressures in the bulk phases. The coordinate axis η is chosen in the direction normal to the dividing surface with the origin $\eta = 0$ at the dividing surface. The coordinates of the boundaries of the interface are denoted as η_o and η_i. The first and the second bending moments entering the expression for the energy (Equation 3) are shown to be given by the first and second moments of distribution of the microscopic pressures over the thickness of the interfaces determined with respect to the dividing surface.

$$C_1 = \int_{\eta_i}^{\eta_o} (P_{i-o} - P_T) \cdot \eta \cdot d\eta \tag{17}$$

$$C_2 = \int_{\eta_i}^{\eta_o} (P_{i-o} - P_T) \cdot \eta^2 \cdot d\eta, \tag{18}$$

where P_{i-o} is a step function equal to P_o for $\eta \geq 0$ and P_i for $\eta < 0$.

To determine the interfacial tension γ, one needs to introduce also the zero moment of distribution of the microscopic pressures,

$$C_0 = \int_{\eta_i}^{\eta_o} (P_{i-o} - P_T) \cdot d\eta. \tag{19}$$

Then, γ is given by

$$\gamma = C_0 + J \cdot C_1 + K \cdot C_2. \tag{20}$$

The tension (Equation 20) depends explicitly on the curvatures J and K and becomes equal to the zero moment C_0 only for a flat interface. Note that different definitions have been introduced in the literature for the interfacial tension γ in terms of the moments C_0, C_1 and C_2; however, only one of them given by Equation 20 is in agreement with the original definition of Gibbs entering Equations 1 – 3 (see, for a review, Murphy 1966).

More recently, the tension and the bending moments of amphiphile monolayers have been expressed in terms analogous to the local thermodynamics and are referred to as the lateral stress profile (see, for a review, Helfrich 1990). Within this approach, one uses, instead of the pressure tensor P_n^m, a tensor of microscopic stresses σ_n^m. The two tensors can be related by $\sigma_n^m = P_{1-2} - P_n^m$, so that both description are equivalent.

It has been noted (Helfrich 1990) that the first moment C_1 is independent of the position of the dividing surface if $C_0 = 0$ and, analogously, the second moment C_2 does not depend on the choice of the dividing surface if $C_0 = 0$ and $C_1 = 0$. A physical basis for the description of amphiphile membranes in terms of microscopic stresses is due to a mean field theory of hydrocarbon chains packing in monolayers, accounting accurately for the statistics of chain conformations (for a review, see Ben-Shaul 1995). Numerical analysis based on this theory allows us to reconstruct the profile of distribution of the microscopic stresses over the hydrophobic part of an amphiphile monolayer.

Analogously to the relationship between the interfacial stresses and the microscopic stress tensor, the elastic moduli of an interface can be expressed through the distribution of the microscopic rigidities across the thickness of the transition region (for reviews, see Petrov and Bivas 1979, Helfrich 1990). Although limiting the generality of the description, such approach permits the analysis of the roles of different part of the interface, such as the polar heads and hydrocarbon tails of amphiphiles monolayers, in determining the resulting elasticity. The first calculations of this kind have been performed for the bending modulus of flat lipid membranes κ, treating it in terms of plausible but speculative models for the distribution of the microscopic rigidities.

Mean field theory (Ben-Shaul 1995) predicted a particular profile of the local rigidities resulting from hydrocarbon chain statistics. Based on these results, the contribution of the hydrocarbon chains to the bending rigidity has been calculated and analysed as function of chain length and other structural parameters of the monolayer. Using Equations 14 and 17, the bending modulus κ can be expressed as a moment of distribution through the monolayer of the local stretching rigidities, which are determined as derivatives of the local stresses with respect to the areas of the corresponding dividing surfaces.

4 Equations of equilibrium and shape of interfaces

The first equation of equilibrium of an interface is due to Laplace (1806) and has a well-known form relating the interfacial tension γ, the total curvature J and the difference of pressure ΔP between the homogeneous phases separated by the interface,

$$\gamma \cdot J = \Delta P. \tag{21}$$

This equation perfectly describes interfaces with considerable tension. However, it proved insufficient for analysis of strongly curved interfaces with small or vanishing tensions.

The equations of equilibrium for an interface with arbitrarily low tension have to account, besides the tension γ, for the bending moments C_1 and C_2. Such equations have been first considered for interfaces between two fluid phases, characterised, independently from the shape of the surface, by a diagonal local pressure tensor (Equation 16, Murphy 1966). The resulting three equations relate the surfaces stresses γ, C_1 and C_2, with the curvatures J, K and the trans-interfacial pressure difference ΔP,

$$\gamma \cdot J - C_1 \cdot (J^2 - 2K) - C_2 \cdot JK = \Delta P \tag{22}$$

$$\nabla_\alpha \gamma - C_1 \nabla_\alpha J - C_2 \nabla_\alpha K = 0 \tag{23}$$

$$\nabla_\alpha C_1 + J \cdot \nabla_\alpha C_2 + b_\alpha^\nu \nabla_\nu C_2 = 0, \tag{24}$$

where ∇_α is the vector of two-dimensional covariant gradient in the plane of the dividing surface and b_α^ν is the tensor of curvature of the dividing surface. Equations 22 – 24 can be seen as determining the equilibrium of an infinitesimal element of the interface with respect to three possible displacements. Equation 22 corresponds to the displacement in the normal direction and is therefore analogous to the Laplace equation. Equation 23 describes the equilibrium with respect to displacement along the surface, and Equation 24 accounts for equilibrium with respect to rotation of the interface element with respect to its initial orientation. Equations 22 – 24 can be derived by variation of the energy (Equation 3) or by direct consideration of conditions of zero forces acting on the element in the normal and tangential directions and of zero resulting torque.

The equation of equilibrium for an amphiphilic membrane, which does not refer to the assumption of diagonal form of the pressure tensor, has been obtained by Helfrich (1973) and then by Evans and Skalak (1979) and Ou-Yang and Helfrich (1989).

Evans and Skalak (1979) derived the equations of mechanical equilibrium for the cases of axisymmetric shapes of the membranes, considering balance of forces and torques acting on an element of the surface. The membrane stresses were expressed in terms of the tension resultants T_m, T_ϕ and the moment resultants M_m, M_ϕ, where the subscripts m and ϕ indicate, respectively, the components parallel and perpendicular to the meridian of the axisymmetric shape. The equilibrium equations in the directions tangential and normal to the surface are, respectively,

$$\frac{\partial r T_m}{\partial s} - T_\phi \frac{\partial r}{\partial s} + c_m \cdot \left[\frac{\partial r M_m}{\partial s} - M_\phi \frac{\partial r}{\partial s} \right] = 0 \tag{25}$$

$$c_m \cdot T_m + c_\phi \cdot T_\phi - \frac{1}{r} \cdot \frac{\partial^2 r M_m}{\partial s^2} + \frac{1}{r} \cdot \frac{\partial}{\partial s} (M_\phi \frac{\partial r}{\partial s}) = \Delta P, \tag{26}$$

where c_m and c_ϕ are the principal curvatures of the surface, s is the arc-length of the meridional curve and r is the distance between a currents point on the surface and the axis of symmetry. The Equations 25 and 26 are valid for membranes with arbitrary elastic properties. They describe the fluid membranes as a particular case of vanishing lateral shear elasticity, where the tension and moment resultants can be related to the bending moments (Equations 5, 6 and 19), by

$$T_m = C_0 + C_1 \cdot c_\phi \tag{27}$$

$$T_\phi = C_0 + C_1 \cdot c_m \tag{28}$$

$$M_m = C_1 + C_2 \cdot c_\phi \tag{29}$$

$$M_\phi = C_1 + C_2 \cdot c_m. \tag{30}$$

Inserting Equations 27 – 30 into Equations 25 and 26, and using Equation 20, one can present Equation 25 in the form

$$\frac{\partial \gamma}{\partial s} - C_1 \cdot \frac{\partial J}{\partial s} - C_2 \cdot \frac{\partial K}{\partial s} = 0, \tag{31}$$

which is a particular case of the Murphy equation (Equation 23).

From the of equilibrium in the normal direction (Equation 26), we obtain

$$\gamma \cdot J - C_1 \cdot (J^2 - 2K) - C_2 \cdot JK - \nabla^2 C_1 - c_\phi \cdot \nabla^2 C_2 - \frac{\partial C_2}{\partial s}\frac{\partial c_\phi}{\partial s} = \Delta P, \tag{32}$$

where ∇^2 is the two-dimensional Laplace operator in the surface plane. Although Equation 32 is limited by consideration of the axisymmetric shapes only, it goes beyond the Murphy equation of equilibrium in normal direction (Equation 22). Indeed, in the cases where the bending moments C_1 and C_2 have constant values along the surface, Equation 32 is a particular case of Equation 22. However, an important difference between the equation of Evans and Skalak and that of Murphy consists in accounting for the contributions related to the changes of the bending moments over the surface, which is of great significance for the analysis of shapes of inhomogeneously curved membranes.

The equilibrium equation derived by Helfrich (1973) and developed by Ou-Yang and Helfrich (1989) is based on the Helfrich model of bending elasticity of fluid membranes and was performed by variational method. Owing to explicitly accounting for the relationships between the membrane stresses and deformations, this equation allows one to calculate the equilibrium shapes of membranes. Therefore, I will refer to it as the shape equation, which is

$$\lambda \cdot J + \kappa(J - J_s) \cdot (\frac{1}{2}J^2 - 2K + \frac{1}{2}J_s \cdot J) + \kappa \nabla^2 J = \Delta P, \tag{33}$$

where λ is called the tensile stress and plays the role of the Lagrange multiplier controlling constancy of the membrane surface.

The Ou-Yang-Helfrich theory refers neither to any particular symmetry of the membrane shape nor to any form of the pressure tensor. Its limitation consists of the consideration of small deviations of the membrane from the flat shape. Therefore, for the cases of small curvatures, the Ou-Yang-Helfrich equation (Equation 33) includes the equations of Murphy and of Evans and Skalak as particular cases and goes beyond both of them. Indeed, inserting into Equations 23 and 31 the relations between the bending moments and the curvatures, Equations 12 and 13, and integrating, we obtain an expression for the tension in terms of the curvatures:

$$\gamma = \lambda + \frac{1}{2}\kappa(J - J_s)^2 + \bar{\kappa}K, \tag{34}$$

where λ is an integration constant. Considering the shapes with homogeneous curvatures $\nabla^2 J = 0$ and using Equations 34, 12 and 13, one derives from Equation 33 the Murphy equation (Equation 22). Substituting Equation 34 for λ in the Ou-Yang-Helfrich, accounting for the relationships 12 and 13, and reducing the Laplacian ∇^2 to the form valid for the axisymmetric shapes, one obtains from Equation 33 the Evans and Skalak equation (Equation 32). This particular form of Equation 33 describing the axisymmetric shapes has been successfully applied to determination of the shapes of lipid vesicles of different topologies (for a review, see Seifert and Lipowsky 1995).

Although much has been done on determinations of equilibrium of the interfaces, the Equations 22 – 24, 31, 32 and 33 derived up to now are not sufficient to analyse the strongly and inhomogeneously curved amphiphile membranes as those building the bicontinuous cubic phases or similar structures. Equations 22 – 24 are exact and are valid for the cases of the arbitrarily high curvatures of the interface, but they do not account for the contributions of the non-diagonal component of the pressure tensor. On the other hand, Equations 31, 32 and 33, which do not refer to any assumptions concerning the pressure tensor, have other limitations. The Evans and Skalak theory considers the axisymmetric shapes only, while the Ou-Yang-Helfrich theory is valid only for the weakly bent membranes.

References

Andelman D, Kozlov M M and Helfrich W, 1994, *Europhys Lett* **25** 231.

Ben-Shaul A, 1995, in *Structure and Dynamics of Membranes*, eds Lipowsky R and Sackmann E (Elsevier, Amsterdam).

Cevc G and Marsh D, 1987, *Phospholipid Bilayers: Physical Principles and Models* (Willey, New York).

Evans E and Skalak R, 1979, *Mechanics and Thermodynamics of Biomembranes* (CRC Boca Raton, Florida).

Gelbart W M, Roux D and Ben-Shaul A, eds, 1993, *Modern Ideas and Problems in Amphiphilic Sciences* (Springer, Berlin).

Gibbs J W, 1876, 1878, On Equilibrium of Heterogeneous Substances, in *The Collected Works of J. Willard Gibbs (1948)* (Yale University Press, New Haven).

Gruner S, 1985, *Proc Natl Acad Sci USA* **82** 3665.

Gruner S, 1989, *Proc Natl Acad Sci USA* **93** 7562.

Helfrich W, 1973, *Z Naturforsch* **28C** 693.

Helfrich W, 1990, Elasticity and Thermal Undulations of Fluid Films of Amphiphiles, in *Liquids and Interfaces. Les Houches XLVIII 1988*, eds Charvolin J, Joanny J F and Zinn-Justin J (Elsevier, Amsterdam).

Israelachvili J N, 1985, *Intermolecular and Surface Forces* (Academic Press, London).

Koynova R and Caffrey M, 1994, *Chem Phys Lip* **69** 1.

Kozlov M M, Leikin S L and Markin V S, 1989, *J Chem Soc Faraday Trans 2* **85** 277.

Kozlov M M and Markin V S, 1990, *J Col Interf Sci* **138** 332.

Kozlov M M and Winterhalter M, 1991a, *J Phys II France* **1** 1077.

Kozlov M M and Winterhalter M, 1991b, *J Phys II France* **1** 1085.

Kozlov M M, Winterhalter M and Lerche D, 1992, *J Phys II France* **2** 175.

Kozlov M M, Leikin S and Rand R P, 1994, *Biophys J* **67** 1603.

Lasic DD, 1995, Applications of liposomes, in *Structure and Dynamics of Membranes*, eds Lipowsky R and Sackmann E (Elsevier, Amsterdam).

Leikin S, Kozlov M M, Fuller N L and Rand R P, 1996, *Biophys J* **71** 2623.

Lichtenberg D, Opatowski E and Kozlov M M, 2000, *Biochim Biophys Acta* **1508** 1.

Lipowsky R and Sackmann E, eds, 1995, *Structure and Dynamics of Membranes* (Elsevier, Amsterdam).

Luzzati V, 1968, The structure of the liquid-crystalline phases of lipid-water systems, in *Biological Membranes*, ed Chapman D (Academic Press, New York).

Mitov M, 1978, *C R Acad Sci* **31** 513.

Murphy C L, 1966, *Thermodynamics of low tension and highly curved interfaces*, Ph.D. Thesis, University of Minnesota, Chemical Engineering.

Nitsche J C C, 1989, *Lectures on minimal surfaces* (Cambridge University Press, Cambridge).

Ou-Yang Z and Helfrich W, 1989, *Phys Rev A* **39** 5280.

Parsegian V A, Rand R P, Fuller N and Rau D C, 1986, *Methods Enzymol* **127** 400.

Petrov A G and Bivas I, 1984, *Prog Surface Sci* **16** 389.

Rand R P and Fuller N, 1994, *Biophys J* **66** 2127.

Rand R P, *Structural parameters of aqueous phospholipid mixtures.* www.brocku.ca/researchers/peterrand/osmotic/osfile.htmldata.

Safran S A, 1994, *Statistical Thermodynamics of Surfaces, Interfaces and Membranes* (Addison-Wesley, New York).

Seddon J M and Templer R, 1993, *Phil Trans R Soc Lond A* **344** 377.

Seifert U and Lipowsky R, 1995, in *Structure and Dynamics of Membranes*, eds Lipowsky R and Sackmann E (Elsevier, Amsterdam).

Tanford C, 1980, *The Hydrophobic Effect — Formation of Micelles and Biological Membranes*, second edition (John Willey and Sons).

Introduction to electrostatics in soft and biological matter

David Andelman

School of Physics & Astronomy, Tel Aviv University
Ramat Aviv 69978, Tel Aviv, Israel

andelman@post.tau.ac.il

1   Introduction

It is hard to overestimate the importance of electrostatic interactions associated with charged objects in soft and biological matter. In aqueous environments, typical to many of these systems, charges tend to dissociate and affect a wealth of functional, structural and dynamical properties. Without attempting to enumerate an exhaustive list, we mention a few examples. Polymers are flexible and elongated one-dimensional objects (see, e.g., chapters by Warren, Podgornik, MacKintosh and Bensimon in this volume). In aqueous solutions they often carry charges, like the naturally occurring DNA or synthetic polyelectrolytes such as polystyrene sulfonate. The charges on the polymer chain and the counter-ions in solution have an important effect on the rigidity of such chains and on inter- and intra-chain interactions leading to interesting phenomena of aggregation and condensation, as is seen most often in the presence of multivalent counter-ions. The process whereby polyelectrolyte chains migrate in external electric fields is called electrophoresis and is another important phenomena with many applications. Other charged structures are biological cell membranes (see, e.g., chapters by Kozlov and Olmsted in this volume). These soft and fluctuating two-dimensional objects are naturally built out of mixtures of phospholipids with or without net charge. Finally, we mention globular proteins with charge groups on their surface (chapter by Elber in this volume), self-assembly of micelles made of charged amphiphiles (chapter by Olmsted in this volume) and charged colloidal particles (see the chapter by Frenkel in this volume) where the charges play a role in stabilising suspensions.

When will the electrostatic interactions influence the structural properties of soft materials? For soft materials the thermal energy $k_B T$ is comparable to the typical energy associated with deformations and structural degrees of freedom. We therefore introduce the length scale at which the thermal energy is equal to the coulombic energy between two unit charges, the so-called *Bjerrum length*

$$l_B = \frac{e^2}{\varepsilon k_B T},\tag{1}$$

which is approximately 7Å in water (dielectric constant $\varepsilon = 80$) at room temperature, $T = 300$ K. An important concept introduced by Debye and Hückel (1923) is the screening of the electrostatic interaction between two charges by the presence of all other cations and anions in the solution. This will be further discussed below.

In this chapter we will briefly review some of the most fundamental concepts related to electrostatic interactions in soft and biological matter. As this is a vast topic, we will restrict the discussion only to static properties of systems in thermodynamic equilibrium excluding the interesting phenomena of dynamical fluctuations and dynamical responses to external fields. Most of the discussion will be restricted to a mean-field approximation of the electric double-layer problem and the solutions of the classical Poisson-Boltzmann equation. Various effects of fluctuations and correlations will only be briefly mentioned toward the end of the present chapter. An excellent reference for the electric double layer is the classical book of Verwey and Overbeek (1948) which explains the DLVO (Deryagin-Landau-Verwey-Overbeek) theory for stabilisation of charged colloidal systems. More recent treatments can be found in many books on colloid science and interfacial phenomena. For example, Evans and Wennerstöm (1994), Israelachvili (1992), and in a review by the present author Andelman (1995). The topic of polyelectrolytes is briefly treated in most polymer books (e.g., De Gennes (1979)), while the classical (and somewhat outdated) book is by Oosawa (1971). For a more recent review on charged polymers, see Netz and Andelman (2003) and references therein.

2   The Poisson-Boltzmann theory

We will now derive the Poisson-Boltzmann (PB) theory for ionic solutions. As a mean-field theory, the PB theory relies on the following assumptions: (1) the only interactions to be considered are coulombic interactions between charged bodies, (2) permanent and induced dipole-dipole interactions are neglected, (3) the charges are taken as point-like objects neglecting any finite size effect and any short-range non-electrostatic interactions, (4) the aqueous solution is modelled as a continuous medium with a dielectric constant ε. For water, the dielectric constant is taken to be $\varepsilon = 80$ and (5) the electrostatic potential $\phi(\mathbf{r})$ that each ion sees is a continuous function that depends in a mean-field way on all the other ions. The charge density profile of all ions $\rho(\mathbf{r})$ is also a mean-field continuous function of the position $\mathbf{r}$.

It is possible to derive the PB equation starting from a field theory and to obtain the PB equation as a first-order term in a systematic expansion (Borukhov *et al* 1998, 2000, Netz and Orland 2000, Burak *et al* 2004a). We will use a simpler and more heuristic approach. Consider an ionic solution with two ionic species having positive and negative charge densities (per unit volume) of ρ_+ and ρ_-, respectively. The total charge density at each point is $\rho = \rho_+ + \rho_-$. Defining $n_\pm$ as the number density (per unit volume) of the two species, then $\rho_\pm(\mathbf{r}) = ez_\pm n_\pm(\mathbf{r})$, where $z_+ > 0$ is the valency of the cations and $z_- < 0$ of the anions.

The ions are assumed to be mobile and in thermodynamic equilibrium. They will adjust to the presence of some fixed electrostatic boundary conditions, which can be either a constant surface potential (*Dirichlet* boundary condition) or a constant surface charge density (*Neumann* boundary condition). At any point $\mathbf{r}$, the relation between the potential ϕ and the charge density ρ is given in terms of the Poisson equation:

$$\nabla^2 \phi = -\frac{4\pi}{\varepsilon}\rho(\mathbf{r}) = -\frac{4\pi e}{\varepsilon}\left[z_+n_+(\mathbf{r}) + z_-n_-(\mathbf{r})\right]. \tag{2}$$

Note that cgs (Gaussian) electrostatic units are used throughout this chapter. However, by using dimensionless energy units and expressing all lengths in terms of the Bjerrum length l_B and the Debye-Hückel screening length (introduced below), the results can be made independent of any specific system of units. Using the above equation one can deduce the electrostatic potential for a *given* ionic distribution. However, in the liquid solution the ions are mobile and will adjust their position according to the local potential they feel. As each ionic species is in thermodynamic equilibrium, its corresponding density has a Boltzmann distribution

$$n_\pm = n_\pm^0 e^{-ez_\pm\phi/k_BT}, \tag{3}$$

where n_i^0 is the reference density of ith species ($i = \pm$) taken at zero potential, $\phi \to 0$. Substituting Equation 3 into Equation 2, we get the Poisson-Boltzmann (PB) equation for the potential ϕ:

$$\nabla^2\phi(\mathbf{r}) = -\sum_{i=\pm} \frac{4\pi e n_i^0 z_i}{\varepsilon} e^{-ez_i\phi(\mathbf{r})/k_BT}. \tag{4}$$

Alternatively, the PB equation (Equation 4) can be derived by requiring that the electro-chemical potential $\mu_\pm$ for the two ionic species is a constant throughout the system

$$\mu_\pm = ez_\pm\phi + k_BT\ln(n_\pm) = \text{constant}. \tag{5}$$

The PB equation is a very useful analytical approximation with many applications. Because the equation is non-linear, it has closed-form analytical solutions only for a limited number of simple charged boundary conditions. On the other hand, by solving it numerically or within some further approximations or limits, we can obtain the ionic profiles as well as the free energy of complex structures, e.g., the free-energy change of a charged globular protein approaching an oppositely charged lipid membrane. Like any approximation, the PB theory has its limits of validity; however, in physiological conditions (electrolyte strength of about 0.1 M), it describes rather well the ionic distributions as long as the surfaces are not too highly charged. The PB theory produces good results for monovalent ions but misses some important features associated with multivalent counterions.

Throughout this chapter we present results for the following two limiting cases:

- The first is the *counter-ion only* case, where there is only one species of ions in solution neutralising the charged surface: $n_-^0 = 0$ and $n_+^0 = n_0$. Then, the PB equation reads

$$\nabla^2\phi = -\frac{4\pi e n_0 z_+}{\varepsilon} e^{-ez_+\phi/k_BT}. \tag{6}$$

- The second is the *added electrolyte* (added salt) case where the system is placed in contact with an infinite reservoir of electrolyte. For simplicity, we treat only the symmetric monovalent electrolyte (*e.g.*, $Na^+ Cl^-$): $z_{\pm} = \pm 1$ and $n_+^0 = n_-^0 = n_0$. Here

$$\nabla^2 \phi = \frac{8\pi e n_0}{\varepsilon} \sinh \frac{e\phi}{k_B T}. \tag{7}$$

We remark that it is rather straightforward to extend the above PB results to any multivalent ionic system, $z_- : z_+$.

The linearised PB equation: Debye-Hückel theory

In the case of low electrostatic potentials a very useful approximation can be used. In this case, the PB equation (Equation 4) can be linearised (as long as $|\phi| < 25\text{mV}$) resulting in the famous Debye-Hückel (DH) theory.

$$\nabla^2 \phi \simeq \frac{8\pi e^2 n_0}{\varepsilon k_B T} \phi(\mathbf{r}) = \lambda_D^{-2} \phi(\mathbf{r}). \tag{8}$$

The new parameter λ_D introduced above has units of length and is known as the *Debye-Hückel screening length*

$$\lambda_D = \sqrt{\frac{\varepsilon k_B T}{8\pi e^2 n_0}} = (8\pi l_B n_0)^{-1/2} \sim n_0^{-1/2}. \tag{9}$$

The screening length varies from about 3Å at the strong ionic strength of 1 M of NaCl to about 1μm in pure water where the ionic strength due to the dissociating OH^- and H^+ ions is 10^{-7} M. A useful formula to remember is that for n_0 measured in molar units, λ_D in Angstroms is given by

$$\lambda_D = \frac{3.05[\text{Å}]}{\sqrt{n_0[\text{M}]}}. \tag{10}$$

The DH treatment gives a simple description to the many-body interactions between ions. It simply states that the interaction between any given pair of ions at distance $r = |\mathbf{r}|$ will decay exponentially due to the screening by all other cations and anions surrounding the ionic pair. Broadly speaking, this screened potential within the DH theory varies like $r^{-1} \exp(-r/\lambda_D)$. To a first approximation, one can say that for $r \leq \lambda_D$ the coulombic interaction ($\sim r^{-1}$) is only slightly screened, whereas for $r > \lambda_D$ it is strongly (exponentially) screened.

In the remainder of this chapter we will consider the PB equation in various simple geometries. We will first discuss solutions of the PB equation in planar geometries and then mention with less detail solutions in cylindrical and spherical geometries.

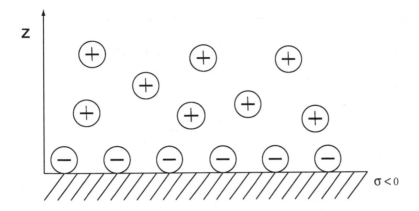

Figure 1. *Schematic illustration of the electric double layer problem for the counter-ion only case. A negative surface with surface charge density σ is placed at $z = 0$, while its counter-ions are released in the solution. The surface is infinite in the (x, y) plane. Counter-ions are attracted to the surface and create a density profile, $n(z)$.*

3 PB equation in planar geometry

3.1 A single charged surface

3.1.1 Counter-ion only

One of the simpler analytical solutions of the PB equation was formulated almost a century ago by Gouy (1910, 1917) and Chapman (1913). The problem they addressed is the profile of a cloud of counter-ions forming a diffusive *electric double-layer* close to a planar surface having a fixed surface charge density, σ. Without loss of generality, the surface charges are taken as anions ($\sigma < 0$) and the counter-ions as monovalent cations ($z_+ = 1$) having a density profile $n(z) = n_+(z)$. The system geometry is depicted on Figure 1. The charged surface is at $z = 0$ and the counter-ions occupy the positive half plane, $z > 0$. As the $z = 0$ charged surface is infinite, the system is translationally invariant in the perpendicular x, y directions and the PB equation reduces to an ordinary differential equation

$$\phi''(z) = -\frac{4\pi e n_0}{\varepsilon} \, e^{-e\phi/k_B T}, \tag{11}$$

with the boundary condition

$$\frac{d\phi}{dz}\bigg|_{z=0} = -\frac{4\pi}{\varepsilon}\sigma > 0. \tag{12}$$

Equation 11 is a second-order differential equation. Using the boundary condition, Equation 12, it can be integrated analytically, yielding the following potential and ionic profile

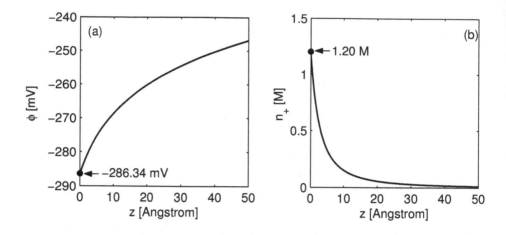

Figure 2. *The electric double layer for a single charged surface in contact with an aqueous solution of monovalent cations. The charged surface is at $z = 0$ with $\sigma = -e/250\,\text{Å}^2$. (a) Potential profile ϕ as function of the distance from the surface, z. The value of the surface potential is $\phi_s = -286.34\,\text{mV}$. (b) Density profile of the counter-ions, n_+ as function of the distance z. The value at the surface is $n_+(0) = 1.2\,\text{M}$ and the Gouy-Chapman length is $b \simeq 5.68\,\text{Å}$.*

$$\phi(z) = \frac{2k_{\mathrm{B}}T}{e}\ln(z+b) + \phi_0,$$

$$n(z) = \frac{1}{2\pi l_{\mathrm{B}}}\frac{1}{(z+b)^2},$$

(13)

where ϕ_0 is a reference potential and the length b is called the *Gouy-Chapman length*

$$b = \frac{\varepsilon k_{\mathrm{B}}T}{2\pi e|\sigma|} = \frac{e}{2\pi|\sigma|l_{\mathrm{B}}} \sim \sigma^{-1}.$$

(14)

Whereas the Bjerrum length is a measure of the electrostatic interactions in units of $k_{\mathrm{B}}T$ and is constant at about 7Å for aqueous solutions at room temperature, the Gouy-Chapman length is inversely proportional to σ, the surface charge density. For strongly charged surfaces, b is only a few angstroms. Although the entire profile is diffusive as it decays algebraically, a simple meaning of b is that counter-ions accumulated in the layer of thickness b close to the surface have an integrated charge (per unit area) of $\frac{1}{2}|\sigma|$, balancing half of the surface charge. Note also that the potential has a logarithmic divergence as $z \to \infty$. This is associated with the infinite extent of the charged surface at $z = 0$. On the other hand, the electric field, $\mathbf{E} = -\nabla\phi$, decays to zero as it should for $z \to \infty$. In Figure 2 we present the potential and ionic profile for a surface density of $\sigma = -e/250\,\text{Å}$. The figure shows clearly the build-up of the diffusive layer of counter-ions attracted to the negatively charged surface, reaching a limiting value of $n_+(0) = 1.2\,\text{M}$. The Gouy-Chapman length is here $b \simeq 5.68\,\text{Å}$.

We discuss next a charged surface placed in contact with an electrolyte bath. Here the potential will decay to zero far away from the surface, even for charged surfaces of infinite extent, because of the screening by the bulk electrolyte reservoir.

3.1.2 Added electrolyte

We look now at another case of experimental interest in which the charged surface at $z = 0$ is placed in contact with an electrolyte bath. On the surface, the same boundary condition, Equation 12, holds. For simplicity, we will consider a monovalent electrolyte $z_\pm = \pm 1$. In the bulk, far away from the surface ($z \to \infty$), we know that $n_\pm(\infty) = n_0$, where n_0 is the electrolyte bulk concentration.

The PB equation (Equation 7) can be integrated for this model system, yielding an analytical solution for the potential and ionic densities,

$$
\phi(z) = -\frac{2k_BT}{e} \ln \frac{1 + \gamma e^{-z/\lambda_D}}{1 - \gamma e^{-z/\lambda_D}},
$$

$$
n_\pm = n_0 \left(\frac{1 \pm \gamma e^{-z/\lambda_D}}{1 \mp \gamma e^{-z/\lambda_D}} \right)^2,
\tag{15}
$$

where the parameter γ is the positive root of a quadratic equation,

$$
\gamma = -\frac{b}{\lambda_D} + \sqrt{\left(\frac{b}{\lambda_D} \right)^2 + 1},
\tag{16}
$$

and the surface potential $\phi_s = \phi(0)$ is related to γ by Equation 15

$$
\phi_s = -\frac{4k_BT}{e} \operatorname{arctanh}(\gamma).
\tag{17}
$$

Once the potential profile is known, the two ionic profiles can be simply calculated from the Boltzmann distribution: $n_\pm(z) = n_0 \exp(\mp e\phi(z)/k_BT)$ as is depicted in Figure 3. The negatively charged surface attracts the counter-ions and repels the co-ions. The ratio b/λ_D is inversely proportional to the surface density. For small surface charge and/or high electrolyte strength, b/λ_D is large, yielding $\gamma \simeq \lambda_D/2b$, and

$$
\phi(z) \simeq \phi_s e^{-z/\lambda_D} \simeq -\frac{2k_BT}{e} \frac{\lambda_D}{b} e^{-z/\lambda_D},
\tag{18}
$$

which coincides with the DH (linearised) limit of the PB equation (Equation 8). Note the difference between the counter-ion case where the potential diverges logarithmically, Equation 13, and the electrolyte-added case, Equation 15, where the potential decays to zero. In the limit of weak surface potential (or weak surface charge) the potential decays exponentially, Equation 18, with the Debye-Hückel screening length, λ_D, as its characteristic length. Within the PB treatment, Equation 15 is the exact solution for any amount of electrolyte and surface charges. It interpolates between these two limits.

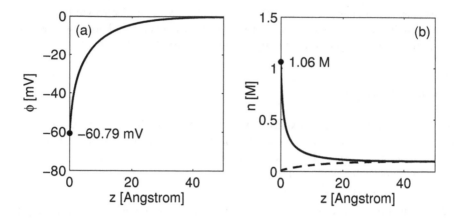

Figure 3. *The electric double layer for a single charged surface in contact with a 1:1 monovalent electrolyte reservoir of concentration $n_0 = 0.1$ M. The charged surface is at $z = 0$ with $\sigma = -e/25$ Å^2. This σ is ten times larger than the value used in Figure 2. (a) Potential profile ϕ as function of distance from the surface, z. The value of the surface potential is $\phi_s \simeq -60.8$ mV. (b) Density profile of the counter-ions, n_+ (solid line) and co-ions (dashed line) n_- are plotted as function of the distance z. The value at the surface is $n_+(0) \simeq 1.06$ M.*

3.1.3   The Grahame equation and the Contact theorem

The PB equation can be integrated once and leads to a relation known as the *Grahame equation* and also as the *Contact theorem* (Grahame 1947, Israelachvili 1992). This is a relation between the surface charge density σ and the limiting value of the ionic density profile at the boundary, $n_\pm(z{=}0)$.

$$\sigma^2 = \frac{\varepsilon k_B T}{2\pi}\left[n_+(0) + n_-(0) - 2n_0\right] \simeq \frac{\varepsilon k_B T}{2\pi}\left[n_+(0) - 2n_0\right],$$

and

$$\sigma^2 = \frac{\varepsilon k_B T}{\pi} n_0 \left[\cosh\frac{e\phi_s}{k_B T} - 1\right].$$

(19)

For large ϕ_s, $n_+(0)/n_-(0) = \exp(2e|\phi_s|/k_B T) \gg 1$, and $n_-(0)$ is neglected in the above equation.

For example, for a surface charge density of one electronic charge per 25 Å^2 (as in Figure 3) and an ionic strength of $n_0 = 0.1$ M, the limiting value of the counter-ion density at the surface is $n_+(0) \simeq 1.06$ M.

3.2   Modified Poisson-Boltzmann equation

As we saw in the preceding section, the density of the accumulated counter-ions at the surface can reach very high, sometimes unrealistic, values. A simple modification of the PB equation allows a remedy of this problem. In its modified form, the only other added

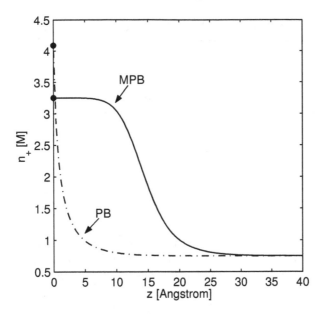

Figure 4. *Comparison of the modified PB profile (MPB, solid line) having $a = 8$ Å, with the regular PB one (PB, dash-dotted line). The surface charge density is $\sigma = -e/25$ Å^2 and the 1:1 electrolyte ionic strength is $n_0 = 0.75$ M. Note that while the PB value at the surface is $n_+(0) \simeq 4.09$ M, the modified PB density saturates at $n_+(0) \simeq 3.26$ M.*

ingredient is the entropy of the solvent in addition to that of the ions. This is especially of importance when the counter-ions have a large size and/or are multivalent.

In the case of a 1:1 electrolyte, the PB equation with the entropy modification results in the following equations for the profile densities and potential (Borukhov *et al* 1997)

$$n_\pm(z) = \frac{n_0 e^{\mp e\phi/k_B T}}{1 - \varphi_0 + \varphi_0 \cosh(e\phi/k_B T)}, \tag{20}$$

$$\nabla^2\phi = -\frac{4\pi e}{\varepsilon}(n_+ - n_-) = \frac{8\pi e n_0}{\varepsilon} \frac{\sinh(e\phi/k_B T)}{1 - \varphi_0 + \varphi_0 \cosh(e\phi/k_B T)}, \tag{21}$$

where $\varphi_0 = \pi a^3 n_0/3$ is the volume fraction of the ions at bulk electrolyte concentration n_0 and a is taken as the molecular size of both the solute and solvent. It is easy to see from Equation 20, that the counter-ions have a Fermi-Dirac like distribution. For small potentials, $e\phi/k_B T \ll 1$, the distribution reduces to the usual Boltzmann one, whereas for high potential, e.g., close to a highly charged surface, the counter-ion density saturates at close packing densities of $1/a^3$. This is very useful for multivalent counter-ions, where the regular PB theory gives unreasonably high values of ionic densities close to charged interfaces. As an example we show in Figure 4 the modified and the regular PB profiles for a 1:1 electrolyte. Large ion size, $a = 8$ Å is chosen to emphasise the saturation effect in the modified PB profile close to the charged surface. Note that the modified PB has a lower limiting value, $n_+(0)$, as well as a saturated accumulated layer of counter-ions close to the surface.

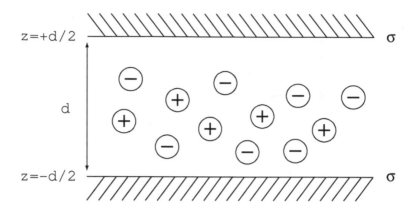

$z=+d/2$ σ

d

$z=-d/2$ σ

Figure 5. *Schematic illustration of the two-surface system. The charged and planar surfaces are located at $z = \pm d/2$, and separated by a distance d. The surface charge is taken to be negative and is neutralised by the ions in solution. In the symmetric case, σ_1 and σ_2 are equal to the same value σ. For counter-ion only case, the two-surface charge is neutralised by the counter-ions. When electrolyte is added, the system is couple with an electrolyte reservoir of density $n_{\pm}^0 = n_0$.*

It is also easy to derive the modified Grahame equation relating the surface charge density σ with the counter-ion density at the surface. Neglecting the contribution of the co-ions at the surface, $n_-(0) \ll 1$, the Grahame equation, Borukhov *et al* (1997), reads

$$\sigma^2 \simeq \frac{\varepsilon k_B T}{2\pi} \frac{1}{a^3} \ln \frac{1 - 2a^3 n_0}{1 - a^3 n_+(0)}. \tag{22}$$

Similarly, the surface charge density can be related with the surface potential ϕ_s by relating $n_+(0)$ to ϕ_s from Equation 20. Note that in the limit of small a, by expanding the logarithm in Equation 22, the Grahame Equation (Equation 19) for the regular PB case is recovered.

3.3 Two planar surfaces

The PB equation can be solved for two planar surfaces. We will restrict ourselves to the case of two equally charged surfaces located at $z = \pm d/2$, each having a charge density $\sigma < 0$ as is depicted in Figure 5. Generalisations to non-equal surface charges exist as well (Parsegian and Gingell 1972). For planar and infinite surfaces, the PB equation reduces to an ordinary differential equation depending only on the coordinate z. We will consider separately the counter-ion only and added-electrolyte cases.

It is instructive to write down the electrostatic free energy for the two-surface problem. It comprises of the electrostatic energy and the entropy of the ions in solution (without considering the modifications of Section 3.2).

$$\mathcal{F} = U - TS = \int f \, d^3\mathbf{r} = \text{Area} \cdot \int f \, dz,$$

$$f = \frac{\varepsilon}{8\pi}(\nabla\phi)^2$$
$$+ k_{\mathrm{B}}T\left(n_+(\mathbf{r})\ln\frac{n_+(\mathbf{r})}{n_0} + n_-(\mathbf{r})\ln\frac{n_-(\mathbf{r})}{n_0} - [n_+(\mathbf{r}) + n_-(\mathbf{r}) - 2n_0]\right). \quad (23)$$

From the free energy per unit area $\mathcal{F}/\text{Area}$, we can calculate the osmotic pressure by taking a variation with respect to the inter-surface spacing d, while keeping the temperature and species chemical potentials fixed,

$$\Pi = -\frac{1}{\text{Area}} \frac{\delta\mathcal{F}}{\delta d}\bigg|_{T,\mu}. \quad (24)$$

For the symmetric case of two equally charged surfaces, the profiles are symmetric about the mid-plane located at $z = 0$, yielding there a zero electric field, $E = 0$. By taking the full variation of the free energy, Equation 23, it can be shown that the pressure is simply equal to the excess of osmotic pressure calculated at the mid-plane with respect to the bulk electrolyte solution

$$\frac{\Pi}{k_{\mathrm{B}}T} = \sum_i \left[n_i(z=0) - n_i^0\right]. \quad (25)$$

The osmotic pressure can be calculated at any point in the solution. Although the expression is different, its value agrees with the above expression calculated at the mid-plane.

3.3.1 Counter-ions only

For a symmetric two-plate system, it is enough to consider the interval $[0, d/2]$ because of the $z \leftrightarrow -z$ symmetry. The boundary conditions are $d\phi/dz|_{z=d/2} = (4\pi/\varepsilon)\sigma$ and $d\phi/dz|_{z=0} = 0$. Let us call $n(z) = n_+(z)$ and denote by n_{m} and ϕ_{m} the values of the density and potential at the mid-plane, respectively. For these boundary conditions, the PB equation can now be solved analytically.

$$\phi(z) = \frac{k_{\mathrm{B}}T}{e}\ln(\cos^2 Kz) < 0,$$
$$\quad (26)$$
$$n(z) = n_{\mathrm{m}}e^{-e\phi(z)/k_{\mathrm{B}}T} = \frac{n_{\mathrm{m}}}{\cos^2 Kz},$$

where the new length in the problem $1/K$ is related to n_{m} by

$$K^2 = \frac{2\pi e^2}{\varepsilon k_{\mathrm{B}}T}n_{\mathrm{m}}. \quad (27)$$

Using the boundary condition at $z = d/2$ we get a transcendental relation for K

$$Kd\tan(Kd/2) = -\frac{2\pi e\sigma}{\varepsilon k_{\mathrm{B}}T}d = \frac{d}{b}, \quad (28)$$

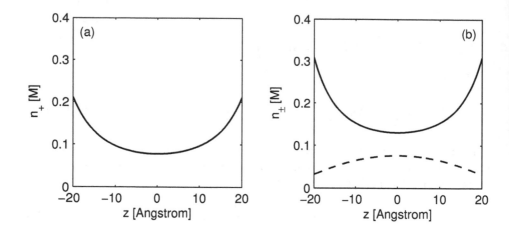

Figure 6. *Ion density profiles between two identical charged surfaces with $\sigma = -e/750 \text{ Å}^2$ each, at separation $d = 40 \text{ Å}$ located at $z = \pm 20 \text{ Å}$. In (a) the $n_+(z)$ profile is plotted from Equations 26-28 for the counter-ion only case, while in (b) n_+ and n_- (solid and dashed lines, respectively) are plotted for 1:1 electrolyte with $n_0 = 0.1 \text{ M}$. See Equations 35-37. As $b \simeq 17 \text{ Å}$, and $\lambda_D \simeq 9.5 \text{ Å}$, we are in between the Intermediate and DH regions of Figure 7 where $d > b > \lambda_D$.*

where b is the Gouy-Chapman length defined in Equation 14. A typical counter-ion profile is shown in Figure 6a for $\sigma = -e/750 \text{ Å}^2$ and $d = 40 \text{ Å}$.

The osmotic pressure, Equation 25, calculated in the counter-ion only case is

$$\frac{\Pi(d)}{k_B T} = \frac{\varepsilon k_B T}{2\pi e^2} K^2 = \frac{1}{2\pi l_B} K^2. \tag{29}$$

Since K depends on other system parameters, we discuss now separately two limits, depending on how strong the surface charge is.

Weak surface charges

For $d/b \ll 1$, the surface charge is weak. From Equation 27, $(Kd)^2 \simeq 2d/b \ll 1$. The pressure is then

$$\frac{\Pi(d)}{k_B T} \simeq -\frac{1}{d}\frac{2\sigma}{e} = \frac{1}{\pi l_B b}\frac{1}{d}. \tag{30}$$

This regime is called the *ideal gas* regime as is apparent from the above pressure expression. The density (per unit volume) of the counter-ions is almost constant between the two plates and is equal to $2|\sigma|/(ed)$. The main contribution to the pressure comes from the ideal-gas-like pressure of the cloud of counter-ions. Note that this regime occurs only for small separations, $d < b$. For weakly charged surfaces, b is relatively large and this regime can be seen for separations in the range of a few angstroms or more.

Strong surface charges

Here, from Equation 27, $d/b \gg 1$ and $Kd \to \pi$. The leading order term in the pressure is then

$$\frac{\Pi(d)}{k_B T} \simeq \frac{\pi}{2l_B} \frac{1}{d^2} = \frac{\pi \varepsilon k_B T}{2e^2} \frac{1}{d^2}. \tag{31}$$

It is interesting to note that the above pressure equation is independent of the surface charge density and is closely related to the *Langmuir equation*, as is discussed in Israelachvili (1992). But note that this equation holds for counter-ions only. As soon as one adds electrolyte, the pressure expression changes as is shown in the next section.

3.3.2 Added electrolyte

The PB equation is now considered for an electrolyte solution between two charged surfaces, restricting ourselves to a 1:1 symmetric and monovalent electrolyte, Equation 7. However, here the exact solution can only be expressed in terms of an elliptic integral. Let us define a dimensionless potential $\eta \equiv -e\phi/k_B T$ so that $\eta > 0$ for $\sigma < 0$. At the mid-plane $\eta_m \equiv \eta(z = 0)$ and on the charged surface $\eta_s \equiv \eta(d/2)$. The PB equation and boundary conditions are now written in terms of η

$$\frac{d^2\eta}{dz^2} = \lambda_D^{-2} \sinh \eta, \tag{32}$$

$$\left. \frac{d\eta}{dz} \right|_{z=d/2} = \frac{2}{b} \quad \text{and} \quad \left. \frac{d\eta}{dz} \right|_{z=0} = 0. \tag{33}$$

First integration from the mid-plane position ($z = 0$) to an arbitrary z gives

$$\lambda_D \frac{d\eta}{dz} = \sqrt{2 \cosh \eta(z) - 2 \cosh \eta_m}. \tag{34}$$

A further definite integration gives an elliptic integral

$$\frac{z}{\lambda_D} = \int_{\eta_m}^{\eta} \frac{d\eta'}{\sqrt{2 \cosh \eta' - 2 \cosh \eta_m}}. \tag{35}$$

The boundary condition (Equation 33) can be inserted in Equation 34 yielding

$$\cosh \eta_s = \cosh \eta_m + \frac{2\lambda_D^2}{b^2}, \tag{36}$$

whereas the second boundary condition at $z = d/2$ is expressed as

$$\frac{d}{2\lambda_D} = \int_{\eta_m}^{\eta_s} \frac{d\eta}{\sqrt{2 \cosh \eta - 2 \cosh \eta_m}}. \tag{37}$$

The last three equations (35)-(37) completely determine the potential $\eta(z)$ and the two species density profiles, $n_\pm(z) = n_0 \exp(\pm\eta(z))$ and their mid-plane values $n_m^\pm = n_0 \exp(\pm\eta_m)$, as function of the three system parameters: the inter-surface spacing d, the surface charge density σ (or equivalently b) and the electrolyte bulk ionic strength

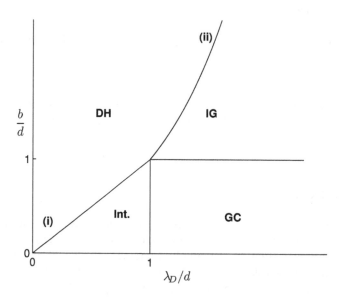

Figure 7. *Schematic representation of the various limits of the PB equation for two flat and equally charged surfaces at separation d. The diagram is plotted in terms of two dimensionless ratios: b/d and λ_D/d, where b, d and λ_D are the Gouy-Chapman length, the inter-surface spacing and the Debye-Hückel length, respectively. The four regions discussed in the text are: the ideal gas (IG), the Gouy-Chapman (GC), intermediate (Int.) and the Debye-Hückel (DH) regions. They are separated by 3 straight lines: $b = \lambda_D$, $b/d = 1$, $\lambda_D/d = 1$ and a parabolic one $b/d = (\lambda_D/d)^2$. The DH region is further divided into two sub-regimes: (i) large d and (ii) small d spacing.*

n_0 (or equivalently λ_D). An example of the counter-ion and co-ion profiles is shown in Figure 6b. Note that in the figure, the three lengths are chosen such that $d > b > \lambda_D$, placing us in between the DH and intermediate regions of Figure 7.

Once the profiles are calculated, the pressure has a simple dependence on the mid-plane properties.

$$\frac{\Pi(d)}{k_B T} = n_m^+ + n_m^- - 2n_0 = 2n_0 \left(\cosh \eta_m - 1 \right). \tag{38}$$

We end the treatment of the 1:1 electrolyte solution between two identically charged surfaces by giving several limiting expressions for the pressure. The exact form can be obtained from the numerical solution of Equations 35-37 as outlined above. Figure 7 summarises the four different regimes in the $(\lambda_D/d, b/d)$ plane. More details about these limiting expressions can be found in Andelman (1995).

Ideal gas region

In the limit of $b/d \gg 1$ and $(\lambda_D/d)^2 \gg b/d$, the pressure reduces to the expression obtained for the counter-ion only case in the limit of small surface charge, Equation 30.

The validity of this *ideal gas* region is for low electrolyte ionic strength and small surface charge.

$$\frac{\Pi(d)}{k_B T} \simeq -\frac{1}{d}\frac{2\sigma}{e} = \frac{1}{\pi l_B b}\frac{1}{d}. \tag{39}$$

Gouy-Chapman region

In the region defined by $\lambda_D/d \gg 1$ and $b/d \ll 1$ where the electrolyte strength is still weak, but for large surface charge density, the expression for the pressure coincides with the other limit, Equation 31. This region is called the *Gouy-Chapman* region

$$\frac{\Pi(d)}{k_B T} \simeq \frac{\pi \varepsilon k_B T}{2e^2}\frac{1}{d^2} = \frac{\pi}{2l_B}\frac{1}{d^2}. \tag{40}$$

Intermediate region

Within the limits of validity $\lambda_D/d \ll 1$ and $b \ll \lambda_D$, in the *intermediate* region, the surface potential is rather large $\eta_s \geq 1$ and $\gamma = \tanh(\eta_s/4) \approx 1$. The PB equation cannot be linearised. On the other hand, the mid-plane potential is small $\eta_m = 8\gamma \exp(-d/2\lambda_D) \ll 1$, and the coupling between the two surfaces is weak.

$$\frac{\Pi(d)}{k_B T} = n_0 \eta_m^2 \simeq \frac{8}{\pi l_B \lambda_D^2}\, e^{-d/\lambda_D}. \tag{41}$$

Debye-Hückel region

The last region is the DH region where the PB equation can be linearised. This region can further be divided into two limits. For large d denoted as case *(i)* in Figure 7, $\lambda_D/d \ll 1$ and $b \gg \lambda_D$, and for small d denoted as case *(ii)* in Figure 7, $\lambda_D/d \gg 1$ and $(\lambda_D/d)^2 \ll b/d$.

The pressure of the linearised DH equation in both limits is given by

$$\frac{\Pi(d)}{k_B T} \simeq \frac{1}{2\pi l_B b^2}\frac{1}{\sinh^2(d/2\lambda_D)}, \tag{42}$$

which reduces in the large d separation, $d \gg \lambda_D$, case (i), to the well-known result

$$\frac{\Pi(d)}{k_B T} \simeq \frac{2}{\pi l_B b^2}\, e^{-d/\lambda_D}, \tag{43}$$

and for small d, $d \ll \lambda_D$, case (ii), to

$$\frac{\Pi(d)}{k_B T} \simeq \frac{2}{\pi l_B}\frac{\lambda_D^2}{b^2 d^2}. \tag{44}$$

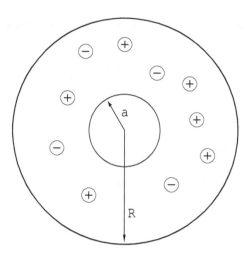

Figure 8. *Schematic illustration of the two-cylinder problem. The aqueous solution is bounded between two concentric cylinders. The inner one at r = a is negatively charged, σ < 0. On the outer cylinder, r = R, the electric field vanishes as the entire cell is electrically neutral.*

4   Poisson-Boltzmann equation in cylindrical coordinates

The PB equation can be solved in cylindrical geometry to model the accumulation of charges around rod-like and charged objects such as rigid polyelectrolytes or elongated colloidal particles. In several situations analytical solutions exist, whereas in others one needs to rely on numerical solutions and approximations.

In the case where we have a solution of rod-like molecules (or cylindrical colloidal particle) we can construct a *cell model*. The model is composed of two concentric cylinders with an ionic solution between them (neglecting any finite size of the cylinder caps), Figure 8. The inner cylinder models a charged and elongated particle, whereas the outer cylinder defines the boundary of the specific volume per charged (cylindrical) particle in case of a multiparticle solution. Because of the cylindrical symmetry it is clear that the potential $\phi(r, \theta, \varphi)$ depends only on the coordinate r, measured from the major axis of the cylinder at the origin. The inner cylinder of radius a has a surface charge density (per unit area) $\sigma < 0$. Alternatively, we can define on the inner cylinder the charge *line density* ρ (number of charges per unit length)

$$\rho \equiv 2\pi a |\sigma/e|. \tag{45}$$

The outer cylinder of radius R defines the total volume of the aqueous solution per charged object (cylinder). From the cylindrical symmetry and the requirement to have charge neutrality within the cell of radius R, the electric field has to vanish at $r = R$.

The boundary conditions then are

$$\frac{d\phi}{dr}\bigg|_{r=a} = -\frac{4\pi}{\varepsilon}\sigma = \frac{2e}{\varepsilon}\frac{\rho}{a},$$

$$\frac{d\phi}{dr}\bigg|_{r=R} = 0.$$

$$(46)$$

4.1 The linearised PB equation: Debye-Hückel theory

We discuss the PB equation in the linear DH limit. As in Section 2, the linear DH gives:

$$\nabla^2\phi = \frac{d^2\phi}{dr^2} + \frac{1}{r}\frac{d\phi}{dr} = \lambda_D^{-2}\phi(r). \tag{47}$$

This equation is the modified Helmholtz equation in cylindrical coordinates and has an analytical solution satisfying the boundary conditions (Equation 46):

$$\frac{e\phi}{k_B T} = -\frac{2}{b\kappa_D}\frac{K_0(\kappa_D r)I_1(\kappa_D R) + I_0(\kappa_D r)K_1(\kappa_D R)}{K_1(\kappa_D a)I_1(\kappa_D R) - I_1(\kappa_D a)K_1(\kappa_D R)}, \tag{48}$$

where $\kappa_D \equiv \lambda_D^{-1}$ is the inverse of the Debye-Hückel screening length, b the Gouy-Chapman length as in Equations 9 and 11, and the functions I_n and K_n are the nth order modified Bessel functions of the first and second kind, respectively.

The limit when the outer cylinder radius goes to infinity, $R \to \infty$, corresponds to the infinite dilution limit of one charged object (cylinder) embedded in an aqueous ionic solution. Then Equation 48 reduces to

$$\frac{e\phi}{k_B T} = -\frac{2}{b\kappa_D}\frac{K_0(\kappa_D r)}{K_1(\kappa_D a)}. \tag{49}$$

The above expression decays exponentially to zero for $\kappa_D r \gg 1$

$$\phi \sim \frac{1}{\sqrt{r}}e^{-\kappa_D r}. \tag{50}$$

4.2 The non-linear PB solution: Counter-ion only and Manning condensation

The non-linear PB equation in cylindrical geometry has an analytical solution for the counter-ion only case (Fuoss *et al.* 1951). We note that rather recently an additional analytical solution has been derived for the added-electrolyte case in a certain limit (Tracy and Widom 1997). This is the limit of infinite dilution ($R \to \infty$) and vanishing inner cylinder radius, $\kappa_D a \to 0$. However, we will restrict ourselves to the counter-ion only case and discuss the interesting phenomena of counter-ion condensation in the infinite dilution limit, known as the Manning condensation (Manning 1969, Oosawa 1971, Le Bret and Zimm 1984). This condensation phenomenon cannot be obtained within the linearised DH regime.

Let us consider again the PB equation for two concentric cylinders, but this time with counter-ions only. The PB equation and the boundary conditions at $r = a$ and $r = R$ is written for the dimensionless potential $\eta = -e\phi/k_B T$

$$\frac{d^2\eta}{dr^2} + \frac{1}{r}\frac{d\eta}{dr} = 4\pi l_B n_0 e^\eta, \tag{51}$$

and

$$\left.\frac{d\eta}{dr}\right|_{r=a} = 4\pi l_B \frac{\sigma}{e} = -\frac{2l_B\rho}{a} \quad \text{and} \quad \left.\frac{d\eta}{dr}\right|_{r=R} = 0. \tag{52}$$

It is possible to map exactly the two-cylinder problem into the simpler two-plate problem discussed earlier in Section 3.3.1 (Burak 2004b). This mapping is an alternative way of looking at the original two-cylinder solution derived by Fuoss *et al* (1951) and detailed in Oosawa (1971).

First we change the distance variable r into u

$$u = \ln\frac{r}{a}, \tag{53}$$

yielding the PB equation

$$\frac{d^2\eta}{du^2} = 4\pi\tilde{n}_0 e^{\eta+2u}, \tag{54}$$

with a renormalised charge density $\tilde{n}_0$ defined as

$$\tilde{n}_0 = l_B a^2 n_0. \tag{55}$$

Making another change of variables for the potential

$$\psi = -\eta - 2u, \tag{56}$$

we obtain an exact mapping of the original cylindrical problem into an equivalent PB for two planar surfaces:

$$\frac{d^2\psi}{du^2} = -4\pi\tilde{n}_0 e^{-\psi} = -4\pi\tilde{n}(\psi), \tag{57}$$

with two boundary conditions

$$\left.\frac{d\psi}{du}\right|_{u=0} = 2(l_B\rho - 1) \quad \text{and} \quad \left.\frac{d\psi}{du}\right|_{u=d} = -2. \tag{58}$$

The mapping is done between the PB equation solved in cylindrical geometry for two concentric cylinders, and the PB equation solved in planar geometry for the counter-ions only case having two planar but *non-identical* charged surfaces. One surface at $u = 0$ has a surface charge density of $(1 - l_B\rho)/2\pi l_B$. This charge can be positive or negative. The second surface at $d = \ln(R/a)$ has a negative surface charge density of $-1/(2\pi l_B)$. The mapping between the two-cylinder problem and the two-plane problem is summarised in the table in the following text.

Before we detail the solution of the two concentric cylinder, let us introduce the concept of the *Manning condensation*. It can be easily understood from this mapping by thinking of the analog planar case in the large d separation. For $l_B\rho < 1$, the surface

	distance	potential	ref. density	inner b.c.	outer b.c.
cylinder	r $a \leq r \leq R$	$\eta = -\dfrac{e\phi}{k_B T}$	n_0	$\dfrac{2 l_B \rho}{a}$	0
planar	$u = \ln \dfrac{r}{a}$ $0 \leq u \leq d = \ln \dfrac{R}{a}$	$\psi = -\eta - 2u$	$\tilde{n}_0 = l_B a^2 n_0$	$\dfrac{1 - l_B \rho}{2\pi l_B}$	$-\dfrac{1}{2\pi l_B}$

Table 1. *Mapping between PB equation in cylindrical and planar geometries*

at $u = 0$ has the same sign as the counter-ions, while the surface at $u = d$ is attractive. When $d \to \infty$ the counter-ions will be repelled from the $u = 0$ surface and will 'run away' to infinity, gaining both entropy and electrostatic attraction with the other surface. However, for $l_B \rho > 1$ the $u = 0$ surface is attractive for the counter-ions. When the other surface is taken to infinity, $d \to \infty$, some of the counter-ions will stay behind balancing entropy and electrostatic attraction, in such a way that the effective charge is always $\rho^* = 1/l_B$. This is the Manning condensation. It states that in infinite dilution (one charged cylinder) and without added salt, the effective charge density of the polyelectrolyte chain never exceeds $1e$ per $l_B \simeq 7\,\text{Å}$.

We now mention the solution for the two concentric cylinders representing a finite concentration of charged rod-like molecules in the solution. Returning to the mapping introduced above, the charges on the two planar surfaces are not identical. The parameter $1 - l_B \rho$ is the charge density on the surface at $u = 0$, and it can be positive ($l_B \rho < 1$) or negative ($l_B \rho > 1$). The non-linear case of no-added electrolyte for two concentric cylinders was solve by Fuoss *et al.* (1951). Here we use the mapping to the planar geometry to get the same results within a different method. The charge density profile expressed in the cylindrical geometry is

$$
n(r) = \frac{1}{2\pi l_B r^2} \times
\begin{cases}
B^2 \left[\sinh(B \ln \frac{r}{R} - \operatorname{arctanh} B) \right]^{-2} & \rho < \rho^* \\[2ex]
\left(\ln(r/R) + 1 \right)^{-2} & \rho = \rho^*, \\[2ex]
B^2 \left[\sin(B \ln \frac{r}{R} - \arctan B) \right]^{-2} & \rho > \rho^*
\end{cases}
\tag{59}
$$

where the critical value ρ^* is given by

$$
\rho^* = \frac{1}{l_B} \frac{\ln(R/a)}{\ln(R/a) + 1},
\tag{60}
$$

and has the limit of $\rho^* \to 1/l_B$ for infinite dilution, $R/a \to \infty$, in agreement with the Manning condensation threshold. The only other parameter in Equation 59 is the integration constant B, which can be obtained from the boundary condition at the inner cylinder $r = a$. Depending on the value of ρ with respect to ρ^*, B can be obtained by inverting the following equation

$$\rho = \frac{1}{l_{\rm B}} \times \begin{cases} 1 - B\coth[B\ln(R/a) + \text{arctanh}\,B] & \rho < \rho^* \\ \\ 1 - B\cot[B\ln(R/a) + \arctan B] & \rho > \rho^* \end{cases} \quad (61)$$

5   Poisson-Boltzmann equation in spherical coordinates: Charged colloids

Dispersion of small (submicron) particles in a liquid solution is called a colloidal suspension. The suspension can be stabilised against van der Waals attractive forces by several means. In aqueous solutions if the particles are charged, the competition between the electrostatic repulsion and the van der Waals attraction can stabilise the suspension. This is the idea behind the famous DLVO theory of Deryagin, Landau, Verwey and Overbeek (Deryagin and Landau 1941, Verwey and Overbeek 1948), where the attractive van der Waals attraction is balanced with screened repulsive electrostatic repulsion, resulting in a secondary minimum for particle-particle interaction. (For a general introduction to colloids, see the chapter by Frenkel in this volume.)

Let us now consider the limit of infinite particle dilution: one spherical charged particle immersed in a solution containing its counter-ions and possibly added salt. The PB equation can be solved in spherical coordinates. For a perfect spherical particle of radius a and charge Qe, the charge density is $\sigma = Qe/4\pi a^2$.

The linearised PB equation (the DH limit) in spherical coordinates is simply written as:

$$\nabla^2\phi = \frac{d^2\phi}{dr^2} + \frac{2}{r}\frac{d\phi}{dr} = \lambda_{\rm D}^{-2}\phi(r). \quad (62)$$

The linearisation can be justified in the presence of high added salt and moderate particle charge density. For the one-sphere problem, we require the potential and the electric field to vanish at infinity, while on the spherical surface, $r = a$, the potential boundary condition is

$$\left.\frac{d\phi}{dr}\right|_{r=a} = -\frac{Qe}{\varepsilon a^2}$$

$$\left.\frac{d\phi}{dr}\right|_{r=\infty} = 0. \quad (63)$$

Clearly the solution of the linearised PB equation has the DH form in spherical coordinates: $\exp(-\kappa_{\rm D}r)/r$, where $\kappa_{\rm D} = 1/\lambda_{\rm D}$. Together with the boundary condition we get

$$\frac{e\phi(r)}{k_{\rm B}T} = \frac{Ql_{\rm B}}{1 + \kappa_{\rm D}a}\frac{e^{-\kappa_{\rm D}(r-a)}}{r}. \quad (64)$$

The linearised PB equation is correct only in the high salt limit, $\kappa_{\rm D}a \gg 1$. However, even at lower salt concentration, it is useful to consider an *effective* particle charge $Q_{\rm eff}$.

Far away from the charged sphere, the non-linear PB solution will have an asymptotic solution behaving just like the linear PB solution, Equation 64, but with an effective charge Q_{eff} replacing the nominal charge Q of the sphere.

We concentrate on the case of a sphere with enough charge such that $a/b = l_{\text{B}}Q/2a \gg 1$, where b is the same Gouy-Chapman length introduced in the planar geometry, Equation 11, and is equal here to $b = 2a^2/l_{\text{B}}Q$. In the very high salt limit, Q_{eff} is about equal to Q. As we lower the amount of salt, the correction to Q_{eff} is found to be

$$Q_{\text{eff}} = Q\left(1 - \frac{1}{4\kappa_{\text{D}}^2 b^2} + \cdots\right). \tag{65}$$

At intermediate salt concentration, $\kappa_{\text{D}}a \simeq 1$, the behaviour is non-monotonic (and will not be detailed here), while for $\kappa_{\text{D}}a \ll 1$ but not smaller than $\exp(-l_{\text{B}}/2a)$, Q_{eff} saturates at a value (Ramanathan 1986, 1988) that does not depend on Q itself:

$$Q_{\text{eff}} = \frac{2a}{l_{\text{B}}} \ln\left[\frac{4}{\kappa_{\text{D}}a} \ln\left(\frac{1}{\kappa_{\text{D}}a}\right)\right]. \tag{66}$$

Taking values of typical colloidal suspensions we have $a \simeq 200\,\text{Å}$ and $Q = 2000$. This gives us $a/b \simeq 35 \gg 1$. For high salt, Q_{eff} is slightly lower than 2000. As $\kappa_{\text{D}}a$ is lowered, Q_{eff} becomes much lower than Q and then reaches an effective value of about 400 for $\kappa_{\text{D}}a = 10^{-2}$, namely, about a fifth of its original value. It is interesting that for a large range of low salinity, Q_{eff} is a weak function of Q, and depends mainly on the salt concentration and particle size a. In other words, different values of Q will give roughly the same Q_{eff} for the same salinity and particle size.

A similar notion of charge renormalisation (Q_{eff}) was introduced by Alexander *et al.* (1984), for a solution containing a finite concentration of particles. In the *absence* of salt (counter-ion only), it was proposed that the potential far away from the charged spherical particles looks like a DH potential with an effective charge that is due to the presence of all other charged spheres (Belloni 1998). Roughly speaking, Q_{eff} is equal to a/l_{B} multiplied by a logarithmic correction that depends on system parameters. However, this logarithmic correction is small, and this charge renormalisation resembles that of Equation 66 above.

6   Beyond the PB treatment

In this chapter we concentrated on the relatively simple Poisson-Boltzmann equation and have shown how its solutions in different geometries (planar, cylindrical, spherical) are related to several interesting physical problems. The PB theory is a mean-field one and as such it neglects fluctuations and correlations. In addition, it neglects the finite size of the ions and the fact that the solvent is not a continuous media. (But see Burak and Andelman (2000, 2001) and references therein for discrete solvent corrections.) At present, a unified theory that takes into account all corrections to PB is not available, but there are a number of attempts where specific corrections to PB have been proposed and studied in detail. We briefly mention some of these corrections (see also the chapter by Podgornik in this volume).

Strong deviations from PB behaviour is seen in the cases where the concentration of ions in solution is very large. There, electrostatic interactions are highly screened and the specificity of ions, the structure of the water shell around them (hydration shell) and the ionic finite size and polarizabilities come into play. Molecular dynamic (MD) simulations have shown that the water shell around ions causes short-range attraction (Guàrdia *et al.* 1991). Another interesting effect for high electrolyte concentration, $n_0 > (2/\pi)l_B^{-3}$ electrolytes, are phase transitions and related critical phenomena as reviewed by Fisher (1994) and Levin (2002).

Several attempts have been made in the past to use liquid-state theories, (Hansen and McDonald 1986, Rosenfeld and Ashcroft 1979, Henderson 1992), to improve upon the PB treatment by calculating the corrections due to correlations. Although these methods involved uncontrolled approximations (unlike perturbative methods), they are quite successful in high salt concentrations. We mention here only one variant called the *anisotropic hypernetted chain* (AHNC) method. The ANHC involved integral equations and can be used in anisotropic charged systems such as ionic profiles close to charged surfaces (Henderson 1992, Kjellander 1996, Kjellander and Marčelja 1984, 1985). For divalent counter-ions (such as Ca^{++}), ANHC calculations, in agreement with Monte-Carlo simulations and experiments, have confirmed attractive interaction between two highly-charged surfaces with inter-surface separation of a few Angstroms. This result has important consequences in the study of clays and zeolites. This attraction clearly goes beyond the PB treatment because it can be shown rigorously that equally charged surfaces always repel each other within the PB formalism (Neu 1999, Sader and Chan 1999, 2000). Another correlation-induced attraction can be found from Monte-Carlo simulations (Moreira and Netz 2002, Guldbrand *et al.* 1984, Kjellander *et al.* 1992, Gronbech-Jensen *et al.* 1997, Deserno *et al.* 2003); and other analytical techniques (Attard *et al.* 1988, Podgornik 1990, Pincus and Safran 1998, Netz and Orland 2000, Burak and Andelman 2001). Corrections to PB can be quite substantial when the counter-ions in solution are multivalent and the surface charges are large.

The phenomenon of DNA condensation and aggregation in presence of multivalent counter-ions is another example where PB fails to provide the full physical picture. More details are given in the chapter of Podgornik in this volume. The attraction that causes the aggregation and condensation is especially strong in presence of trivalent and tetravalent counter-ions such as spermidine and spermine (Anderson and Record 1980, Raspaud *et al.* 1998) and is a topic of numerous investigations (Bloomfield 1991, Rau and Parsegian 1992a, 1992b, Ha and Liu 1997, Gelbart *et al.* 2000, Grosberg *et al.* 2002, Burak *et al.* 2003, 2004c).

7   Concluding remarks

In this chapter we reviewed some of the underlying principles behind the behaviour of charges in solution. In particular, we considered the way ions in solution will react to the presence of charged boundary conditions, like a charged surface or particle. Considerable insight can be gained from simple models of one charged surface immersed in an ionic solution, or the forces that exist between two such surfaces as mediated by the ionic solution. Other geometries are also useful to consider. Charged cylinders can

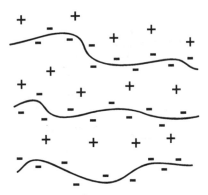

Figure 9. *A schematic illustration of a lamellar stack of membranes. Each membrane is made of a charged bilayer. The entire stack is neutral. The membrane undulations depends on $k_{\mathrm{B}}T$ and the elastic bending modulus.*

be thought of as models of long, rod-like charged molecules, whereas spheres model colloidal particles.

The chapter mainly describes results obtained within the Poisson-Boltzmann formalism that is a mean-field approximation. Corrections to this theory are due to correlations and fluctuations of charge densities, and may play a substantial role.

The simple geometries of a plane, cylinder and sphere may be too simplified in some applications. The geometrical shape of charged membranes and macromolecules is often more complex. In addition, the shape often is not rigid but can deform at room temperature. Hence, the flexibility of the objects and its charge contributions has to be considered.

We close this chapter by mentioning two such examples. Biological membranes are two-dimensional flexible objects (see the chapter by Kozlov in this volume). A stack of membranes forming a lamellar system is shown on Figure 9. When the membranes are charged, the most straightforward effect is a stiffening of the elastic constants. The exact expression of the bending modulus depends on the amount of salt and membrane thickness and charge. A simple result can be obtained in the linearised DH limit where the elastic bending modulus is increased by an amount proportional to

$$\frac{\sigma^2 \lambda_{\mathrm{D}}^3}{\varepsilon} \sim \frac{\lambda_{\mathrm{D}}^3}{l_{\mathrm{B}} b^2}. \tag{67}$$

When the charged membranes are composed of mixtures of charged and neutral lipids, the charges can rearrange themselves laterally on the membrane. This can lead to lateral phase separation and nucleation of charged domains.

Another example is related to the flexibility of charged polymers, depicted in Figure 10. Polymers have a persistence length ℓ_p above which long chains behave as a random walk, while for lengths smaller than ℓ_p, the chain is nearly rigid, rod-like (see the chapters by Warren and Mackintosh in this volume). As the chains become charged, the main effect is the rigidifying of the chains resulting in an increase in the persistence length. Using the linearised PB equation, the electrostatic persistence length was calcu-

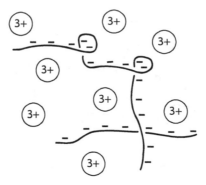

Figure 10. *A schematic representation of a polyelectrolyte solution. The chains are negatively charged and the counter-ions are trivalent.*

lated by Odijk (1977) and independently by Skolnick and Fixman (1977)

$$\ell_p = \frac{l_B \rho^2}{4\kappa_D^2}, \tag{68}$$

where ρ is the line charge density (number of charges per unit length) on the chain. For highly charged chains such as DNA and, in particular, in presence of multivalent counterions, this result has corrections that lead to possible chain instabilities and collapse. Such a phenomena has been observed for DNA as well as for synthetic polyelectrolytes.

Acknowledgements

I am indebted to Y. Burak for comments and to E. Mar-Or for technical assistance. The discussion of the PB solution in cylindrical geometry is based on the PhD thesis of Y. Burak (Tel Aviv University, 2004). This chapter was completed during a stay at the Isaac Newton Institute, University of Cambridge. I thank the Institute for its hospitality. Support from the U.S.-Israel Binational Science Foundation (B.S.F.) under grant No. 287/02 and the Israel Science Foundation under grant No. 210/01 is gratefully acknowledged.

References

Alexander S, Chaikin P M, Grant P, Morales J and Pincus P, 1984, *J Chem Phys* **80** 5776.
Andelman D, 1995, Electrostatic Properties of Membranes: The Poisson-Boltzmann Theory, chapter 12 in *Handbook of Biological Physics*, volume 1, eds Lipowsky R and Sackmann E (Elsevier Science, Amsterdam).
Anderson C F and Record Jr M T, 1980, *Biophys Chem* **11** 353.
Attard P, Mitchell D J, and Ninham B W, 1988, *J Chem Phys* **88** 4987.
Belloni L, 1998, *Colloid Surf A* **140** 227.
Bloomfield V A, 1991, *Biopolymers* **31** 1471.
Borukhov I, Andelman D and Orland H, 1997, *Phys Rev Lett* **79** 435.
Borukhov, I, Andelman D and Orland H, 1998, *Eur Phys J B* **5** 869.
Borukhov I, Andelman D and Orland H, 2000, *Electrochim. Acta* **46** 221.

Burak Y and Andelman D, 2000, *Phys Rev E* **62** 5296.
Burak Y and Andelman D, 2001, *J Chem Phys* **114** 3271.
Burak Y, Ariel G and Andelman D, 2003, *Biophys J* **85** 2100.
Burak Y, Andelman D and Orland H, 2004a, *Phys Rev E* **70** 016102.
Burak Y, 2004b, *Ph.D Thesis* (Tel Aviv University), unpublished.
Burak Y, Ariel G and Andelman D, 2004c, *Curr Opin Colloid Interface Sci* **9** 53.
Chapman, D L, 1913, *Philos Mag* **25** 475.
Evans D F and Wennerström, H, 1994, *The Colloidal Domain* (VCH Publishers, New-York).
Debye P and Hückel E, 1923, *Physik Z* **24** 185. An English translation is published in: *The collected works of Peter J W Debye*, 1954 (Interscience, New York).
De Gennes P G, 1979, *Scaling Concepts in Polymer Physics* (Cornell Univ., Ithaca).
Deryagin B V and Landau L D, 1941, *Acta Physicochim URSS* **14** 633.
Deserno M, Arnold A and Holm C, 2003, *Macromolecules* **36** 249.
Fisher M E, 1994, *J Stat Phys*, **75** 1, and references therein.
Fuoss R M, Katchalsky A and Lifson S, 1951, *Proc Natl Acad Sci (USA)* **37** 579.
Le Bret M and Zimm B H, 1984, *Biopolymers* **23** 287.
Gelbart W M, Bruinsma R F, Pincus P A and Parsegian V A, 2000, *Physics Today* **53** 38.
Gouy G, 1910, *J Phys (Paris)* **9** 457.
Gouy G, 1917, *Ann Phys* **7** 129.
Grahame D C, 1947, *Chem Rev* **41** 441.
Grønbech-Jensen N, Mashl R J, Bruinsma R F and Gelbart W M, 1997, *Phys Rev Lett* **78** 2477.
Grosberg A Y, Nguyen T T and Shklovskii B, 2002, *Rev Mod Phys* **74** 329.
Guàrdia E, Rey R and Padró A, 1991, *J Chem Phys* **95** 2823.
Guldbrand L, Jönsson B, Wennerström H and Linse P, 1984, *J Chem Phys* **80** 2221.
Ha B Y and Liu A J, 1997, *Phys Rev Lett* **79** 1289.
Hansen J P and McDonald I R, 1986, *Theory of Simple Liquids*, second edition (Academic Press, London).
Henderson D, 1992, chapter 6 in *Fundamentals of Inhomogeneous Fluids* 177, ed Henderson D (Marcel Dekker, New York).
Israelachvili J N, 1992, *Intermolecular and Surface Forces* (Academic Press, London).
Kjellander R and Marčelja S, 1984, *Chem Phys Lett* **112** 49; 1985, **114** 124(E).
Kjellander R and Marčelja S, 1985, *J Chem Phys* **82** 2122.
Kjellander R, Åkesson T, Jönsson B and Marčelja S, 1992, *J Chem Phys* **97** 1424.
Kjellander R, 1996, *Ber Bun Phys Chem* **100** 894.
Levin Y, 2002, *Rep Prog Phys* **65** 1577.
Manning, G S, 1969, *J Chem Phys* **51** 924.
Moreira A G and Netz R R, 2002, *Eur Phys J E* **8** 33.
Netz R R and Andelman D, 2003, *Phys Rep* **380,** 1.
Netz R R and Orland H, 2000, *Eur Phys J E* **1** 203.
Neu J C, 1999, *Phys Rev Lett* **82** 1072.
Odijk T, 1977, *J Polym Sci Part B: Polym Phys* **15** 477.
Oosawa F, 1971, *Polyelectrolytes* (Marcel Dekker, New York).
Parsegian V A and Gingell D, 1972, *Biophys J* **12** 1192.
Pincus P A and Safran S A, 1998, *Europhys Lett* **42** (1998) 103.
Podgornik R, 1990, *J Phys A* **23** 275.
Ramanathan G V, 1986, *J Chem Phys* **85** 2957.
Ramanathan G V, 1988, *J Chem Phys* **88** 3887.
Raspaud E, de la Cruz M O, Sikorav J L and Livolant F, 1998, *Biophys. J* **74** 381.
Rau D C and Parsegian A, 1992a, *Biophys J* **61** 246.
Rau D C and Parsegian A, 1992b, *Biophys J* **61** 260.
Rosenfeld Y and Ashcroft N W, 1979, *Phys Rev A* **20** 1208.

Sader J E and Chan D Y C, 1999, *J Coll Interface Sci* **213** 268.

Sader J E and Chan D Y C, 2000, *Langmuir* **16** 324.

Skolnick J and Fixman M, 1977, *Macromolecules* **10** 944.

Tracy C A and Widom H, 1997, *Physica A* **244** 402.

Verwey E J W and Overbeek J Th G, 1948, *Theory of the Stability of Lyophobic Colloids* (Elsevier, New York).

Thermal Barrier Hopping in Biological Physics

Tom C B McLeish

Department of Physics and Astronomy, University of Leeds
Leeds LS2 9JT, UK

t.c.b.mcleish@leeds.ac.uk

1   Introduction

The rate at which things happen in life is important. Too fast risks losing control, and lost control means lost coordination in a complex world in which any one process depends on others. Panic replaces coordinated flight and as a result you get eaten. Too slow — and you just get eaten. Yet the requirement of rate control would appear unrealisable within the nanoscale world of biomolecular machines and cell membranes that control metabolism. Here, Brownian motion is always a large component of dynamics. Every molecule is buffeted continuously by the stochastic and unpredictable thermal forces from its environment. What scope is there here for clocks? The answer is a deep and interesting one: if care is taken to design an 'energy landscape' in such a way that the ensemble of states that define the start and end points of a biochemical process are separated by a third ensemble that must be visited on the way but requires significantly higher internal energy (on a scale of $k_B T$) than the initial state, then the mean transition rate over this 'energy barrier' is controlled and relatively narrowly distributed.

This well-known process is schematically represented in Figure 1. In the following, and at other points in this the rest of this volume, we have needed the rate for the escape time of a single degree of freedom (e.g., diffusing particle) over a barrier. We usually write the rate in terms of a 'Bolzmann activation energy' as

$$\tau_{esc} \simeq \tau_0 e^{E^{\neq}/k_B T}, \tag{1}$$

where $E^{\neq}$ is the barrier height or activation energy and τ_0 some local microscopic hopping timescale. For many qualitative purposes, this is sufficient. A few examples in molecular biology and biological physics will remind us how often we need this result:

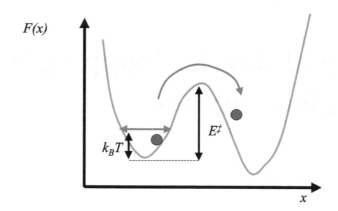

Figure 1. *A schematic representation of a barrier-crossing problem. The particle spends most of its time within k_BT of the bottom of its initial free-energy well, occasionally experiencing a larger fluctuation in energy that carries it over the barrier.*

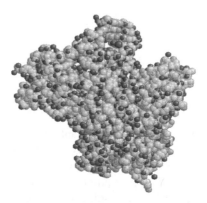

Figure 2. *Topoisomerase I enzyme in a space-filling representation.*

- Enzyme activity is regularly described this way, where the first minimum represents the reactants and the second the products (Alberts *et al.* 2002). Figure 2 displays one of the topoisomerase family of proteins that are needed to bind, cleave and restore a section of DNA. (For more details on topoisomerases, see Bensimon's chapter in this volume.)

  Figure 1 in this case refers to the initial and final topological state of the DNA and enzyme. But the example also illustrates the many simplifications implicit in looking at this complex process in this way. The DNA and protein are complex objects with many internal degrees of freedom. It is not obvious that a one-dimensional 'reaction pathway' will emerge as a consistent model of enzyme action.

- Linear molecular motors such as myosin on actin and kinesin on microtubules have been modelled at the simplest level as 'Brownian ratchets' (Jülicher *et al.* 1997). The energy landscape contains a series of minima that represent the binding sites

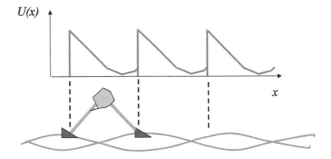

Figure 3. *The one-dimensional energy landscape models a myosin-actin system.*

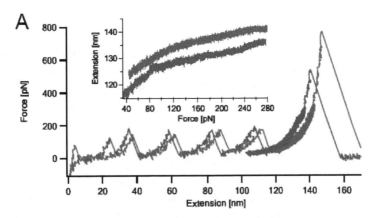

Figure 4. *The extension-force trace of two concatamers consisting of five repeated I27 domains of titin. The inset shows an expanded region of the response of the final domain to unfold. (Courtesy of M. Kawakami, University of Leeds.)*

of the motors along their molecular 'track' (Figure 3).

The energy-consuming process of ATP hydrolysis is modelled by changing the landscape itself so that the motor finds itself continually on an energy gradient that drives procession. However, Brownian noise means that reverse steps are always possible, and are given by expressions analogous to Equation 1.

- Single-molecule force spectroscopy explores the resistance of biomolecules such as globular proteins to tensile forces applied by atomic force microscopy, laser tweezers or biomembrane probes.

Figure 4 gives an example of the well-known sawtooth function of pulling force with distance that arises from the successive unfolding of repeated protein domains concatenated into a single chain. The very complex process of mechanical denaturing of these protein domains is also typically modelled as a two-state process

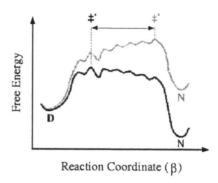

Figure 5. *A one-dimensional barrier-crossing model of protein folding from the denatured state (D) to the folded "native" state (N).*

along a single reaction coordinate, where the first minimum is the folded state, and the second is the unfolded and stretched protein. One simple consequence of applying rate formulae such as Equation 1 to this problem is the prediction (also seen in experiment) that the mean unfolding force varies logarithmically with the imposed unfolding rate.

- Protein folding itself is often modelled by diffusion on a one-dimensional landscape, in spite of the huge number of degrees of freedom actually involved.

  In this case, the landscape may exhibit more complex structure arising from 'intermediate states' illustrated in Figure 5. Adding chemical denaturant is modelled as a change of the overall gradient as in the figure. In complex landscapes this may cause the transition state itself to change.

For these and similar examples, the simple treatment embodied in Equation 1 is often good enough, but there are times when the prefactors to the expression will be important. It is also worth keeping in mind that barrier crossing is an example of an extremal problem (crossing events are rare and not representative of the equilibrium ensemble). In extremes our usual thinking about mean behaviour can let us down badly. In the following we will cover some of the very basic ground in actually deriving mean first passage times for different landscapes, in particular, so that approximations may be avoided where they are misleading.

2   A preliminary: Diffusion on a flat landscape

To illustrate the non-intuitive behaviour that can emerge from the extemal nature of first passage problems, we will first treat a simple case in which the potential is flat (see Figure 6). Diffusers with diffusion constant D are introduced at the origin at a steady rate J and absorbed when they reach $x = -x_0$. It turns out that we also need to keep the problem finite in $x > 0$, so a reflecting boundary at some distant point $x = L$ is placed.

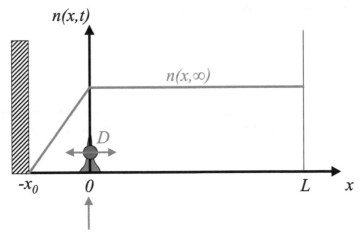

Figure 6. *Diffusing particles are introduced at the origin, reflected at $x = L$ and absorbed at $x = -x_0$. The steady-state concentration distribution is indicated.*

What is the mean time of absorption at $x = -x_0$ following injection of a diffuser at the origin? A naïve guess would be to invert the standard result for the mean-square diffusion with time and estimate $\langle \tau \rangle \simeq x_0^2/D$; but is this true? First, we prove a preliminary result that connects the steady-state density distribution $n_\infty(x)$ with the mean lifetime $\langle \tau \rangle$. If we write the probability that a diffuser introduced at $t = 0$ is still present somewhere in the domain at $t = t'$ as $P(t')$ (so that the probability that the diffuser is absorbed exactly at t' is $-dP(t')/dt'$), we can write the total number of diffusers present at steady state N_∞ in two different ways, firstly as an integral over the steady-state distribution $n(x, t \to \infty)$ and then also as a time history over the birth rate and survival probability:

$$
\begin{aligned}
N_\infty &= \int_{-x_0}^{L} n(x, t \to \infty)\, dx = \int_0^\infty J.P_s(t')\, dt' \\
&= J\left\{ [t' P_s(t')]_0^\infty - \int_0^\infty t' \left(\frac{dP_s(t')}{dt'} \right) dt' \right\} \\
&= J\left\{ 0 + \int_0^\infty t' p_s(t')\, dt' \right\} \\
&= J\langle \tau \rangle.
\end{aligned}
\tag{2}
$$

So the mean first passage time is, in general, just given by

$$
\langle \tau \rangle = \frac{N_\infty}{J} = \frac{1}{J} \int_{-x_0}^{L} n(x, t \to \infty)\, dx.
\tag{3}
$$

Note that, as the problem is a linear one, we shall expect that the final result be independent of J. Now, solving the determining equations

$$\frac{\partial n\,(x,t)}{\partial t} = D\frac{\partial^2 n\,(x,t)}{\partial x^2} \tag{4}$$

$$J = -D\frac{\partial n}{\partial x}$$

at steady state for the boundary conditions of Figure 6, we find zero curvature everywhere, giving the piecewise linear solution for $n\,(x,\infty)$ shown in the figure:

$$\frac{\partial^2 n}{\partial x^2} = 0 \tag{5}$$

$$n_\infty\,(x) = \begin{cases} \left(\frac{J}{D}\right)(x+x_0) & x<0 \\ \frac{Jx_0}{D} & x>0 \end{cases}.$$

Integrating over the particle density gives the total number of diffusers at steady state,

$$N_\infty = \int_{-x_0}^{L} n_\infty(x)dx = \frac{Jx_0}{D}\left(\frac{x_0}{2}+L\right), \tag{6}$$

and finally for the mean lifetime,

$$\langle\tau\rangle = \frac{N_\infty}{J} = \frac{x_0}{D}\left(\frac{x_0}{2}+L\right) \simeq \frac{x_0 L}{D}. \tag{7}$$

This is a surprise: the mean first passage time depends not only *linearly* on the distance to the absorbing boundary x_0 but also on the distance to the reflecting boundary. Physically, this is because the first passage time is actually very broadly distributed and dominated by particles that set off in the 'wrong' direction initially.

3 First passage times: an exact result

Now, we set about using the tools developed in the last section to tackle a more interesting potential. Consider a potential $U(x)$ with a single minimum at $x=0$ (the mean position of the diffusers). We want to calculate the average first passage time of a particle through position $s>0$, given that it is introduced at $x=0$ at $t=0$. This is equivalent to the mean lifetime of the particle if an absorbing barrier is placed at $x=s$. To solve this problem, we consider introducing, as before, a steady current $J\delta(x)$ of diffusers at the origin and waiting until a steady-state number density $n_\infty(x)$ of diffusers has been established. Then, the total number of particles in the distribution is just the 'supply' current multiplied by the mean survival time τ_{esc}. So

$$\tau_{esc} = \frac{1}{J}\int_{-\infty}^{s} n_\infty(x)dx. \tag{8}$$

We assume that the diffusion constant is D, and we will work with units of k_BT for the energy. We first need to specify contributions to the local current of particles $j(x)$,

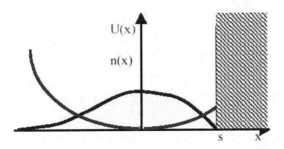

Figure 7. *The steady state distribution of diffusers introduced at the origin into a potential $U(x)$ with a minimum there, and an absorbing boundary at $x = s$.*

for if we do, then the general expression for the rate of change of local number density is just the 'divergence' of the current

$$\frac{\partial n}{\partial t} = -\frac{\partial j}{\partial x}. \tag{9}$$

Contributions to the current are of two types. One we have seen earlier, the diffusive contribution

$$j_{\text{diff}} = -D\frac{\partial n}{\partial x} \tag{10}$$

that yields the well-known diffusion equation again when substituted into Equation 9. The second contribution is from the force on the particles when they are moving in a potential. The force per particle is just

$$f = -\frac{\partial U}{\partial x}. \tag{11}$$

This is opposed by the drag coefficient per particle ζ to give a drift velocity for each particle at x of

$$v = -\frac{1}{\zeta}\frac{\partial U}{\partial x} \tag{12}$$

and a final forced current due to the motion of all particles at x of

$$j_f = -\frac{1}{\zeta}\frac{\partial U}{\partial x}n\left(x,t\right). \tag{13}$$

A very important relation exists between the two apparently unrelated constants D and ζ. Due to Einstein, it arises from the balancing of the two currents j_{diff} and j_f at equilibrium, when there is no net current. The equilibrium condition is just set by the Boltzmann distribution (writing $k_B T$ explicitly for a moment)

$$n\left(x,t\right) = n_0 e^{-U(x)/kT}. \tag{14}$$

Now, by setting the total current at equilibrium to zero,

$$j_{\text{diff}} + j_f = 0 \qquad (15)$$

$$-D\frac{\partial n}{\partial x} - \frac{1}{\zeta}\frac{\partial U}{\partial x}n(x,t) = 0$$

$$\frac{D}{k_B T}\frac{\partial U}{\partial x}n_0 e^{-U(x)/kT} - \frac{1}{\zeta}\frac{\partial U}{\partial x}n_0 e^{-U(x)/kT} = 0$$

and cancelling terms $\frac{\partial U}{\partial x}n_0 e^{-U(x)/kT}$, we find that this only holds if

$$D = \frac{k_B T}{\zeta}, \qquad (16)$$

which is the Stokes-Einstein relation. Diffusion and drag turn out to be deeply connected. We can now write the dynamical equation from the divergence relation, the two currents and the Stokes-Einstein relation. The form is made even simpler if we work from now on again in units such that $k_B T = 1$. Then, $n(t,x)$ satisfies

$$\frac{\partial n}{\partial t} = -\frac{\partial}{\partial x}D\left(-\frac{\partial n}{\partial x} - n\frac{\partial U}{\partial x}\right) + j\delta(x) = 0 \qquad (17)$$

at steady state. In the $x > 0$ regions we may integrate this once directly to give

$$\frac{\partial n}{\partial x} + \frac{\partial U}{\partial x} = \frac{-j}{D}, \qquad (18)$$

and once more by using the integrating factor e^U throughout:

$$\frac{\partial}{\partial x}\left(ne^U\right) = \frac{-je^U}{D} \implies n(x) = \frac{j}{D}e^{-U(x)}\int_x^s e^{U(x')}dx' \text{ for } x > 0. \qquad (19)$$

In $x < 0$, there is no current (all the introduced flux ends up at the absorbing boundary at $x = s$), so the population distribution is just proportional to the equilibrium distribution, with the prefactor chosen to match with the solution in $x > 0$. So:

$$n(x) = \frac{j}{D}e^{-U(x)}\int_0^s e^{U(x')}dx' \text{ for } x < 0. \qquad (20)$$

Now, integrating $n(x)$ over all x (it is sensible to reverse the order of integration between x and x' so that the two pieces in $x < 0$ and $x > 0$ can be joined in one initial integration from $-\infty$ to x') and using the relation for the mean lifetime we find

$$\tau(s) = \frac{1}{D}\int_0^s dx' e^{U(x')}\int_{-\infty}^{x'} dx e^{-U(x)}. \qquad (21)$$

This is the exact solution for the mean lifetime. There is usually no need to approximate it further, as in nearly all cases, the integrals are cheap to compute, but it can (and very often is) approximated when the barrier is high ($U(s) \gg 1$). For now the inner integral is dominated by the contribution near the origin where U is at its minimum — a good approximation is the consequent Gaussian integral arising from the approximation

$$U(x) \simeq \frac{1}{2}U''(0)x^2. \qquad (22)$$

The outer integral is likewise dominated by the contribution near the upper limit. This may be expanded as

$$U(x) \simeq U(s) - (x - s)U'(s) \tag{23}$$

for a cusp-shaped barrier as in the figure, or, for smooth barriers where the gradient of the potential vanishes at the top of the barrier, as

$$U(x) \simeq U(s) - \frac{1}{2}(x - s)^2 U''(s) \tag{24}$$

to give an exponential or another Gaussian integral respectively. The final results are, for a cusp-shaped barrier,

$$\tau(s) \simeq \frac{1}{DU'(s)} \sqrt{\frac{2\pi}{U''(0)}} e^{U(s)} \tag{25}$$

and for a smooth barrier,

$$\tau(s) \simeq \frac{\pi}{D} \sqrt{\frac{1}{U''(s)U''(0)}} e^{U(s)}. \tag{26}$$

These last two approximations occur very frequently in the literature, and unfortunately not exclusively when the barrier is high. The exact result and the approximations were most famously recorded in the literature by Kramers (1940).

4   Landscapes and intermediate states

The results of the last section tell us that the prefactor of the dominant $e^{E^{\neq}/k_B T}$ is not necessarily close to unity, especially when the potential is large (and so, by inheritance, its local curvature and gradients). In both cases of smooth and sharp barriers, the result for the mean first passage time at $x = s$ can be interpreted as

$$\tau(s) \simeq \frac{l_0 l_b}{D} e^{U(s)},$$

where the lengths l_0 and l_b are the effective widths of the potential at the minimum and at the barrier (by "effective widths" we mean the lateral scales in x over which the potential changes by $k_B T$, or by order 1 in our dimensionless energy). Clearly, strategies that minimise these lengths (they may be very much shorter than the length scale of s itself, for example) may radically speed up barrier crossing without altering the (free) energy of the barrier at all. For example, the potentials (1) and (2) of Figure 8 have minima and maxima whose widths are characterised by the very different lengthscales l_1 and l_2.

In consequence, they have barrier-crossing times that follow the ratio

$$\frac{\langle \tau_1 \rangle}{\langle \tau_2 \rangle} = \left(\frac{l_1}{l_2} \right)^2,$$

which in this case is of the order 10^2, even though their barrier heights are identical. Although potential landscapes such as this one are rather artificial, a more natural way

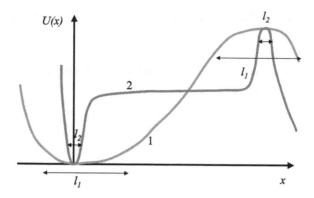

Figure 8. *Two potential landscapes with identical activation energies but widely different first passage times. Potential (2) has faster barrier crossing by virtue of the smaller effective widths at the origin and at the barrier.*

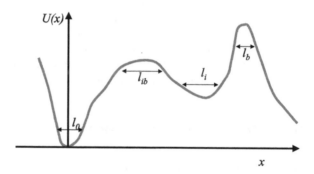

Figure 9. *A barrier with an intermediate state, whose presence may actually increase the rate of barrier crossing.*

of creating the same short effective length scales at the minimum and at the barrier is to introduce "intermediate states". The corresponding landscape now looks like Figure 9.

In the case of such states, we may write down the integral expression for the exact mean passage time, and then approximate it when the barrier is high in the spirit of the results for simple barriers. Terms in $e^{U(x')}$ within the integral will be dominated by local maxima of the potential, while terms in $e^{-U(x)}$ will pick up behaviour from the local minima. So,

$$
\begin{aligned}
\tau(s) &= \frac{1}{D} \int_0^s dx' e^{U(x')} \int_{-\infty}^{x'} dx\, e^{-U(x)} \\
&\simeq \frac{1}{D} \left[l_{ib} e^{U_{ib}} l_0 + l_b e^{U_b} \left(l_0 + e^{-U_i} l_i \right) \right] \\
&= \frac{1}{D} \left[l_{ib} l_0 e^{U_{ib}} + l_b l_i e^{(U_b - U_i)} + l_b l_0 e^{U_b} \right].
\end{aligned}
$$

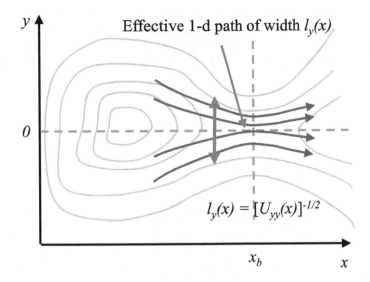

Figure 10. *A two-dimensional barrier-crossing problem, with the potential surface represented by contours. The highest flux of diffusers is restricted to a path of the effective width of the saddle, l_y.*

The terms in this approximation bear an interesting physical meaning. The final term is just the result for the simple barrier: it represents a diffuser whose extremal trajectory begins near the origin and finishes at the barrier without any re-equilibration along the way. The first two terms describe a second pathway that the intermediate state opens up: the first is the mean first passage time to the first barrier and the second the mean activated time to the main barrier, taking the intermediate state as the starting point. Adding those terms together gives the mean time for the new pathway that spends sufficient time at the intermediate state to explore the metastable configurations there, before making another activated crossing of the main barrier. A recent discussion of the role of complex intermediate states in barrier crossing has been made in the context of protein folding (Sanchez and Kiefhaber 2003). The main point here is again that the structure of the landscape, even in one dimension, can make a big difference to the first passage times, without changing the barrier height (and the Boltzmann factor that it induces).

5 Higher-dimensional barrier crossing

In nearly all physical cases, barrier crossing takes place in a space of several active degrees of freedom. This is true of all the biological examples of the introduction, and supremely true in the case of protein folding, which constitutes a search in very large (of order several hundred) degrees of freedom. How is the barrier-crossing rate modified when other degrees of freedom are active? We begin by taking a two-dimensional system with a basin and a saddle (see Figure 10).

The central result, Equation 3, applies in the natural generalisation of integration over the two-dimensional steady-state distribution of diffusers. However, the effect of introducing the new degree of freedom in this case can be seen by taking, as before, the simpler case of large potentials when the barrier at x_b is a deep minimum in the new co-ordinate, y, so that it becomes a saddle. Now the current equation in two dimensions,

$$\mathbf{j} = D\left(-n\nabla U - \nabla n\right), \tag{27}$$

can be evaluated along the saddle line at x_b where, by definition, $j_y = 0$ and so

$$\frac{1}{n}\frac{\partial n}{\partial y} = -\frac{\partial U}{\partial y} \qquad \Rightarrow \qquad n\left(x_b, y\right) = n\left(x_b, 0\right) e^{[U(x_b,0) - U(x_b,y)]}. \tag{28}$$

The distribution of diffusers is limited effectively to a path over the saddle of an effective width l_y given by the curvature of the sides of the saddle. We can map approximately the two-dimensional problem in $n_2(x, y)$ onto the one-dimensional one above in $n(x)$ by writing $n(x) = n_2(x, 0).l_y(x)$ where we define the variable width $l_y(x) = \ln\left[U_{yy}(x)/2\pi\right]$ (here we use the notation $\partial^2 U/\partial y^2 \equiv U_{yy}$ etc.). The final result at the level of the high barrier approximation is

$$\langle \tau \rangle \simeq \frac{\pi}{D}\sqrt{\frac{1}{U_{xx}\left(0\right)U_{xx}\left(x_b\right)}}\sqrt{\frac{U_{yy}\left(x_b\right)}{U_{yy}\left(0\right)}}e^{U(x_b)}. \tag{29}$$

Note that the additional term may be absorbed in to the definition of the (one-dimensional) potential so that

$$U(x, y) \to U_1(x) = U(x, 0) + \left[\frac{1}{2}\ln U_{yy}\left(x, 0\right) - \frac{1}{2}\ln U_{yy}\left(0, 0\right)\right]. \tag{30}$$

The new term results from the projection of the full potential down to an effective potential in one degree of freedom, and looks remarkably like an expression for entropy (except, of course, we are dealing with non-equilibrium processes here). It is very tempting to extend the definition of *free energy* so that $U_1(x)$ becomes the effective one-dimensional free energy of the projected system. The obvious generalisation of Equation 29 to more degrees of freedom applies with the curvatures in all the saddle dimensions occurring as a sequence of products (or sums in the entropic terms of the renormalised projected potential). These results are discussed in greater detail by Hanggi *et al.* (1990).

Not all high-dimensional spaces contain neat divisions of the degrees of freedom into saddle dimensions and path dimensions. For example, I have recently discussed strategies for designing the energy surface in the high degrees of freedom that arise in protein folding (McLeish 2005). New complexities arise when the effective diffusion constants in different degrees of freedom differ widely. Low barriers in slow degrees of freedom may have to compete with higher barriers in faster degrees of freedom. A physical example of this kind arises in the entangled dynamics of melts of star-shaped polymers (McLeish 2003). Clearly the range of internal modes available to proteins, from fast local distortion to much slower diffusion or conformational changes, bring questions such as these into the analysis of enzyme function.

References

Alberts B, Johnson A, Lewis J, Raff M, Roberts K and Walter P, 2002, *Molecular Biology of the Cell*, fourth edition (Garland Science, New York).

Hanggi P, Talkner P and Borkovec M, 1990, *Rev Mod Phys* **62** 251.

Jülicher F, Ajdari A and Prost J, 1997, *Rev Mod Phys* **69** 1269.

Kramers H A, 1940, *Physica* **7** 284.

McLeish T C B, 2003, *J Rheol* **47** 177.

McLeish T C B, 2005, *Biophys J* **88** 172.

Sanchez I E and Kiefhaber T, 2003, *J Mol Biol* **325** 367.

References

Part II

Biological Applications

Part II

Biological Applications

Elasticity and dynamics of cytoskeletal filaments and their networks

Fred C MacKintosh

Free University of Amsterdam
De Boelelaan 1081, 1081HV Amsterdam, The Netherlands
fcm@nat.vu.nl

1   Introduction

Biological cells are often subject to a variety of external stresses and forces. They also exert forces on their surroundings, for instance, in cell locomotion. There are many important and outstanding problems in cell biology concerning the origins and regulation of cell mechanical properties (Alberts *et al.* 1994, Boal 2002). Such mechanical factors determine how a cell maintains and modifies its shape, how it moves, and even how cells adhere to one another. Mechanical stimulus of cells can also result in changes in gene expression.

As materials or mechanical systems, cells exhibit rich composite structures on the nanometre to micrometre scale, involving rather soft membranes and rather rigid filamentous proteins or biopolymers. Most plant and animal cells, in fact, possess a complex network structure of biopolymers and associated proteins and enzymes for bundling, cross-linking, and active force generation. This *cytoskeleton* is often the principal determinant of cell elasticity and mechanical stability. Over the last few years, the single-molecule properties of many of the important constituents of the cytoskeleton have been studied in detail by biophysical techniques such as high-resolution microscopy, scanning force microscopy, and optical tweezers. At the same time, numerous *in vitro* experiments aimed at understanding some of the unique mechanical and dynamic properties of solutions and networks of cytoskeletal filaments have been performed. In parallel with these experiments, theoretical models have emerged that have both served to explain many of the essential material properties of these networks, as well as motivate quantitative experiments to determine, e.g., concentration dependence of shear moduli and the effects of cross links. In this chapter, we focus on the state of the theoretical modelling of both cytoskeletal solutions and networks, as well as the properties of synthetic semiflexible polymer systems with similar underlying properties.

One of the principal components of the cytoskeleton, and in fact one of the most prevalent proteins in the cell, is actin, which exists in both monomeric/globular (G-actin) and polymeric/filamentary (F-actin) forms. In animal cells especially this often forms a network of entangled and cross-linked filaments known as the actin cortex, which is frequently found near the periphery of cells, and appears to be strongly associated with the outer membrane. *In vivo*, this network is far from passive, with both active motion and force generation during cell locomotion, and with a strong coupling to membrane proteins that appears to play a key role in the ability of cells to sense and respond to external stresses.

Fundamental to understanding any of these complex structures is a quantitative model for the structure, interactions, and response of networks such as the actin cortex. Unlike conventional polymer networks and gels, however, these networks have been clearly shown to possess properties that cannot be modelled by existing polymer theories. These properties include anomalously large shear moduli, strong signatures of non-linear response (in which, for example, the shear modulus can increase by a full factor of 10 or more under modest strains of only 10% or so) (Janmey *et al.* 1994), and unique dynamics. In a very close and active collaboration between theory and experiment over the past decade or so, a standard model of sorts for the material properties of such networks has emerged. Central to these models has been the semiflexible nature of the constituent filaments, which is both a fundamental property of almost any filamentous protein, as well as a clear departure from conventional polymer physics, which has focused on flexible or rod-like limits. In contrast, biopolymers such as F-actin are nearly rigid on the scale of a micrometre, while at the same time showing significant thermal fluctuations on the cellular scale of a few microns.

We shall begin with an introduction to the models of single-filament response and dynamics. In fact, we shall spend most of our time on a detailed understanding of these single-filament properties. Because cytoskeletal filaments are the most important structural components in cells, a quantitative understanding of their mechanical response to bending, stretching, and compression is crucial for any model of the mechanics of networks of these filaments. We see how the fundamental properties of the individual filaments can explain many of the properties of solutions and networks.

2 Single-filament properties

Biopolymers such as those that make up the cytoskeleton consist of aggregates of large globular proteins that are bound together rather weakly, as compared with most synthetic, covalently-bonded polymers. Nevertheless, they can in many cases form filaments of surprising stability and strength. Being rather bulky, however, with diameters as large as 10-25 nm, they are far from flexible to bending. In fact, the most fundamental aspect in determining their long wavelength behaviour can be said to be their bending rigidity. In many cases, however, the molecular weight or contour length of these filaments is still

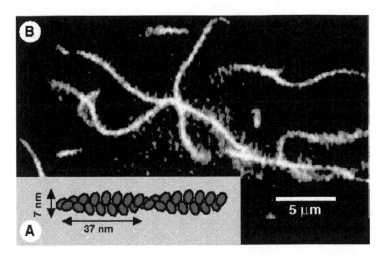

Figure 1. *Entangled solution of semiflexible actin filaments. (A) In physiological conditions, individual monomeric actin proteins (G-actin) polymerise to form double-stranded helical filaments known as F-actin. These filaments exhibit a polydisperse length distribution of up to about 70 μm in length. (B) A solution of 1.0 mg/ml actin filaments, approximately 0.03% of which have been labelled with rhodamine-phalloidin in order to visualise them by florescence microscopy. The average distance ξ between chains in this figure is approximately 0.3 μm. Reprinted with permission from MacKintosh F C, Käs J, and Janmey P A,* Physical Review Letters, *75 4425 (1995). Copyright 1995 by the American Physical Society.*

large enough that they may begin to show significant thermal bending fluctuations. Thus they are said to be *semi-flexible* or *worm-like*.

The length at which significant bending fluctuations occur is a convenient characterisation of the mechanical stiffness of such polymers. This thermal bending length or *persistence length* ℓ_p is more precisely defined in terms of the the angular correlations (e.g., of the local tangent along the polymer backbone), which decay exponentially with a characteristic length ℓ_p. The persistence lengths of a few important biopolymers are given in Table 1, along with their approximate diameter and length. The significance of the persistence length is nicely illustrated in Figure 1, showing fluorescently labelled F-actin filaments on the micrometre scale.

2.1 Worm-like chain

When modelling such a polymer that is rigid on the molecular scale of a few nanometres, it is natural to think of the chain as a continuous line with finite resistance to bending. This is the essence of the so-called worm-like chain model. This can be described by an energy or Hamiltonian of the form,

$$H_{\text{bend}} = \frac{\kappa}{2} \int ds \left| \frac{\partial \mathbf{t}}{\partial s} \right|^2, \qquad (1)$$

Type	Approximate diameter	Persistence length	Contour length
DNA	2 nm	50 nm	$\lesssim 1$ m
F-actin	7 nm	17 μm	$\lesssim 50$ μm
Microtubule	25 nm	$\sim$ 1-5 mm	10s of μm

Table 1. *Persistence lengths and other parameters of various biopolymers (Howard 2001, Gittes* et al. *1993).*

where κ is the *bending modulus* and t is a (unit) tangent vector along the chain. Here, the chain position $\mathbf{r}(s)$ is described by an arc-length coordinate s along the chain backbone. Hence, the tangent vector is given by

$$\mathbf{t} = \frac{\partial \mathbf{r}}{\partial s}.$$

The bending modulus κ has units of energy times length. A natural energy scale for a rod subject to Brownian fluctuations is $k_B T$, where T is the temperature and k_B is Boltzmann's constant. Thus, $\ell_p = \kappa/(k_B T)$ is a length. This is a fundamental length, known as the *persistence length*, which has a very simple physical interpretation. It is the typical length scale over which the polymer forgets its orientation, i.e., the maximum length over which the polymer will appear straight in the presence of the constant Brownian forces it experiences in a medium at finite temperature.

More precisely, for a homogeneous rod of diameter $2a$ consisting of a homogeneous material, the bending modulus should be proportional to the Young's modulus E, which has units of energy per volume. Thus, on dimensional grounds, we expect that $\kappa \sim Ea^4$. In fact (Landau and Lifshitz 1986),

$$\kappa = \frac{\pi}{4} Ea^4.$$

This is often expressed as $\kappa = EI$, where I is the moment of inertia of the cross-section. For a rod-like object consisting of a homogeneous elastic material with modulus E, the bending modulus can only depend linearly on E, with a purely geometric prefactor depending only on the cross-section. The factor $\pi a^4/4$ happens to be the right one for a solid rod or radius a. For a hollow tube, the prefactor would be different, but still of order a^4, where a is the (outer) radius.

In general, for bending in 3D, there are two independent directions for deflections of the rod or polymer transverse to its local axis. It is often instructive to consider a simpler case of a single transverse degree of freedom, i.e., motion confined to a plane. Here, the integrand in Equation 1 becomes $(\partial\theta/\partial s)$, where θ is simply the local angle that the chain axis makes relative to any fixed axis. A discrete approximation to the corresponding energy is

$$H \simeq \frac{\kappa}{2} \sum_i (\Delta\theta_i)^2 / \Delta s, \tag{2}$$

where $\Delta\theta_i = \theta_i - \theta_{i-1}$.

Thus, from the equipartition theorem,

$$(\Delta\theta_i)^2 = \frac{k_B T \Delta s}{\kappa}. \tag{3}$$

This can be used to determine the correlations of orientations from one point along the chain to another. Specifically, we note that

$$
\begin{aligned}
\cos(\theta_m - \theta_n) &= \cos(\Delta\theta_m + \theta_{m-1} - \theta_n) \tag{4}\\
&= \cos(\Delta\theta_m)\cos(\theta_{m-1} - \theta_n) - \sin(\Delta\theta_m)\sin(\theta_{m-1} - \theta_n),
\end{aligned}
$$

and thus,

$$
\begin{aligned}
\langle\cos(\theta_m - \theta_n)\rangle &= \langle\cos(\Delta\theta_m)\rangle\langle\cos(\theta_{m-1} - \theta_n)\rangle\\
&\quad \cdots \tag{5}\\
&= \langle\cos(\Delta\theta_m)\rangle^{m-n-1}.
\end{aligned}
$$

Here, we have used the independence of the various $\Delta\theta_i$, and the fact that $\langle\sin(\Delta\theta_m)\rangle = 0$. From this, we can find the correlation function

$$\langle \mathbf{t}(s) \cdot \mathbf{t}(s')\rangle \simeq \langle\cos(\Delta\theta)\rangle^{|s-s'|/\Delta s} \simeq e^{-k_B T|s-s'|/2\kappa}. \tag{6}$$

Here, we have used the fact that $\langle\cos(\Delta\theta)\rangle \simeq 1 - \frac{1}{2}\langle\Delta\theta^2\rangle$, for small Δs and $\Delta\theta$. Of course, this is all for motion confined to a plane. Taking into account the two independent transverse directions for thermal fluctuations, one finds in 3D that

$$\langle \mathbf{t}(s) \cdot \mathbf{t}(s')\rangle = e^{-|s-s'|/\ell_p}, \tag{7}$$

where ℓ_p is the persistence length above. This persistence length provides a *geometric* measure of the *mechanical* stiffness of the rod, provided that it is in equilibrium at temperature T.

This provides, in principle, a way to measure the persistence length, and thus the bending modulus of filaments by imaging the angular correlations along a filament. As discussed in the following text, however, one must also be careful to account for the dynamics of filaments.

2.2 Force extension

At the heart of any model for network properties there must be at least two things: a detailed model for the response at a single molecule level and a characterisation of the way in which the single polymers are connected or otherwise interact with each other. The latter, being inherently a collective property, we reserve for later discussion.

A single filament can respond to forces in at least two ways. It can respond to both transverse and longitudinal forces by either bending or stretching/compressing. On length scales shorter than the persistence length, bending can be described in mechanical terms, as for elastic rods. By contrast, stretching and compression can involve both a purely elastic or mechanical response (again, as in the stretching, compression, or even buckling of macroscopic elastic rods), as well as an entropic response. The latter comes from the thermal fluctuations of the filament. Perhaps surprisingly, as we shall see below, the longitudinal response can be dominated by entropy even on length scales small compared with the persistence length. Thus, it is incorrect to think of a filament as truly rod-like, even on length scales short compared with ℓ_p.

The longitudinal single filament response is often described in terms of a so-called force-extension relationship. Here, the force required to extend the filament is measured or calculated in terms of the degree of extension along a line. At any finite temperature, there is a resistance to such extension due to the presence of thermal fluctuations that make the polymer deviate from a straight conformation. This has been the basis of mechanical studies, for example, of long DNA (Bustamante *et al.* 1994). In the limit of large persistence length, this can be calculated as follows (MacKintosh *et al.* 1995).

We consider a filament segment of length ℓ that is short compared with the persistence length ℓ_p. It is then nearly straight, with small transverse fluctuations. We let the x-axis define the average orientation of the chain segment, and let u and v represent the two independent transverse degrees of freedom. These can then be thought of as functions of x and time t in general. For simplicity, we shall mostly consider just one of these coordinates, $u(x, t)$. The bending energy is then

$$H_{\text{bend}} = \frac{\kappa}{2} \int dx \left(\frac{\partial^2 u}{\partial x^2} \right)^2 = \frac{\ell}{4} \sum_q \kappa q^4 u_q^2, \qquad (8)$$

where we have represented $u(x)$ by a Fourier series

$$u(x, t) = \sum_q u_q \sin(qx). \qquad (9)$$

This is appropriate for the case of fixed boundary conditions $u = 0$ at the ends, $x = 0, \ell$. For this case, the wave vectors $q = n\pi/\ell$, where $n = 1, 2, 3, \ldots$.

We assume that the chain has no compliance in its contour length, i.e., that the total arc length

$$\int dx \sqrt{1 + |\partial u/\partial x|^2} \qquad (10)$$

is fixed. Then, the contraction of the chain relative to its full contour length in the presence of thermal fluctuations in u is

$$\Delta \ell = \int dx \left(\sqrt{1 + |\partial u/\partial x|^2} - 1 \right) \simeq \frac{1}{2} \int dx \, |\partial u/\partial x|^2. \qquad (11)$$

The integration here is actually over the projected length of the chain. But, to leading (quadratic) order in the transverse displacements, we make no distinction between projected and contour lengths here, and above in H_{bend}.

Thus, the contraction

$$\Delta \ell = \frac{\ell}{4} \sum_q q^2 u_q^2. \qquad (12)$$

Conjugate to this variable is the tension τ in the chain. Thus, we consider the effective energy

$$H = \frac{1}{2} \int dx \left[\kappa \left(\frac{\partial^2 u}{\partial x^2} \right)^2 + \tau \left(\frac{\partial u}{\partial x} \right)^2 \right] = \frac{\ell}{4} \sum_q \left(\kappa q^4 + \tau q^2 \right) u_q^2. \qquad (13)$$

Under a constant tension τ therefore, the equilibrium amplitudes u_q must satisfy

$$\langle |u_q|^2 \rangle = \frac{2k_B T}{\ell \left(\kappa q^4 + \tau q^2 \right)}, \tag{14}$$

and the contraction

$$\langle \Delta \ell \rangle = k_B T \sum_q \frac{1}{(\kappa q^2 + \tau)}. \tag{15}$$

There are, of course, two transverse degrees of freedom, and so this last answer incorporates a factor of two appropriate for a chain fluctuating in 3D.

Semi-flexible filaments exhibit a strong suppression of bending fluctuations for modes of wavelength less than the persistence length ℓ_p. More precisely, as we see from Equation 14 above, the mean-square amplitude of shorter wavelength modes are increasingly suppressed as the fourth power of the wavelength. This has important consequences for many of the thermal properties of such filaments. In particular, it means that the longest unconstrained wavelengths tend to be dominant in most cases. This allows us, for instance, to anticipate the scaling form of the end-to-end contraction $\Delta \ell$ between points separated by arc length ℓ in the absence of an applied tension. We note that it is a length and it must vary inversely with stiffness κ and must increase with temperature. Thus, since the dominant mode of fluctuations is that of the maximum wavelength, ℓ, we expect the contraction to be of the form $\langle \Delta \ell \rangle_0 \sim \ell^2 / \ell_p$. More precisely, for $\tau = 0$,

$$\langle \Delta \ell \rangle_0 = \frac{k_B T \ell^2}{\kappa \pi^2} \sum_{n=1}^{\infty} \frac{1}{n^2} = \frac{\ell^2}{6\ell_p}. \tag{16}$$

Similar scaling arguments to those above lead us to expect that the typical transverse amplitude of a segment of length ℓ is approximately given by

$$\langle u^2 \rangle \sim \frac{\ell^3}{\ell_p}. \tag{17}$$

in the absence of applied tension. The precise coefficient for the mean-square amplitude of the midpoint between ends separated by ℓ (with vanishing deflection at the ends) is $1/24$.

For a finite tension τ, however, there is an extension of the chain (toward full extension) by an amount

$$\delta \ell = \langle \Delta \ell \rangle_0 - \langle \Delta \ell \rangle_\tau = \frac{k_B T \ell^2}{\kappa \pi^2} \sum_n \frac{\phi}{n^2 \left(n^2 + \phi \right)}, \tag{18}$$

where $\phi = \tau \ell^2 / (\kappa \pi^2)$ is a dimensionless force. The characteristic force $\kappa \pi^2 / \ell^2$ that enters here is the critical force in the classical Euler buckling problem (Landau and Lifshitz 1986). Thus, the force-extension curve can be found by inverting this relationship. In the linear regime, this becomes

$$\delta \ell = \frac{\ell^2}{\ell_p \pi^2} \phi \sum_n \frac{1}{n^4} = \frac{\ell^4}{90 \ell_p \kappa} \tau, \tag{19}$$

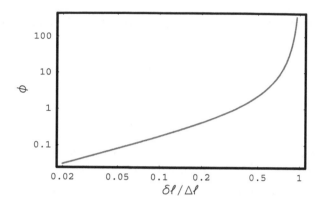

Figure 2. *The dimensionless force ϕ as a function of extension $\delta\ell$, relative to maximum extension $\Delta\ell$. For small extension, the response is linear.*

i.e., the effective spring constant for longitudinal extension of the chain segment is $90\kappa\ell_p/\ell^4$. The scaling form of this could also have been anticipated, based on very simple physical arguments similar to those above. In particular, given the expected dominance of the longest wavelength mode (i.e., ℓ), we expect that the end-to-end contraction scales as $\delta\ell \sim \int |\partial u/\partial x|^2 \, dx \sim u^2/\ell$. Thus, $\langle\delta\ell^2\rangle \sim \ell^{-2}\langle u^4\rangle \sim \ell^{-2}\langle u^2\rangle^2 \sim \ell^4/\ell_p^2$, which is consistent with the effective (linear) spring constant derived above. The full nonlinear force-extension curve can be calculated numerically by inversion of the expression above. This is shown in Figure 2. Here, one can see both the linear regime for small forces with the effective spring constant given above as well as a divergent force near full extension. In fact, the force diverges in a characteristic way, as the inverse square of the distance from full extension: $\tau \sim |\delta\ell - \Delta\ell|^{-2}$ (Fixman and Kovac 1973).

Before concluding our discussion of the longitudinal response of semiflexible polymers, it is worth asking about another obvious contribution to their response. This, we can think of as the *zero-temperature* or *purely mechanical* response. After all, we are treating semiflexible polymers as little bendable rods. To the extent that they behave like rigid rods, we might expect them to respond to longitudinal stresses by, e.g., increasing/decreasing their arc length. Based on the arguments above, it seems that the persistence length ℓ_p determines the length below which filaments behave like rods, and above which they behave like flexible polymers with significant thermal fluctuations. Perhaps surprisingly, however, even for short segments of semiflexible polymers of length much less than the persistence length, their longitudinal response can be dominated by the entropic force-extension described above, i.e., in which the response is due to transverse thermal fluctuations.

In order to examine this, we consider a simple model of a semiflexible polymer as a homogeneous elastic rod of radius a. We have already seen that the bending modulus is $\kappa \sim Ea^4$. Likewise, the (linear) stretching/compression of such an elastic rod is described by the Hamiltonian

$$H_{\text{stretch}} = \frac{1}{2}\mu \int ds \left(\frac{d\ell(s)}{ds}\right)^2, \tag{20}$$

where $d\ell/ds$ gives the relative change in length (strain) along the filament. The stretch modulus $\mu \sim Ea^2$. The effective (mechanical) spring constant of a segment of length ℓ is thus $k_{\mathrm{M}} \sim \mu/\ell \sim Ea^2/\ell$, compared with the effective (thermal) spring constant $k_{\mathrm{T}} \sim \kappa\ell_p/\ell^4 \sim a^2\ell_p/\ell^3$. Since the system will respond primarily according to the softer effective spring constant, the dominant response will be thermal if $\ell^3 \gtrsim a^2\ell_p$, and will be mechanical only if $\ell^3 \lesssim a^2\ell_p$. Thus, even segments of length much less than ℓ_p can still respond according to the thermal response described above. For F-actin, for example, even filament segments as short as a fraction of 1 μm in length may be dominated in their longitudinal compliance by the thermal response arising from bending fluctuations.

2.3 Dynamics

In the preceding text, we have considered only static properties at the single-chain level. The dynamics of individual chains exhibit rich behaviour that can have important consequences even at the level of bulk solutions and networks. The principal dynamic modes come from the transverse motion, i.e., the degrees of freedom u and v. Thus, we must consider time dependence of these quantities. The transverse equation of motion of the chain can be found from H_{bend}, together with the hydrodynamic drag of the filaments through the solvent. This is done via a Langevin equation describing the net force per unit length on the chain at position x,

$$0 = -\zeta\frac{\partial}{\partial t}u(x,t) - \kappa\frac{\partial^4}{\partial x^4}u(x,t) + \xi_{\perp}(x,t), \tag{21}$$

which is, of course, zero within linearised, inertia-free (low Reynolds number) hydrodynamics that we assume here. The first term represents the hydrodynamic drag per unit length of the filament. Here, we have assumed a constant transverse drag coefficient that is independent of wavelength. In fact, the actual (low Reynolds number) drag per unit length on a rod of length L is $\zeta = 4\pi\eta/\ln(AL/a)$, where L/a is the aspect ratio of the rod and A is a constant of order unity that depends on the precise geometry of the rod. For a filament fluctuating freely in solution, a weak logarithmic dependence on wavelength is thus expected. In practice, the presence of other chains in solution gives rise to an effective screening of the long-range hydrodynamics beyond a length of order the separation between chains, which can then be taken in place of L above. The second term in the Langevin equation above is the restoring force per unit length due to bending. It has been calculated from $-\delta H_{\mathrm{bend}}/\delta u(x,t)$ with the help of integration by parts. Finally, we include a random force $\xi_{\perp}$.

A simple force balance in the Langevin equation above leads us to conclude that the characteristic relaxation rate of a mode of wavevector q is (Farge and Maggs 1993)

$$\omega(q) = \kappa q^4/\zeta. \tag{22}$$

The fourth-order dependence of this rate on q is to be expected from the appearance of a single time derivative along with four spatial derivatives in Equation 21. This relaxation rate determines, among other things, the correlation time for the fluctuating bending modes. Specifically, in the absence of an applied tension,

$$\langle u_q(t)u_q(0)\rangle = \frac{2k_BT}{\ell\kappa q^4}e^{-\omega(q)t}. \tag{23}$$

That the relaxation rate varies as the fourth power of the wavevector q has important consequences. For example, while the time it takes for an actin filament bending mode of wavelength 1μm to relax is of order 10 ms, it takes about 100 s for a mode of wavelength $10\ \mu$m.

The very strong wavelength dependence of the relaxation rates in Equation 22 has important consequences, for instance, for imaging of the thermal fluctuations of filaments, as is done in order to measure ℓ_p and the filament stiffness (Gittes *et al.* 1993). This is the basis, in fact, of most measurements to date of the stiffness of DNA, F-actin, and other biopolymers. Using Equation 23, for instance, one can both confirm thermal equilibrium and determine ℓ_p by measuring the mean-square amplitude of the thermal modes of various wavelengths. However, in order to resolve the various modes and to establish that they behave according to the thermal distribution, one must sample over times long compared to $1/\omega(q)$ for the longest wavelengths $\lambda \sim 1/q$. At the same time, one must be able to resolve fast motion on times of order $1/\omega(q)$ for the shortest wavelengths. Given the strong dependence of these relaxation times on the corresponding wavelengths, for instance, a range of order a factor of 10 in the wavelengths of the modes corresponds to a range of 10^4 in observation times.

Another way of looking at the result of Equation 22 is that a bending mode of wavelength λ relaxes (i.e., fully explores its equilibrium conformations) in a time of order $\zeta\lambda^4/\kappa$. Since it is also true that the longest (unconstrained) wavelength bending mode has by far the largest amplitude, and thus dominates the typical conformations of any filament, Equations 15 and 23, we can see that in a time t, the *typical* or dominant mode that relaxes is one of wavelength $\ell_\perp(t) \sim (\kappa t/\zeta)^{1/4}$. As we have seen in Equation 17, the mean-square amplitude of transverse fluctuations increases with filament length ℓ as $\langle u^2 \rangle \sim \ell^3/\ell_p$. Thus, in a time t, the expected mean-square transverse motion is given by (Farge and Maggs 1993, Amblard *et al.* 1996)

$$\langle u^2(t) \rangle \sim (\ell_\perp(t))^3 /\ell_p \sim t^{3/4}, \tag{24}$$

because the typical and dominant mode contributing to the motion at time t is of wavelength $\ell_\perp(t)$.

The dynamics of longitudinal motion can be calculated similarly. Here, however, we must account for the fact that the mean-square longitudinal fluctuations $\langle \delta\ell^2(t) \rangle$ of a long filament involve the sum (in quadrature) of independently fluctuating segments along a full filament of length ℓ. The typical size of such independently fluctuating segments at time t is $\ell_\perp(t)$, of which there are $\ell/\ell_\perp(t)$. As shown above, the mean-square amplitude of longitudinal fluctuations of a fully relaxed segment of length $\ell_\perp(t)$ is of order $\ell_\perp^4(t)/\ell_p^2$. Thus, the longitudinal motion is given by (Granek 1997, Gittes and MacKintosh 1998)

$$\langle \delta\ell(t)^2 \rangle \sim \frac{\ell}{\ell_\perp(t)} \times \frac{\ell_\perp(t)^4}{\ell_p^2} \sim t^{3/4}, \tag{25}$$

where the mean-square amplitude is smaller than that for the transverse motion by a factor of order ℓ/ℓ_p. Thus, both for the short-time fluctuations as well as the static fluctuations of a filament segment of length ℓ, a filament end explores a disk-like region with longitudinal motion smaller than perpendicular motion by this factor.

For the problem as stated above, i.e., for an isolated fluctuating filament in a quiescent solvent, there is a potential problem with the analysis above, which includes only the effect of drag for motion perpendicular to the filament (Everaers *et al.* 1999). In fact, there is a finite propagation of tension along a semiflexible filament, expressed by yet another length (Morse 1998b)

$$\ell_{\parallel}(t) \sim \sqrt{\ell_{\perp}(t)\ell_p} \sim t^{1/8}. \tag{26}$$

This represents, for instance, the range along the filament over which the tension has spread from a point of disturbance. At very short times, it is possible to observe a $t^{7/8}$ motion in some cases rather than the $t^{3/4}$ in Equation 25 (Everaers *et al.* 1999). For the high-frequency rheology of semiflexible polymer solutions, however, only a dynamical regime corresponding to Equation 25 is observed (Gittes *et al.* 1997, Schnurr *et al.* 1997) and expected (Morse 1998a, Gittes and MacKintosh 1998), at least for long filaments (Pasquali *et al.* 2001). This illustrates a key difference between the dynamics of single polymers in a quiescent solvent and the dynamics of sheared solutions.

3 Solutions of semi-flexible polymer

Given the rigidity of semi-flexible polymers at scales shorter than their contour length, it is not surprising that in solutions they interact with each other in very different ways than flexible polymers would, e.g., at the same concentration. In addition to the important characteristic lengths of the molecular dimension (say, the filament diameter $2a$, the material parameter ℓ_p, and the contour length of the chains), there is another important new length scale in a solution, the *mesh size*, or typical spacing between polymers in solution, ξ. A simple estimate (Schmidt *et al.* 1989) shows how ξ depends on the molecular size a and the polymer volume fraction ϕ. In the limit that the persistence length ℓ_p is large compared with ξ, we can approximate the solution on the scale of the mesh as one of rigid rods. Hence, within a cubical volume of size ξ, there is of order one polymer segment of length ξ and cross-section a^2, which corresponds to a volume fraction ϕ of order $(a^2\xi)/\xi^3$. Thus, we have

$$\xi \sim a/\sqrt{\phi}. \tag{27}$$

While the mesh size characterises the typical spacing between polymers within a solution, it does not entirely determine the way in which they interact with each other. For instance, for a random static arrangement of rigid rods, it is not hard to see that polymers will not touch each other on average except on a much larger length: imagine threading a random configuration of rods at small volume fraction with a thin needle. An estimate of the distance between typical interactions (entanglements) of semiflexible polymers must account for their thermal fluctuations (Odijk 1983). As we have seen above, the transverse range of fluctuations δu a distance ℓ away from a fixed point grows according to $\delta u^2 \sim \ell^3/\ell_p$. Along this length, such a fluctuating filament explores a narrow cone-like volume of order $\ell \delta u^2$. An entanglement that leads to a constraint of the fluctuations of such a filament occurs when, with probability of order unity, another filament crosses through this volume, in which case it will occupy a volume of order $a^2 \delta u$, since $\delta u \ll \ell$. Thus, the volume fraction and the contour length ℓ between constraints is of order $\phi \sim a^2/(\ell \delta u)$. Taking the corresponding length as an entanglement length, we find

$$\ell_e \sim (a^4 \ell_p)^{1/5} \phi^{-2/5}, \tag{28}$$

which is larger than the mesh size ξ in the semiflexible limit $\ell_p \gg \xi$.

These transverse entanglements, separated by a typical length ℓ_e, govern the elastic response of solutions in a way first outlined by Isambert and Maggs (1996). A more complete discussion of the rheology of such solutions can be found in Morse (1998b) and Hinner *et al.* (1998). The basic result for the rubber-like plateau shear modulus for such solutions can be obtained by noting that the number density of entropic constraints (entanglements) is thus $n\ell/\ell_c \sim 1/(\xi^2 \ell_e)$, where $n = \phi/(a^2\ell)$ is the number density of chains of contour length ℓ. In the absence of other energetic contributions to the modulus, the entropy associated with these constraints results in a shear modulus of order $G \sim k_B T/(\xi^2 \ell_e) \sim \phi^{7/5}$. This has been well established in experiments such as those of Hinner *et al.* (1998).

With increasing frequency, or for short times, the macroscopic shear response of solutions is expected to show the underlying dynamics of individual filaments. One of the main signatures of the frequency response of polymer solutions in general is an increase in the shear modulus with increasing frequency. In practice, for high molecular weight F-actin solutions of approximately 1 mg/ml, this is seen for frequencies above a few Hertz. Initial experiments measuring this response by imaging the dynamics of small probe particles have shown that the shear modulus increases as $G(\omega) \sim \omega^{3/4}$ (Gittes *et al.* 1997, Schnurr *et al.* 1997), which has since been confirmed in other experiments and by other techniques.

This behaviour can be understood in terms of the dynamic longitudinal fluctuations of single filaments, as shown above (Morse 1998a, Gittes and MacKintosh 1998). Much as the static longitudinal fluctuations $\langle \delta \ell^2 \rangle \sim \ell^4/\ell_p^2$ correspond to an effective longitudinal spring constant $\sim k_B T \ell_p^2/\ell^4$, the time-dependent longitudinal fluctuations shown in Equation 25 correspond to a time- or frequency-dependent compliance or stiffness in which the effective spring constant increases with increasing frequency. This is because, on shorter timescales, fewer bending modes can relax, which makes the filament less compliant. Accounting for the random orientations of filaments in solution results in a frequency-dependent shear modulus

$$G(\omega) \sim \frac{1}{15} \rho \kappa \ell_p \left(-2i\zeta/\kappa \right)^{3/4} \omega^{3/4} - i\omega\eta, \tag{29}$$

where ρ is the polymer concentration measured in length per unit volume.

4 Network elasticity

In a living cell, there are many different specialised proteins for binding, bundling, and otherwise modifying the network of filamentous proteins. In fact, more than 100 actin associated proteins alone have been identified. Not only is it important to understand the mechanical roles of, e.g., cross-linking proteins, but as we shall see, these can have a much more dramatic effect on the network properties than is the case for flexible polymer solutions and networks.

The introduction of cross-linking agents into a solution of semiflexible filaments introduces yet another important and distinct length scale, which we shall call the cross-link distance ℓ_c. As we have just seen, in the limit that $\ell_p \gg \xi$, individual filaments may only

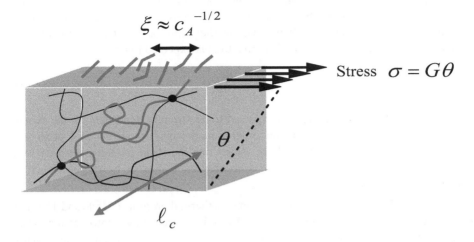

Figure 3. *The macroscopic shear stress depends on the mean tension in each filament, and on the area density of such filaments passing any plane.*

interact with each other infrequently. That is to say that, in contrast with flexible polymers, the distance between interactions of one polymer with its neighbours (ℓ_e in the case of solutions) may be much larger than the typical spacing between polymers. Thus, if there are biochemical cross links between filaments, these may result in significant variation of network properties even when ℓ_c is larger than ξ.

Given a network of filaments connected to each other by cross links spaced an average distance ℓ_c apart along each filament, the response of the network to macroscopic strains and stresses may involve two distinct single-filament responses: (1) bending of filaments and (2) stretching/compression of filaments. Models based on both of these effects have been proposed and analysed. Bending-dominated behaviour has been suggested both for ordered (Satcher and Dewey 1996) and disordered (Kroy and Frey 1996) networks. That individual filaments bend under network strain is perhaps not surprising, unless one thinks of the case of uniform shear. In this case, only rotation and stretching/compression of individual rod-like filaments are possible. This is the basis of so-called *affine* network models (MacKintosh *et al.* 1995). In contrast, bending of constituents involves *non-affine* deformations.

It has recently been shown (Head *et al.* 2003a, Wilhelm and Frey 2003, Head *et al.* 2003b) that which of these behaviours is expected depends, e.g., on filament length and cross-link concentration. Non-affine behaviour is expected either at low concentrations or for short filaments, whereas the deformation is increasingly affine at high concentration or for long filaments. For the first of these responses, the network shear modulus (Non-Affine) is expected to be of the form

$$G_{\mathrm{NA}} \sim \kappa/\xi^4 \sim c^2 \tag{30}$$

when the density of cross links is high (Kroy and Frey 1996, Satcher and Dewey 1996)

For affine deformations, the modulus can be estimated using the effective single-filament longitudinal spring constant for a filament segment of length ℓ_c between cross-links, $\sim \kappa\ell_p/\ell_c^4$, as derived above. Given an area density of $1/\xi^2$ such chains passing through any shear plane (see Figure 3), together with the effective tension of order $(\kappa\ell_p/\ell_c^3)\epsilon$, where ϵ is the strain, the shear modulus is expected to be

$$G_{AT} \sim \frac{\kappa\ell_p}{\xi^2\ell_c^3}. \tag{31}$$

This shows that the shear modulus is expected to be strongly dependent on the density of cross links. Recent experiments on *in vitro* model gels consisting of F-actin with permanent cross links, for instance, have shown that the shear modulus can vary from less than 1 Pa to over 100 Pa at the same concentration of F-actin, by varying the cross-link concentration (Gardel *et al.* 2004).

In the preceding derivation we have assumed a thermal/entropic (Affine and Thermal) response of filaments to longitudinal forces. As we have seen above, however, for shorter filament segments (e.g., for small enough ℓ_c), one may expect a mechanical response characteristic of rigid rods that can stretch and compress. This would lead to a different expression (Affine, Mechanical) for the shear modulus

$$G_{AM} \sim \frac{\mu}{\xi^2} \sim \phi, \tag{32}$$

which is proportional to concentration. The expectations for the various mechanical regimes is shown in Figure 4 (Head *et al.* 2003b).

5   Nonlinear response

One of the characteristics of the elastic response of many biopolymer networks is their non-linear behaviour (Janmey *et al.* 1994). In particular, these networks have been shown to exhibit significant (up to a factor of 10 or more) stiffening under strain. In fact, many biological tissues show the same sort of strain stiffening. These materials are compliant, while being able to withstand a wide range of shear stresses.

This strain stiffening behaviour can be understood in terms of the characteristic force-extension behaviour of an individual semiflexible filament, as described above (MacKintosh *et al.* 1995). As can be seen in Figure 2, for small extensions or strains, there is a linear increase in the force. As the strain increases, however, the force is seen to grow more rapidly. In fact, in the absence of any compliance in the arc length of the filament, the force strictly diverges at a finite extension. This suggests that for a network, the macroscopic stress should diverge, while in the presence of high stress, the macroscopic shear strain is bounded and ceases to increase. In other words, after being compliant at low stress, such a material will be seen to stop responding even under high applied stress.

This can be made more quantitative by calculating the macroscopic shear stress of a strained network, including random orientations of the constituent filaments. Specifically, for a given shear strain γ, the tension in a filament segment of length ℓ_c is calculated based on the force-extension relation above. This is done within the affine approximation, in which the microscopic strain on any such filament segment is determined precisely by the macroscopic strain, as well as the filament's orientation with respect to the shear. The

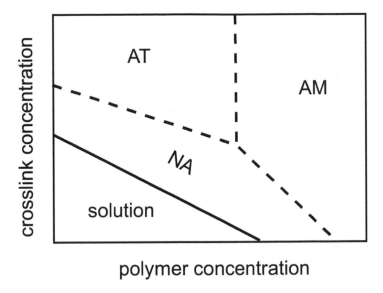

Figure 4. *A sketch of the expected diagram showing the various elastic regimes in terms of crosslink density and polymer concentration. The solid line represents the rigidity percolation transition where rigidity first develops from a solution at a macroscopic level. The other, dashed lines indicate crossovers (not thermodynamic transitions). NA indicates the non-affine regime, while AT and AM refer to affine thermal (or entropic) and mechanical, respectively.*

contribution of such a filament's tension to the macroscopic stress, i.e., in a horizontal plane in Figure 3, also depends on its orientation in space. Finally, the concentration or number density of such filaments crossing this horizontal plane is a function of the overall polymer concentration and the filament orientation.

The full non-linear shear stress is calculated as a function of γ, the polymer concentration, and ℓ_c, by adding all such contributions from all (assumed random) orientations of filaments. This can then be compared with macroscopic rheological studies of cross-linked networks, such as done recently by Gardel *et al.* (2004). These experiments measured the differential modulus, $\partial\sigma/\partial\gamma$, versus applied stress σ and found good agreement with the predicted increase in this modulus with increasing stress. In particular, given the quadratic divergence of the single-filament tension shown above (Fixman and Kovac 1973), this modulus is expected to increase as $\partial\sigma/\partial\gamma \sim \sigma^{3/2}$, which is consistent with the experiments by Gardel *et al.* (2004). This provides a strong test of the underlying mechanism of network elasticity.

In addition to good agreement between theory and experiment for densely crosslinked networks, these experiments have also shown evidence of a lack of strain stiffening behaviour of these networks at lower concentrations (of polymer or cross links), which may provide evidence for a non-affine regime of network response described above.

6 Discussion

Cytoskeletal filaments are not only essential structural elements in cells, but they also have proven remarkable model systems for the study of semiflexible polymers. Their size alone makes it possible to visualise individual filaments directly. They are also unique in the extreme separation of two important lengths, the persistence length ℓ_p and the size of a single monomer. In the case of F-actin, ℓ_p is more than a thousand times the size of a single monomer. This makes for not only quantitative but qualitative differences from most synthetic polymers. We have seen, for instance, that the way in which semiflexible polymers entangle is very different. This makes for a surprising variation of the stiffness of these networks with only changes in the density of cross links, even at the same concentration.

In spite of the molecular complexity of filamentous proteins as compared with conventional polymers, a quantitative understanding of the properties of single filaments provides a quantitative basis for modelling solutions and networks of filaments. In fact, the macroscopic response of cytoskeletal networks quite directly reflects, e.g., the underlying dynamics of an individual semiflexible chain fluctuating in its Brownian environment.

In developing our current understanding of cytoskeletal networks, a crucial role has been played by *in vitro* model systems, such as the one in Figure 1. Major challenges, however, remain for understanding the cytoskeleton of living cells. In the cell, the cytoskeleton is hardly a passive network. Among other differences from the model systems studied to date is the presence of active contractile or force-generating elements such as motors that work in concert with filamentous proteins. Nevertheless, *in vitro* systems may soon permit a systematic and quantitative study of, e.g., various actin-associated proteins for cross-linking and bundling (Gardel *et al.* 2004).

References

Alberts B *et al.*, 1994, *Molecular Biology of the Cell*, third edition (Garland Press, New York).
Amblard F, Maggs A C, Yurke B, Pargellis A N, and Leibler S, 1996, *Phys Rev Lett* **77** 4470.
Boal D, 2002, *Mechanics of the Cell* (Cambridge University Press, Cambridge).
Bustamante C, Marko J F, Siggia E D, Smith S, 1994, *Science* **265** 1599.
Everaers R, Jülicher F, Ajdari A, and Maggs A C, 1999, *Phys Rev Lett* **82** 3717.
Farge E and Maggs A C, 1993, *Macromolecules* **26** 5041.
Fixman M and Kovac J, 1973, *J Chem Phys* **58**, 1564.
Gardel M L, Shin J H, MacKintosh F C, Mahadevan L, Matsudaira P, Weitz D A, 2004, *Science* **304** 1301.
Gittes F, Mickey B, Nettleton J, and Howard J, 1993, *J Cell Biol* **120** 923.
Gittes F, Schnurr B, Olmsted P D, MacKintosh F C, and Schmidt C F, 1997, *Phys Rev Lett* **79** 3286.
Gittes F and MacKintosh F C, 1998, *Phys Rev E* **58** R1241.
Granek R, 1997, *J Phys II (France)* **7** 1761.
Head D A, Levine A J, and MacKintosh F C, 2003a, *Phys Rev Lett* **91** 108102.
Head D A, Levine A J, and MacKintosh F C, 2003b, *Phys Rev E* **68** 061907.
Hinner B, Tempel M, Sackmann E, Kroy K, and Frey E, 1998, *Phys Rev Lett* **81** 2614.
Howard J, 2001, *Mechanics of Motor Proteins and the Cytoskeleton* (Sinauer Press, Sunderland, MA).

Isambert H, Maggs A C, 1996, *Macromolecules* **29** 1036.

Janmey P A, Hvidt S, Käs J, Lerche D, Maggs, A C, Sackmann E, Schliwa M, Stossel T P, 1994, *J Biol Chem* **269**, 32503.

Kroy K and Frey E, 1996, *Phys Rev Lett* **77** 306.

Landau L D and Lifshitz E M, 1986, *Theory of Elasticity* (Pergamon Press, Oxford).

MacKintosh F C, Käs J, and Janmey P A, 1995, *Phys Rev Lett* **75** 4425.

Morse D C, 1998a, *Phys Rev E* **58** R1237.

Morse D C, 1998b, *Macromolecules* **31** 7044.

Odijk T, 1983, *Macromolecules* **16** 1340.

Pasquali M, Shankar V, and Morse D C, 2001, *Phys Rev E* **64** 020802(R).

Satcher R and Dewey C, 1996, *Biophys J* **71** 109.

Schmidt C F, Baermann M, Isenberg G, and Sackmann E, 1989, *Macromolecules* **22**, 3638.

Schnurr B, Gittes F, MacKintosh F C, and Schmidt C F, 1997, *Macromolecules* **30** 7781.

Wilhelm J and Frey E, 2003, *Phys Rev Lett* **91** 108103.

Twisting and stretching DNA: Single-molecule studies

D Bensimon, G Charvin, J-F Allemand, T R Strick and V Croquette

LPS, ENS, UMR 8550 CNRS
24 rue Lhomond, 75231 Paris Cedex 05, France
&
Cold Spring Harbor Laboratory
1 Bungtown road, Cold Spring Harbor NY11724, US

David.Bensimon@lps.ens.fr

1 Introduction

When in 1953 Watson and Crick proposed their famous double helix structure for desoxyribonucleic acid (DNA) (Watson and Crick 1953a), DNA was already known to be the support of genetic heredity (Avery *et al.* 1944, Herschey and Chase 1952). However, this major discovery significantly changed the way of thinking about cellular processes such as the replication of DNA (during cell mitosis) by providing a much needed molecular and structural basis (Watson and Crick 1953b). Ever since, it gradually transpired that the study of molecular interactions within the cell was a necessary step in understanding its function. In the thirty years following the discovery of the double helix, numerous techniques have emerged to advance that study, which by now constitute the bulk of 'molecular biology'. These techniques allow one to transform, synthesise and sequence DNA molecules and also to study and quantify the interactions between biomolecules (protein/DNA interactions, for example). The climax of this so-called genomic era was reached when the human genome sequencing program reached its goal two years ago.

In parallel biophysicists have developed in the past ten years a variety of single-molecule micromanipulation techniques and used them to monitor the mechanical response of different biopolymers, such as double-stranded DNA (dsDNA), single-stranded DNA (ssDNA), ribonucleic acid (RNA), and proteins. From a physical point of view, the study of single-molecule elasticity has provided an ideal testing ground for models of polymer elasticity (see Warren's chapter in this volume for an introduction). It has also stimulated theorists to take into account the interactions (e.g. electrostatic interaction, base pairing, self-avoidance) existing within more realistic polymer chains. These studies might also be helpful in understanding the huge complexity of protein folding, which is one of the great challenges of the 'post-genomic' era.

In biology, and especially in enzymology, single-molecule micromanipulation techniques have demonstrated that elasticity and force generation at the nanometre scale are deeply involved in the mechanism of enzymes and proteins that interact with DNA (such as polymerases or transcription regulation factors). In contrast to bulk assays that measure the average activity of an ensemble of proteins, some of which may be inactive, single-molecule techniques allow the measurement of the distribution of activities of individual enzymes. Moreover, by allowing the measurement of the rate, processivity and step size of single enzymes as a function of force and ATP concentration, these novel techniques have shed a new light on their mechanism.

In this chapter, we first review the various physical and biological reasons that have justified the investigation of DNA mechanics. We then introduce the techniques used to stretch and twist single DNA molecules. We show how such experiments yield a direct observation of the behaviour of DNA under torsional stress, thus allowing for a precise determination of its torsional modulus.

1.1   DNA structure and its biological implications

1.1.1   Primary structure

DNA is composed of two anti-parallel strands wrapping around each other with a right-handed helical symmetry. Each strand consists of a polymer chain of deoxyribose rings linked by phosphodiester bonds between their 3'-OH groups and the phosphate group on the 5' carbon of their neighbour (see Figure 1a). One of four possible different chemical groups: adenine (A), guanine (G), cytosine (C), and thymine (T) is linked to the 1' carbon of the sugar ring. The adenine (guanine) group on one strand is paired with a thymine (cytosine) group on the complementary strand.

1.1.2   Secondary structure

Pairing between A and T and G and C (by hydrogen bonds) yield a stable and almost regular right-handed helix of complementary and anti-parallel strands (see Figure 1(b)) that are to each other as a film is to its negative. In addition to base-pairing, there is a strong hydrophobic attraction between successive base pairs (bp; distance of 3.2 Å) known as 'stacking' interactions. These interactions contribute to the exceptionally high bending rigidity of DNA as compared to artificial polymers such as polyethylene (see the following text).

The geometrical properties of the standard DNA double helix (known as B-DNA: pitch $\sim$ 34, effective diameter $\sim$ 20 Å) (Saenger 1988) vary with environmental parameters such as temperature, solvent, ionic strength and depend on the DNA sequence. Under physiological conditions, the averaged helical repeat is 10.5 base pairs.

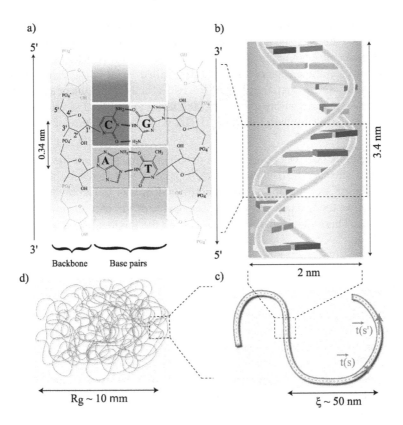

Figure 1. *Primary, secondary and tertiary structure of the DNA molecule. (a) Each of the two strands of the DNA molecule is made of a ribose-phosphate backbone. Chemical groups (Adenine (A), Guanine (G), Cytosine (C) and Thymine (T)) are attached on the ribose sugar. The distance between basepairs is 3.4 Å for the standard B-form of DNA. Note that the two strands are anti-parallel: orientation can be defined using the anchoring points of the phosphate group (3' and 5') on the sugar. (b) Secondary structure of DNA: each strand wraps around each other in a non-symmetric helical fashion. The pitch of the helix is 10.5 base pairs for the standard B-DNA, which means that the twist angle between bases is $\approx 36°$. (c) At larger scales, DNA behaves as a worm-like chain whose orientation decorrelates over a typical length scale of $\xi = 50$ nm. (d) Long DNA molecules ($L >> \xi$) look like random coils with a typical gyration radius (de Gennes 1979) $R_g = \sqrt{2\xi L} \approx 10 \ \mu m$ in the case of a E.* coli *chromosome (4 million base pairs).*

The fact that the bases are buried inside the double helix confers to the molecule a very low reactivity and reduces drastically the interactions of the polymer with itself (in contrast to single-stranded DNA or RNA, which form complex tertiary structures such as hairpins). This feature is essential to preserve the genetic code from being damaged or modified easily, but it presents a challenge to the protein machinery that has to regulate, edit, copy and transcribe the molecule. This machinery must access the bases and therefore open the DNA molecule.

As originally mentioned by Watson and Crick (1953a) and proved by Meselsohn and Stahl (1958), the existence of two copies of the genetic information allows for a simple mechanism of DNA replication: each mother strand is used as a template for the synthesis of its complementary strand, the so-called 'semi-conservative' replication model of DNA. Another essential feature of the two strands complementarity is to permit DNA mismatches and mispairing to be corrected using the original strand as a template. This ensures the extremely high fidelity of DNA replication (an error rate of one base per 10^9).

DNA replication, however, poses a major problem: if the molecule has to be unwound to be duplicated, how are the daughter strands separated 'without everything getting tangled' (Watson and Crick 1953b)? This crucial problem was solved by the discovery of enzymes known as topoisomerases, which are capable of unknotting and untangling the DNA molecule. These topoisomerases control both the torsional stress in DNA and its overall topology.

1.1.3   Tertiary structure

The primary and secondary structures of DNA confer upon the molecule a large bending stiffness, so that the typical length scale, denoted by the persistence length ξ, over which thermal fluctuations are able to bend the axis of the molecule under physiological conditions is about 50 nm (as compared to 1 nm for usual polymers). In other words, if $\mathbf{t}(s)$ and $\mathbf{t}(s')$ are two independent vectors tangent to double helix's axis (parametrised by s), ξ is the typical decorrelation length (see Figure 1(c)):

$$\langle \mathbf{t}(s)\mathbf{t}(s')\rangle = e^{-|s-s'|/\xi}. \tag{1}$$

DNA can thus be modelled as a flexible rope whose random coil geometry constitutes *de facto* the tertiary structure of the DNA. Given the crystallographic (or 'contour') length L of the DNA, one can estimate the typical radius of gyration of the molecule using the model of a 'semi-flexible' polymer in solution (de Gennes 1979):

$$R_g = \sqrt{2\xi L}. \tag{2}$$

For the *E. coli* chromosome, $L \approx 1$ mm so $R_g = 10~\mu\text{m}$ (to be compared with the typical size of the bacteria $\approx 1~\mu\text{m}$), (see Figure 1(d) and 2). Therefore, in order to fit into the cell, DNA has to be packed. The way this packaging is achieved implies major changes in the organisation of DNA. In bacteria the space occupied by the circular chromosomal DNA is minimised by maintaining the molecule in a highly supercoiled state. In eukaryotes (organisms possessing a nucleus) a protein octamer complex known as a histone has about 145 bps of DNA wrapped around it to form a nucleosome core. Such nucleosomes occur every 200 bps along the DNA forming the bead on a string structure of chromatine. Chromatine itself is further compactified into supercoiled structures of higher order, eventually resulting in the packaging of a total DNA about 1 m long (in Humans) into a cell nucleus a few microns in diameter. One of the major challenges of biology is to understand how this extremely compacted molecule is nevertheless regulated, copied, transcribed, repaired and disentangled, as it must be prior to cell division.

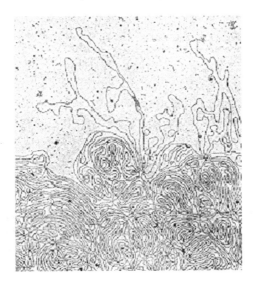

Figure 2. *Protein-depleted chromosome during metaphase. This electron microscopy picture shows the DNA molecule of a chromosome after depletion of the proteins. Assuming 3 billion base pairs for the human genome, each chromosme contains a few hundred million base pairs, i.e., its crystallographic length is about a few centimetres. However, thermal fluctuations tend to entangle the DNA, so that the typical length scale of the DNA coil is a few tens of microns.*

1.2 DNA editing

Compared to the chemical synthesis of artificial polymers (whose length and sequence are poorly controlled), the biochemical synthesis of DNA is a marvel of accuracy and control. Due to the importance of DNA as the support of genetic information, nature has evolved a whole panoply of enzymes that act on DNA to perform a number of text-editing functions: copy (DNA polymerases), cut (restriction enzymes), paste (DNA ligases), correct spelling (mismatch/repair system), ... Although the 'fonts' typically used are A, G, C and T, other 'fonts' are also found in nature: for example, uracil (U) for thymine (T) in messenger RNA or methylated cytosine and adenine in DNA. Artificial 'fonts' have also been introduced by researchers for specific purposes: for example, labelling the four bases with different fluorescent groups for their sequencing (Sanger 1981) or with certain chemical moities (biotine, digoxigenine (DIG)) for their specific binding to certain biomolecules (streptavidin or an antibody against digoxigenine, anti-DIG).

The availability of these extremely powerful molecular editing tools gives scientists unprecedented control on the synthesis and tailoring of DNA for their particular purposes. Thus, a technique known as the polymerase chain reaction (PCR) allows for the generation of faithful copies (with appropriate 'fonts') of a given DNA sequence, whose number increase exponentially with the number of PCR cycles. The use of restriction enzymes and ligases permit a controlled modification of the underlying DNA text and is widely used by molecular biologists, for example, to insert or delete genes. Finally, the availability of off-the-shelf, well-characterised (i.e., sequenced) DNAs from natu-

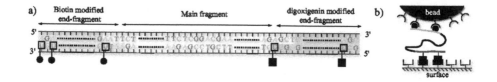

Figure 3. *DNA construct used for micromanipulation. (a) The DNA molecule used for micromanipulation (Strick et al. 1996, 1998a,b) experiments is composed of three parts: the main part, which is a few kilobases long; two fragments (a few hundred base-pairs) synthesised by the polymerase chain reaction (PCR) using biotin or digoxigenin labelled uracile nucleotides (uracile, as thymine, is complementary to adenine). These two fragments prepared separately from the main part of the DNA are cut using appropriate restriction enzymes and ligated using a ligase enzyme. (b) The modification of the small fragments allows for binding on appropriate surfaces (coated with streptavidin or anti-DIG).*

ral sources (plasmids or phages) allows one to use a number of standard 'texts' without having to write one's own from scratch.

To micromanipulate DNA molecules, we use this panoply of enzymes to prepare a DNA molecule (~ 10000 bps long) ligated at its extremities with properly PCR-labelled small fragments (see Figure 3): one fragment multiply tagged with biotin so that it sticks to a streptavidin coated surface, an other multiply labelled with DIG to stick on an anti-DIG-coated bead or surface.

2   Micromanipulation techniques

Many techniques have been developed in the past ten years to micro-manipulate single biomolecules. Common to all these set-ups is the binding of the molecule of interest to a fixed surface at one extremity and to a force sensor at the other. The sensor (or the surface) is often displaced and the resulting force and molecular extension are measured. Different force sensors have been implemented: atomic force (AFM) cantilevers (Florin *et al.* 1994) and microfibers (Ishijima *et al.* 1991, Cluzel *et al.* 1996), optical traps (Smith *et al.* 1996) and optical tweezers (Simmons *et al.* 1996), magnetic traps (Strick *et al.* 1996, Smith *et al.* 1992) and magnetic tweezers (Amblard *et al.* 1996, Gosse and Croquette 2002).

The stiffness of the sensor varies from 10^{-7} N/m for magnetic tweezers to 10^{-2} N/m for AFM cantilevers, so that the range of forces available to stretch biomolecules is large. These different techniques are thus generally complementary.

2.1 Orders of magnitude

Before describing some of these set-ups, let us discuss the order of magnitude of the forces relevant to the physics of biomolecules:

- Covalent binding forces. The largest force encountered, F_{max}, is responsible for breaking a covalent bond, characterised by a typical length ≈ 1 Å and energy ≈ 1 eV: $F_{max} \sim 1\,\text{eV}/1\,\text{Å} = 1.6 \times 10^{-9}\,\text{N} = 1.6\,\text{nN}$. This sets the upper bound for single-molecule-stretching experiments as investigated by the group of H. Gaub using AFM cantilevers (Rief 1997).

- Non-covalent binding forces. Many of the chemical bonds involved in our bodies are non-covalent. They imply a combination of hydrogen binding, hydrophobic and electrostatic interactions. All these weak bonds have a binding energy of a few $k_B T$ while they act on a nanometre scale. Since $1 k_B T = 4 \times 10^{-21}\,\text{J} = 4\,\text{pN nm}$ at room temperature, the tensile force required to break a single weak bond is ~ 4 pN. The stability of the 3D structure of proteins or DNA and their mutual interactions usually involve several weak bonds in a cooperative manner. Thus, forces in the range of 10 to 100 pN will be required to denature protein and DNA or to disrupt their specific binding. One of the strongest non-covalent bond is the one binding biotin and streptavidin (dissociation constant $k_d = 10^{-15}$ M). The force required to break that bond is about 160 pN (Florin *et al.* 1994). In fact, as described by Merkel *et al.* (1997), since thermal fluctuations favour the breaking of a bond under tension, this force is smaller the longer it is applied.

- Entropic forces. In the case of double-stranded DNA, we have seen that thermal fluctuations of energy $1 k_B T$ cause the molecule to bend on the scale of the persistence length $\xi = 50$ nm. Therefore, the force F_e required to stretch DNA to a significant portion of its total length is: $F_e = k_B T/\xi \approx 0.1$ pN. Notice that the *stiffer the polymer*, i.e., the longer the persistence length ξ, the lower the force F_e and thus the *easier it is to stretch it*. This particular feature of polymers is known as the paradox of 'entropic elasticity' and can be explained as follows: the applied force is needed to reduce the number of configurations of the polymer (i.e., reduce its entropy). For a polymer of given length, the stiffer it is, the smaller the number of its possible configurations, the smaller the loss of entropy during stretching and thus the smaller the force required to stretch it.

- Langevin forces. The smallest detectable force on an object of size d in a viscous fluid (viscosity η) is the Langevin force, which models the random collisions on the object by molecules of the surrounding fluid. On a timescale τ the average force felt by the object is:

$$F_L = \sqrt{4\pi k_B T \gamma / \tau}, \tag{3}$$

where γ characterises the dissipation (for a spherical object of diameter d: $\gamma = 3\pi\eta d$). For a $d = 2\,\mu$m bead in water ($\eta = 10^{-3}$ Poise), the intensity of the average force over $\tau = 1$ s is 10 fN. In optimum micro-manipulation conditions, this random force is the major source of experimental noise. Since the viscosity of water cannot be lowered and the timescale τ is set by the biological process studied, the signal-to-noise ratio can only be improved by reducing the size of the sensor.

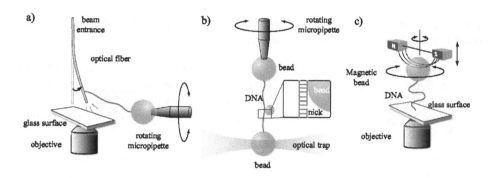

Figure 4. *Principle of the set-ups used to stretch and twist DNA molecules. (a) An optical microfibre forms a force sensor whose deflection is proportional to the force applied on the molecule by pulling a micropipette (Cluzel et al. 1996, Léger et al. 1998). The rotation of the micropipette about its axis provides the ability to twist DNA. (b) A similar set-up where the force sensor is a latex bead trapped between two counter-propagating laser beams. The force applied on the DNA is deduced from the displacement of the bead from its equilibrium position. An extra bead can be used to actually detect the rotation of the molecule (Bryant et al. 2003). The rotation speed of this bead provides an elegant means to directly measure the torque applied on the DNA. (c) Using a small superparamagnetic bead, one can pull on the DNA by placing magnets in the vicinity (few mm) of the bead (Strick et al. 1996). The magnetic moment of the bead follows the direction of the magnetic field and can provide a very strong torque. Thus, rotating the magnets induces the rotation of the bead and the twisting of the DNA.*

2.2   Description of twisting and stretching setups

While all the micro-manipulation techniques cited previously can be used to measure a tensional stress, some of them can also be used to apply a torsional stress on the molecule (see Figure 4). In this review, we shall focus on these set-ups only.

2.2.1   Microfibres

The set-up described in Cluzel *et al.* (1996) and Léger *et al.* (1998) (see Figure 4(a)) uses the deflection of an optical fibre to measure the force applied on a DNA molecule by a micropipette that pulls on it. This device allows the micropipette to rotate so that one can twist the DNA molecule. The typical stiffness of this set-up is about 10^{-5} N/m (Essevaz-Roulet *et al.* 1997, Léger 1999) and its spatial resolution is about 10 nm. However, because of the large size of the fibre, the large Langevin force density ($\approx$ 5 pN) associated to high-bandwith (100 Hz) measurements implies that the minimal force resolution is only a few piconewtons.

2.2.2 Optical trap

An optical trap uses the radiation pressure exerted by two counter-propagating laser beams on a small transparent bead to maintain it at a fixed spatial position. The tension on a DNA molecule tethered at one extremity to such a trapped particle can be deduced from its displacement from equilibrium, knowing the trap stiffness (or alternatively from the transfer of momentum to the diffracted beam). In the set-up shown in Figure 4(b) (Smith *et al.* 1996), a micropipette is used to pull and twist a DNA molecule anchored at its other end to a bead held in that pipette. Finally, a third (reporter) bead is sometimes used to estimate the torque exerted on the DNA molecule. That torque can be deduced from the Stokes drag exerted on the reporter bead as it rotates under the action of the torque in the molecule. On a 1 μm trapped bead, the Langevin force density is about ≈ 10 fN/Hz$^{1/2}$. Assuming a typical measurement bandwith of 100 Hz, the smallest measurable force is therefore ~ 0.1 pN.

2.2.3 Magnetic trap

Another set-up used for stretching and twisting DNA molecules consists of attaching a DNA at one extremity to a superparamagnetic bead and at the other to the surface of a glass capillary (see Figure 4(c)) (Strick *et al.* 1996). By approaching strong permanent magnets (0.5 T intensity, with a gradient on the millimetre scale) one can apply a force of a few pN on the bead and the DNA molecule that anchors it to the surface. One can also twist the DNA molecule by rotating the magnets, since the magnetic moment of the bead follows the direction of the magnetic field like the needle of a compass. Magnetic traps using beads of similar size as optical traps have similar noise limitations.

Optical traps and microfibres are extension clamps. They rely on the stiffness of the sensor to deduce by Hooke's law the exerted force from the displacement of the sensor from its equilibrium position. This typically limits their accuracy to forces measured over one or two orders of magnitude. In contrast, magnetic traps are force clamps. The force varies on a scale of the order of 1 mm, much larger than any variation in molecular extension. Consequently, the force is set by the position of the magnets with respect to the sample and the extension of the molecule is measured.

The force itself can be deduced from the transverse fluctuations of the tethered bead. The magnetic trap is equivalent to a damped pendulum of length $\langle z \rangle$ (the average molecular extension) for which gravity has been replaced by a magnetic field gradient (see Figure 5(a)).

Brownian fluctuations tend to displace the bead from its equilibrium position by an amount δx, thus generating a restoring force $\delta F_x \approx F \delta x / \langle z \rangle \equiv k_x \delta x$, where $k_x \doteq F / \langle z \rangle$ is the effective stiffness of the damped pendulum. By the equipartition theorem: $k_x \langle \delta x^2 \rangle / 2 = k_B T / 2$. Consequently, we can deduce the force applied by the magnets by measuring the averaged fluctuations $\langle \delta x^2 \rangle$ and the average extension $\langle z \rangle$ of the molecule:

$$F = \frac{k_B T \langle z \rangle}{\langle \delta x^2 \rangle}. \tag{4}$$

Figure 5(b) and (c) display typical recordings of z and x as a function of time. His-

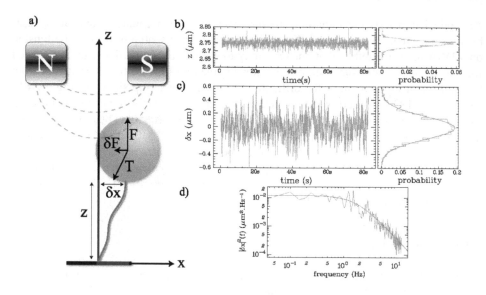

Figure 5. *Principle of force measurement. (a) Thermal energy displaces the bead from its equilibrium position, so that a restoring force $\delta F \approx F\delta x/z$ due to the magnets tends to put it back to equilibrium position. (b) and (c) Typical recordings and histograms of x and z position of the bead as a function of time. From these recordings, we deduce the force applied on the bead using the equipartition theorem. Here $\langle z \rangle = 2.75\mu m$ $\langle \delta x^2 \rangle = 170\,nm$ give $F = 0.38$ pN. (d) Power spectrum of the displacement along x and fitted using a Lorentzian curve $|\delta x^2|(f) = A/(f_0^2 + f^2)$. The cut-off in the spectrum is related to the friction of the bead. Using Fourier analysis we obtain the averaged transverse fluctuations $\langle \delta x^2 \rangle$ integrated over a finite spectrum and corrected for aliasing. We also derive the cut-off frequency f_0 of the system (here $f_0 = 1.8Hz$). This provides us with a test to check that the signal is not undersampled (i.e., to verify if $f_{video} = 25Hz \gg f_0$). We also verify that the recording time T_r is sufficiently long ($f_0 T_r \gg 1$) to yield statistically relevant estimates.*

tograms of position yield $\langle \delta x^2 \rangle = 170$ nm (so that $k_x = k_B T / \langle x^2 \rangle = 1.4 \dot{1}0^{-7}$ N/m) and $\langle z \rangle = 2.75$ μm so that $F = k_x . \langle z \rangle = 0.38$ pN.

3   Stretching DNA

As mentioned earlier, the forces of interest span a range between tens of femtonewtons to hundreds of piconewtons. The combination of the different set-ups described earlier allows one to study the elastic behaviour of DNA over this wide range of forces.

The first measurements of the entropic elasticity of a single DNA molecule were

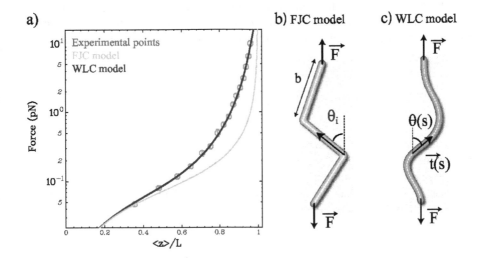

Figure 6. *Force vs. extension for dsDNA and polymer models for DNA. (a) Experimental force extension curve for one dsDNA of 16μm (squares). The fit using the worm-like chain (WLC) model (upper curve) gives ξ = 51.6 nm and L = 15.6μm. The plot of the freely jointed chain (FJC) model (lower curve) using the same value of ξ shows an obvious discrepancy with the experimental data. (b) The FJC model: the DNA chain is modelled as a sequence of segments of size b = 2ξ; θ_i is the angle between the i-segment and the direction of force. (c) The WLC model: the DNA chain is treated a continuous curves with a given bending rigidity B = k_BTξ.*

reported by Smith *et al.* (1992) and were extended to higher forces a few years later (Smith *et al.* 1996, Cluzel *et al.* 1996, Léger *et al.* 1999). Smith *et al.* used a combination of magnetic fields and hydrodynamic drag to pull on a superparamagnetic bead tethered to a surface by a single DNA molecule. Figure 6(a) displays a typical force extension curve obtained on a 16 μm long molecule.

The simplest model to explain the force extension behaviour of a polymer is the freely jointed chain (FJC), which models the polymer as a chain of uncorrelated discrete segments. The energy of such a system is simply given by:

$$\frac{E_{FJC}}{k_B T} = -\sum_i Fb\cos\theta_i, \qquad (5)$$

where θ_i is the angle between segment i and the direction of the force, and $b = 2\xi$ is the length of the segment, the so-called Kuhn length (see Figure 6(a)). The averaged orientation is deduced from a balance between the alignment along the force (that reduces the enthalpy) and the entropic disorientation of the segments. This model, which is equivalent to the Langevin model of paramagnetism, fails to account for the experimental data (see Figure 6). Although it fits the DNA elasticity at very low forces ($F < F_e \sim 0.1$ pN), it deviates significantly at higher forces.

In 1994, Marko and Siggia (Marko and Siggia 1994, 1995) solved a more relevant model that better describes the entropy of a DNA molecule under stretch: a continuous flexible (worm-like) chain with a bending modulus $B = k_B T \xi$. (See the chapters by Warren and MacKintosh in this volume for introductions to the worm-like chain.) The energy of such system is given by:

$$\frac{E_{WLC}}{k_B T} = \xi \int_0^L \left(\frac{d\mathbf{t}}{ds}\right)^2 ds - \frac{F}{k_B T} \int_0^L \cos\theta(s) ds. \qquad (6)$$

This model is formally analogous to the quantum mechanics of a dipole in an electric field, where the free energy of the polymer (in the limit $L \gg \xi$) is given by the ground state energy of the QM problem. This energy can be computed analytically at low and high forces. An analytic formula that fits the elastic behaviour to 0.1% for all forces has been proposed by Bouchiat *et al.* (1999):

$$\frac{F\xi}{k_B T} = \frac{z}{L} - \frac{1}{4} + \frac{1}{4(1 - z/L)^2} + \sum_{i=2}^{7} a_i (z/L)^i, \qquad (7)$$

with $a_2 = -0.5164228$, $a_3 = -2.737418$, $a_4 = 16.07497$, $a_5 = -38.87607$, $a_6 = 39.49944$, $a_7 = -14.17718$. The very good agreement between the model and the data is shown in Figure 6. These experiments provide the most precise measurement of the persistence length of DNA.

4   DNA under torsion

Most polymers are insensitive to torsion because their monomers are linked by single covalent bonds about which they are free to rotate. This property is lost when the polymer possesses no single covalent bond about which torsion can be relaxed. This is the case of the double-helical structure of a DNA molecule with no nicks (no break in one of the strands). This particular feature has very important biological implications. First, from a structural point of view, twisted DNA provides an efficient way to compact the molecule so that it fits the cell size, which may be the reason that *in vivo* all DNAs are twisted. Second, a negatively twisted (underwound) DNA may locally denature, thus facilitating its interactions with a variety of proteins (RNA polymerases, regulation factors, etc.). On the other hand, positively coiled DNA is more stable at high temperature (it denatures less). Thus, thermophilic bacteria that live close to the boiling point of water have enzymes (reverse gyrases) that overwind DNA. Because the topology of DNA plays such as essential role in the cell life, nature has evolved a family of enzymes, known generally as topoisomerase (the just mentioned reverse gyrase is one of them), that control the torsion and entanglement of DNA molecules.

In the following we first introduce the formalism used to describe twisted tubes and molecules. We shall then present some of the biological problems that motivate the study of DNA under torsion. We will finally discuss the various structural transitions observed upon twisting DNA and the theoretical model appropriate for the description of DNA under torsion.

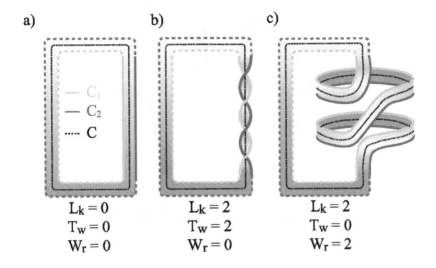

Figure 7. *Linking number L_k, twist T_w and writhe W_r. (a) We define a ribbon by the two curves C_1 and C_2 that define its edges. (b) By wrapping C_1 and C_2 around each other, we change the twist T_w but also the linking number of the two curves. (c) For a given linking number, we can change the geometry so that $T_w = 0$ (no wrapping of C_1 and C_2) and $W_r = 2$, the ribbon crosses itself twice.*

4.1 Topological formalism

The torsional state of DNA can be described using two geometrical variables: the *twist* (T_w) and the *writhe* (W_r) and one topological variable, the *linking number (L_k)*.

The twist T_w is a geometrical variable that measures the number of times the two strands wrap around each other. For example, Figure 7(b) displays two curves C_1 and C_2 wrapped around each other by two turns, so that $T_w = 2$. The natural twist T_{w0} of a DNA molecule equals the total number of base pairs (bps) n divided by the average number of bps in a helical pitch $n_p = 10.5$ bps: $T_{w0} = n/n_p$.

The second geometrical variable, the writhe W_r, is related to the global path adopted by the ribbon (denoted by C in Figure 7): W_r is the number of times C crosses itself. W_r equals zero in the cases shown in Figure 7(a) and (b), but equals two in Figure 7(c). The two canonical writhing structures are solenoids and plectonemes, as seen in Figure 8. Both structures are observed *vivo*.

The linking number L_k characterises quantitatively the fact that the two strands of a circular DNA molecule are intertwined and cannot be separated without breaking one strand and passing the other one through the break. Intuitively, we can define L_k as the number of passages we have to execute in order to unlink the two strands. L_k is a topological variable, i.e., it is not affected by changing the geometry (bending, stretching, etc.) of a circular DNA, as long as no strand is cut. For example, in Figure 7(a) the two curves C_1 and C_2 are not intertwined and have $L_k = 0$. On the other hand, in Figure 7(b)

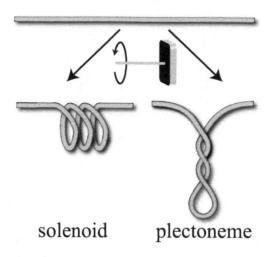

solenoid   plectoneme

Figure 8. *Shape of a supercoiled string. Twisting a string leads to the formation of either solenoids or plectonemes.*

two strand passages (C_1 through C_2) are required to unlink the curves C_1 and C_2 and reach the situation shown in Figure 7(a), so that $L_k = 2$.

Notice that we may extend the definition of the linking number to a linear molecule by joining its ends, as shown in Figure 7(a): the linking number of two plain curves $\mathcal{C}_1$ and $\mathcal{C}_2$ equals 0.

A mathematical theorem due to Calugareanu (1959) and generalised by White (1969) states that:

$$L_k = T_w + W_r. \tag{8}$$

A DNA molecule without intrinsic curvature and under no torsional stress has $W_r = 0$ so that its natural linking number is: $L_{k0} = T_{w0}$. A DNA molecule is supercoiled when its linking number $L_k \neq L_{k0}$. One defines the degree of supercoiling σ by:

$$\sigma \equiv \frac{\Delta L_k}{L_{k0}} \equiv \frac{L_k - L_{k0}}{L_{k0}} = \frac{\Delta T_w + W_r}{L_k}, \tag{9}$$

where $\Delta T_w = T_w - T_{w0}$.

4.2   Biological motivations

As mentioned earlier, DNA supercoiling plays a major role in the cell. Plasmids (small circular DNA molecules, see Figure 9) extracted from bacteria are usually underwound (or negatively supercoiled) and form plectonemic structures with $\sigma \approx -0.06$. Namely, the two strands in a plasmid of 5000 bases wind about each other by 30 turns less than the expected ~ 500 turns.

The whole *E. coli* chromosome (which is a 4 million bps circular molecule) is also negatively supercoiled. This unwinding is generated by an enzyme known as gyrase

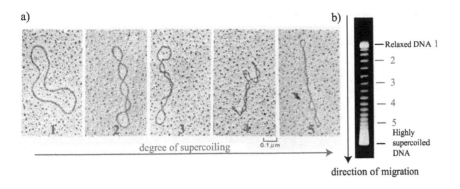

Figure 9. *Electron micrograph images of supercoiled plasmids (5 kbp). The degree of supercoiling increases from left to right.*

(which belongs to the family of topoisomerases), which actively modifies the linking number by passing one strand through an other. Eukaryotic chromosomes are also negatively supercoiled. As previously described, this is achieved by wrapping every 200 bps of DNA around a histone core, thus forming a solenoidal structure with a negative writhe, $W_r = -1$, and a decrease in twist, $\Delta T_w = -1$, for each 200 bps.

A consequence of the intertwining of the two strands of the DNA molecules is that the replication process should in principle generate daughter dsDNA that are catenated L_k times (which in *E. coli* could be as much as 400 000 times). This does not happen because of the action of topoisomerases that actively and completely disentangle the DNA molecules prior to cell division. These enzyme act locally, but systematically reduce the topology of the molecules (a global property) by a mechanism that is not yet fully elucidated. Figure 10 displays an electron micrograph and the corresponding drawing of a partially replicated plasmid . The molecule can be divided in two parts: the unreplicated region (drawn in bold) in front of the replication fork, which is positively supercoiled, and the replicated daughter strands behind the replication fork that remain catenated. At the end of replication, the action of topoisomerases has completely unlinked the two daughter strands.

In bulk assays, the principal tool for the study of topoisomerases and the analysis of the topology of DNA is gel electrophoresis (see Figure 9). Circular DNA molecules (plasmids) with different degrees of supercoiling migrate differently in an electrophoretic gel because of their various degree of compaction. This permits the analysis of the activity of enzymes that modify the topology of DNA. Since it is difficult to artificially generate supercoiled substrate of arbitrary and controlled degree of supercoiling, one often uses as an initial substrate (prior to action of topoisomerases) plasmids extracted from bacteria that are negatively supercoiled (with $\sigma = -0.06$). Bulk assays are thus essentially limited in the range of accessible degrees of supercoiling and irreversible. The advent of single-molecule manipulation techniques and the possibility to reversibly twist single DNA (and braid pairs of molecules) by unlimited number of turns are opening new opportunities for the study of topoisomerases.

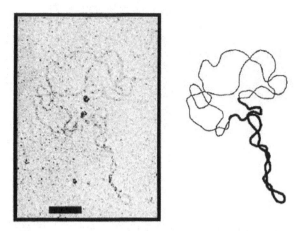

Figure 10. *Electron micrograph of a replicating plasmid. The replication process induces supercoiling in front of the replication fork (bold region). Links between the mother strands implies that the daughter molecules remain catenated. The cell requires the action of topoisomerases to separate the newly synthesised molecules.*

4.3   Twisting DNA at the single-molecule level

All the set-ups previously described permit the change of the linking number of a stretched single DNA molecule by rotating one extremity of the molecule, assuming that the DNA can support a torsional constraint, namely, that it is multiply bound at its ends and is not nicked (i.e., no single-strand breakage in its sugar-phosphate backbone). One is thus able to study the change in the extension of a DNA molecule at a given force as it is twisted by a controlled number of turns n ($\sigma = L_{k0}/n$); see Figure 11(a) (at $F = 1.2$ pN). One observes two different regimes : while for n smaller than a certain value n_b (that varies with F) the extension varies little, it drops almost linearly with n for $n > n_b$.

4.3.1   Buckling instability of DNA

The curve in Figure 11(a) can be easily understood by considering the twisting by n turns of a rubber tube under an applied force F (see Figure 11(b)). Twisting a tube of torsional modulus C (usually normalised by $k_B T$ in the DNA context: $\mathcal{C} = k_B T C$) results first in an increase in the torque Γ: $\Gamma = 2\pi n \mathcal{C}/L$, leaving its extension L almost unchanged. The twist energy, therefore, increases quadratically with n. After a certain number of turns n_b, the associated torque Γ_b becomes so large that it costs less energy to form a loop of size R and pay the price of its bending energy, rather than to increase the torsional energy: the tube buckles.

The formation of a loop involves two different energy terms. First, the bending energy required to create a loop of size $2\pi R$: $E_{bending} = 2\pi R B/(2R^2)$. Second, there is the work done against the force: $W = 2\pi R F$. Balancing the torsional work done upon adding one turn at the buckling transition against the energy of formation of a loop yields:

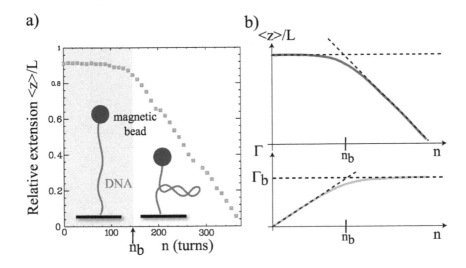

Figure 11. *Extension vs. σ for one dsDNA. (a) Experimental squares show the normalised extension vs. σ behaviour of one single DNA molecule (50 kbp) at F=1.2 pN. We distinguish two regimes: at low n, change in DNA extension is small, while the molecule stores torsional energy; at $n = n_b \approx 140$, the molecule buckles and starts forming plectonemes. After the buckling, the extension decreases almost linearly with the addition of further twisting. (b) As can be understood for an elastic rubber tube, initial twisting does not change the system's extension, but the torque stored in the tube increases linearly with the number of turns n applied. At a given $n = n_b$, forming a loop (plectonemes) cost less energy than increasing the torsional energy. Each additional turn leads to the formation of another loop, so that the extension drops down linearly with n, but the torque $\Gamma = \Gamma_b$ remains constant.*

$$2\pi\Gamma_b = 2\pi R\frac{B}{2R^2} + 2\pi RF. \tag{10}$$

The radius that minimises the bending energy is given by $R = \sqrt{B/2F}$, so that $\Gamma_b = \sqrt{2BF}$, $n_b = L\sqrt{2BF}/(2\pi\mathcal{C})$. Thus, the greater the force, the larger the critical buckling torque Γ_b and the number of turns n_b and the smaller the radius of the loop. Upon further twisting, the tube coils around itself but the torque Γ_b no longer increases. Figure 12 displays the dependence of n_b and the slope $d<z>/dn = 2\pi R$ on the force F for a 48 kbps DNA molecule.

This simple model describes at least qualitatively the behaviour of DNA under torsion (for a more accurate description, see the following text). The buckling transition is not as sharp as expected for a macroscopic transition because of thermal fluctuations, but increasing the force tends to make the transition sharper.

Just as stretching a twisted piece of DNA increases the torque in the molecule, twisting a circular plasmid results in an increase in the tension in the molecule. This tension can be estimated from our measurements. It is given by the critical force F_0

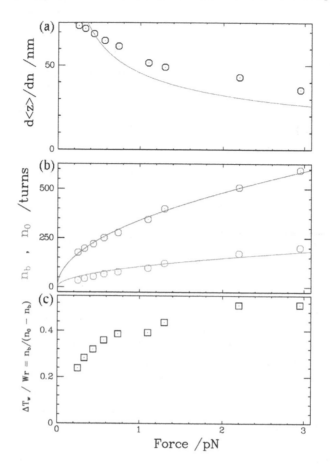

Figure 12. *Variation with the force of n_b, n_0, the ratio $\Delta T_w/W_r$ and $d<z>/dn$ after the buckling instability. (a) The slope $s \equiv d<z>/dn$ after the buckling instability (circles) decreases with the force applied. The fit to the simple model of buckling ($s = AF^{-0.5}$ with $A = 45$) gives a semi-quantitative agreement with the data. Note that $A = 45$ is of the same order of magnitude as the value predicted by the model: $2\pi\sqrt{B/2} \approx 30$; (b) Variation of n_b (experimental lower points) and n_0 (experimental upper points) with the force. The data were fitted using powerlaws ($n = AF^{0.5}$), as predicted by the buckling model. For the lower and upper curves, A equals 340 and 104, respectively. As the buckling transition is not sharp (especially at low forces), the estimation of n_b from extension versus number of turns could not be done with an accuracy better than 20%, whereas errors bars for n_0 is smaller than the size of the symbol. (c) Estimation of the $\Delta T_w/W_r$ ratio from the measurement of n_b and n_0 at different forces.*

required to stretch a supercoiled linear DNA. Alternatively, for a given force F_0, one can measure the number of turns n_0 required to reduce the molecular extension to zero (see Figure 11). The simple model described previously can be used to estimate n_0:

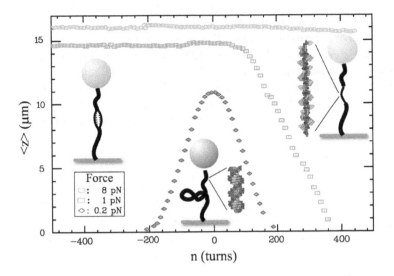

Figure 13. *Extension vs. supercoiling curves obtained for DNA (50 kb) at three different stretching forces. Diamond points: F = 0.2 pN. The system's extension decreases whether over- or underwound. Square points: F = 1 pN. The system contracts when n > 0 but remains almost constant at n < 0. Circular points: F = 8 pN. The DNA extension does not vary much in the plotted range of turns. Sketches are the interpretation of the data; for F = 0.2 pN: formation of plectonemes at low forces of the natural B-DNA (see numerically simulated structure); for F = 1 pN: denaturation of DNA and formation of a single strand bubble; for F = 8 pN: generation of a local DNA structure called P-DNA with exposed basepairs (as can be seen from the numerically simulated structure).*

$\langle z(n_0) \rangle = L - 2\pi R(n_0 - n_b) = 0$, from which we deduce:

$$n_0 = \frac{L}{2\pi}(\frac{1}{B} + \frac{1}{C})\sqrt{2BF}. \tag{11}$$

Notice the fair agreement with the data (see Figure 12(b)). This simple mechanical model predicts that the ratio of twist to writhe in a plasmid is a constant: $T_w/W_r = n_b/(n_0 - n_b) = B/C \sim 0.5$ (see the following text), which is not too far from the ratio (0.33) estimated from electron micrographs (Boles *et al.* 1990).

4.3.2 Torque-induced transitions

The simple mechanical description is essentially correct (though not very precise due to the neglect of thermal fluctuations) at low forces ($F < 0.5$ pN in 10 mM phosphate buffer), where the behaviour of DNA under twist does not depend on the sign of n, namely when the chiral structure of DNA is not reflected in its elastic response to torque (in Figure 13 the lowest curve (diamonds) is symmetric with respect to $n \to -n$).

At higher forces one does observe an asymmetry in the behaviour of DNA under twist that may be understood by invoking changes in the double-helical structure of the DNA. Thus, above 0.5 pN, undertwisting DNA ($n < 0$) leads to the denaturation of the double helix rather than to formation of supercoils (see Figure 13). Because at these forces, the critical torque $\Gamma_d \sim 9$ pN nm (Strick *et al.* 1999, Bryant *et al.* 2003) for denaturation is smaller than the critical torque for buckling Γ_b, instead of buckling the strand unpairs by about 10.5 base pairs for each extra unwinding turn added (Allemand *et al.* 1998, Strick *et al.* 1998a). As a result, some denaturation bubbles appear inside the DNA molecule, especially in A-T rich regions that melt more easily than G-C rich ones (Strick *et al.* 1998a,b).

A similar behaviour occurs when one overtwists DNA at still higher forces ($F > 3$ pN): overwinding induces (at a critical torque $\Gamma_p \sim 34$ pN nm (Bryant *et al.* 2003), see the following text) the local formation of a novel structure of DNA locally called P-DNA(Allemand *et al.* 1998), characterised by a much smaller pitch than B-DNA (about 2.6 bps per turn). Numerical simulations performed by Lavery *et al.* showed that P-DNA exhibits a right-handed helical structure with the phosphate backbone winding at the centre and the bases exposed out in solution (see Figure 13).

These results show that the DNA structure and geometry are greatly affected by the torsion applied on it. However, in physiological conditions it seems that DNA behaves elastically in a reasonable force-torque domain.

4.3.3 The rod-like chain (RLC)

As mentioned previously, thermal fluctuations cause the buckling transition observed in DNA to be smoother than the one for a rubber tube, especially at low forces ($F < 0.5$ pN). To get a better understanding of the behaviour of DNA under twist, one needs to extend the statistical mechanics analysis of the worm-like chain model to a polymer that is not free to rotate about its axis. That description, known as the rod-like chain model (RLC), requires the introduction of an additional torsional energy term in the energy functional for the worm-like chain:

$$E_{RLC} = E_{WLC} + \frac{C}{2} \int_0^L \Omega^2(s) ds, \tag{12}$$

where $\Omega(s)$ is the local twist of the chain. To simplify the calculation the helical structure of the DNA is assumed to decouple from that description. Therefore, this model cannot account for the structural transitions to denatured DNA or P-DNA at higher forces or degrees of supercoiling. The solution of this model can be found in Bouchiat and Mézard (1998) and Moroz and Nelson (1998). One of the major difficulties encountered in this solution is how to account for the topological constraint of fixed linking number. This constraint is a global property of the chain, which imposes non-local interactions in the statistical mechanics of the chain that are intractable. Akin to this problem is the fact that self-avoiding interactions are completely neglected. However, a chain that can freely pass through itself cannot be twisted!

One way out of that conundrum is to study the behaviour of a twisted chain at high forces where the chain is almost straight and unlikely to writhe (Moroz and Nelson 1998, Moroz and Nelson 1997). A problem with that approach is that at high forces the experimental data are skewed by the presence of the aforementioned structural transitions, which makes the comparison with the theory less reliable. Nevertheless, using that approach Moroz and Nelson have derived a best fit value of $C = 110$ nm (see Figure 14).

Another approach (Bouchiat and Mézard 1998, Bouchiat and Mézard 1999) is to use a local approximation for the linking number (the so-called Fuller formula, which is equivalent modulo 2π to the exact (but non-local) Gauss integral) and regularise the resulting theory by the introduction of a cut-off length (of the size of the DNA pitch). Although this model does not take into account the self-avoidance of the chain, as can be seen in Figure 14(a), it nevertheless fits our data remarkably well at low forces ($F < 0.4$ pN) using a value of $C = \mathcal{C}/k_{\mathrm{B}}T = 86$ nm (Bouchiat and Mézard 1998).

The validity of the approximation used by Bouchiat and Mézard (i.e., the use of the Fuller formula) was recently addressed by Rossetto-Giaccherino (2002). By numerical MC simulations of a fluctuating twisted chain, they find that in the experimental domain studied ($F > 0.1$ pN) the fluctuations of writhe are correctly estimated using Fuller's formula (although there are large discrepancies at lower forces). This explains why this local approximation to the linking number yields such a good agreement with the experimental data.

5 DNA-protein interactions

The experiments described earlier show that DNA under twist displays a rich phase diagram with a significant number of transitions induced either by the force or the torque or both. The order of magnitude of the torques involved in these transitions are compatible with those occurring during biological processes a few $k_{\mathrm{B}}T$. While it is known that the twist-induced denaturation of DNA is used by the cell to control the interactions of various enzymes with DNA (e.g., RNA-polymerase, transcription factors), there is as yet no evidence for the utility of a P-DNA structure in cellular processes.

As already mentioned, the role of topoisomerases is to regulate the topological state of the DNA. Although most of these enzymes generally reduce the torsional stress on the molecule, some (such as gyrase or reverse gyrase) catalyse the increase of σ within the DNA and therefore increase its writhe. An understanding of the behaviour of DNA under torsional stress is thus a pre-requisite for a comprehension of their mechanism of action. Other enzymes, such as the helicases of the replication complex, are able to translocate along the DNA and processively melt it, possibly generating a positive torque upstream. The interplay between this torque and the function of these enzymes is an important and still largely untouched issue in molecular biology. More generally, most of the enzymes that interact with DNA locally modify its structure, geometry or topology. To investigate the action of these enzymes one needs first to understand the elastic response of the DNA molecule, a program largely achieved by now. One may then expect that the single-molecule manipulation tools that have contributed so much to that understanding will also play a role in the study of these molecular machines, helping us to address issues such as their rate, processivity, efficiency, step size and interaction with the various DNA states described here.

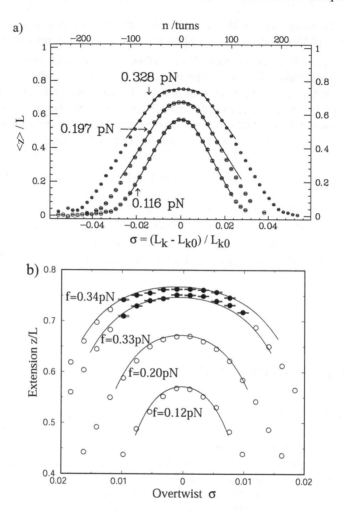

Figure 14. *Fit of the extension vs. number of turns (or degree of supercoiling) curve. (a) Using the local Fuller formula for the writhe, Bouchiat and Mézard obtained a best fit value of C = 86 ± 10 nm. (b) Using a high force expansion of a twisted DNA, Moroz and Nelson obtained a best fit value of C = 110 nm (plot taken from Moroz and Nelson (1998), Figure 1(a)).The solid black symbols were fits to the model; the curves are predictions based on this fit. These predictions agree with other data not used in the fit (open symbols).*

Acknowledgements

We acknowledge useful discussions and exchanges with C. Bustamante, N. Cozzarelli, A. Vologodskii, J. Marko, C. Bouchiat, M. Mézard, P. Nelson and D. Chatenay. This work was supported by grants from CNRS, ARC and the CEE under the 'MolSwitch' program.

References

Allemand J F, Bensimon D, Lavery R and Croquette V, 1998, *Proc Natl Acad Sci USA* **95** 14152–14157.

Amblard F, Yurke B, Pargellis A and Leibler S, 1996, *Rev Sci Instrum* **67(2)** 1–10.

Avery O T, MacLeod C M, and MacCarty M, 1944, *J Exp Med* **79** 137–158.

Boles T C, White J H and Cozzarelli M R, 1990, *J Mol Biol* **213** 931–951.

Bouchiat C and Mézard M, 1998, *Phys Rev Lett* **80** 1556–1559.

Bouchiat C, Wang M D, Block S M, Allemand J F, Strick T R and Croquette V, 1999, *Biophys J* **76** 409–413.

Bouchiat C and Mézard M, 1999, *cond-mat/9904018*.

Bryant Z, Stone M D, Gore J, Smith S B, Cozzarelli N R and Bustamante C, 2003, *Nature* **424** 338–341.

Calugareanu G, 1959, *Rev Math Pures Appl* **4** 5–20.

Cluzel P, Lebrun A, Heller C, Lavery R, Viovy J L, Chatenay D and Caron F, 1996, *Science* **271** 792 –794.

de Gennes P G, 1979, *Scaling concepts in polymer physics* (Cornell University Press, Cornell).

Essevaz-Roulet B, Bockelmann U and Heslot F, 1997, *Proc Natl Acad Sci (USA)* **94** 11935–11940.

Florin E L, Moy V T and Gaub H E,1994, *Science* **264** 415–417.

Gosse C and Croquette V, 2002, *Biophys J* **82** 3314–3329.

Herschey A D and Chase M, 1952, *J Gen Physiol* **36** 39–56.

Ishijima A, Doi T, Sakurada K and Yanagida T, 1991, *Nature* **352** 301–306.

Léger J F, Robert J, Bourdieu L, Chatenay D and Marko J F, 1998, *Proc Natl Acad Sci USA* **95** 12295–12296.

Léger J F, Romano G, Sarkar A, Robert J, Bourdieu L, Chatenay D and Marko J F, 1999, *Phys Rev Lett* **83** 1066–1069.

Marko J F and Siggia E D, 1994, *Science* **265** 506–508.

Marko J F and Siggia E D, 1995, *Macromolecules* **28(26)** 8759–8770.

Merkel R, Nassoy P, Leung A, Ritchie K and Evans E, 1997, *Nature* **397** 50–53.

Meselsohn M and Stahl F, 1958, *Proc Natl Acad Sci (USA)* **44** 671–682.

Moroz J D and Nelson P, 1998, *Macromolecules* **31** 6333–6347.

Moroz J D and Nelson P, 1997, *Proc Natl Acad Sci (USA)* **94** 14418–14422.

Rief M, Oesterhelt F, Heymann B and Gaub H E, 1997, *Science* **275** 1295–1297.

Rossetto-Giaccherino V, 2002, *Mécanique statistique de systèmes sous contraintes: topologie de l'ADN et simulations électrostatiques*, PhD thesis, Université Pierre et Marie Curie Paris VI.

Saenger W, 1988, *Principle of Nucleic Acid Structure* (Springer-Verlag, New York).

Sanger F, 1981, *Science* **214** 1205–1210.

Simmons R M, Finer J T, Chu S and Spudich J A, 1996, *Biophys J* **70** 1813–1822.

Smith S B, Cui Y and Bustamante C, 1996, *Science* **271** 795–799.

Smith S B, Finzi L and Bustamante C, 1992, *Science* **258** 1122–1126.

Strick T, Allemand J F, Bensimon D, Bensimon A and Croquette V, 1996, *Science* **271** 1835–1837.

Strick T, Allemand J F, Bensimon D and Croquette V, 1998a, *Biophys J* **74** 2016–2028.

Strick T R, Bensimon D and Croquette V, 1999, *Genetica, proceedings of NATO ARW on "Structural Biology and Functional Genomics"* **106** 57–62.

Strick T, Croquette V and Bensimon D, 1998b, *Proc Nat Acad Sci (USA)* **95** 10579–10583.

Watson J D and Crick F H C, 1953a, *Nature* **171** 737–738.

Watson J D and Crick F H C, 1953b, *Nature* **171** 964–967.

White J H, 1969, *Am J Math* **91** 693–728.

References

Interactions and conformational fluctuations in DNA arrays

Rudolf Podgornik

Department of Physics, University of Ljubljana
Jadranska 19, SI-1000 Ljubljana, Slovenia
&
Department of Theoretical Physics, J. Stefan Institute
Jamova 39, SI-1000 Ljubljana, Slovenia

rudolf.podgornik@fmf.uni-lj.si

1   Introduction

In this chapter, I will give a broad overview of recent work on the equation of state of DNA in aqueous *monovalent* salt solutions. The picture that I will develop shows that at non-negligible monovalent salt concentrations the direct electrostatic interactions between DNA molecules are almost always masked by the thermal conformational fluctuations of the DNA chains. These thermal fluctuations act to boost the magnitude as well as the spatial range of the electrostatic interactions. This renormalisation of the bare electrostatic interactions is a salient feature of dense systems composed of flexible polyelectrolyte molecules.

DNA can adopt a variety of mesophases in 0.5 M NaCl aqueous solution when the density, as measured, e.g., by the effective interaxial spacing R (Figure 1), is high enough (Podgornik *et al.* 1998). These extend from the melting of the crystal phase at $R \approx 24\,\text{Å}$ to the nematic-isotropic transition at $R \approx 120\,\text{Å}$. In these mesophases, the DNA is orientationally ordered, or even shows long-range hexatic order perpendicular to the long axes of the molecules, but is positionally still a liquid with only short-range positional order (Podgornik *et al.* 1998). This is in particular valid for long $\sim \mu$m fragments of DNA. The local structure of these DNA arrays is presented in Figure 1. In this part of the phase diagram, the equation of state, i.e., the dependence of the osmotic pressure on the macromolecular (DNA) concentration, has been studied very carefully (Strey *et al.* 1998). In this chapter I will give a comprehensive introduction to the equation of state

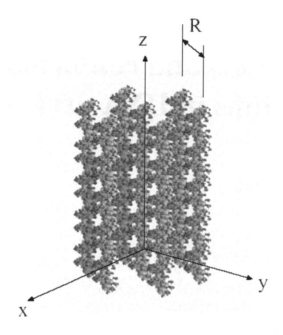

Figure 1. *The schematic geometry of an array of DNA molecules with long-range orientational order and short-range positional order. Drawn approximately to scale. R is the average interaxial separation between the molecules.*

for DNA under various solution conditions with monovalent salts and counter-ions, and make an attempt to explain in some detail the theoretical underpinning of its equation of state in the range $24\,\text{Å} \lesssim R \lesssim 120\,\text{Å}$.

2   Electrostatic interactions

Electrostatic interaction between charged macromolecules is one of the two pillars of the DLVO theory of colloid stability (Israelachvili 1998; Verwey and Overbeek 1948). The other one being the Lifshitz-van der Waals electromagnetic fluctuation forces (Mahanty and Ninham 1976). In DLVO theory both contributions are assumed to be additive (see also Frenkel's chapter in this volume). In general, however, the van der Waals forces and the electrostatic interactions are coupled through the self consistent zero-order term of the Lifshitz formula (Mahanty and Ninham 1976). This coupling is important in cases where counter-ion correlations make a large attractive contribution to the total interaction as, e.g., in the case of polyvalent counter-ions.

In what follows I will dwell exclusively on the electrostatic component of the DLVO theory, assuming that the van der Waals forces are negligible in the context of DNA interactions, their magnitude being always much smaller then the magnitude of the electrostatic interactions. I will also not discuss the water-mediated structural interactions that make their mark on the equation of state at very high DNA densities and/or high salt concentrations.

2.1 Poisson-Boltzmann theory

In our investigation of the electrostatic interactions between charged macromolecules we will start by writing down the expression for the non-equilibrium mean-field free-energy density (f) of a gas of mobile charged particles, i.e., counter-ions and salt ions. It can be written as a difference of the electrostatic field energy density (w) and the ideal entropy density of the mobile charge carriers (s) (Andelman 1995). The electrostatic field energy density can be written as

$$w(\mathbf{E}(\mathbf{r}), \rho_i(\mathbf{r})) = -\tfrac{1}{2}\epsilon\epsilon_0 \boldsymbol{\nabla}\phi(\mathbf{r})^2(\mathbf{r}) + \sum_i z_i e_i n_i(\mathbf{r})\phi(\mathbf{r}), \qquad (1)$$

where $\mathbf{E}(\mathbf{r}) = -\boldsymbol{\nabla}\phi(\mathbf{r})$ is the local electrostatic field, ϵ is the static dielectric constant and n_i is the density of the charged species i of charge $z_i e_i$. The total charge density is thus $\sum_i \rho_i(\mathbf{r}) = \sum_i z_i e_i n_i(\mathbf{r})$. After minimising this expression with respect to $\phi(\mathbf{r})$, one remains with the standard Poisson equation and the usual form of the electrostatic field energy density.

Let us assume furthermore that there are N mobile ionic species with charges $z_i e_i$, whose densities are n_i, while n_i^0 is the particle density of the same mobile charged species in the bulk with which the system is in chemical equilibrium. Then, the entropy difference between the volume under consideration and the bulk is given by

$$s(n_i(\mathbf{r})) = -k_B \sum_i \left(n_i(\mathbf{r}) \log\left(\frac{n_i(\mathbf{r})}{n_i^0}\right) - (n_i(\mathbf{r}) - n_i^0) \right), \qquad (2)$$

where k_B is the Boltzmann constant. The complete non-equilibrium free-energy difference is then defined as a volume integral:

$$\mathcal{F} = \int d^3\mathbf{r}\, f(\mathbf{E}(\mathbf{r}), n_i(\mathbf{r})) = \int d^3\mathbf{r}\, (w(\mathbf{E}(\mathbf{r}), n_i(\mathbf{r})) - Ts(n_i(\mathbf{r}))). \qquad (3)$$

By minimising the free energy (Equation 3) with respect to the electrostatic potential, one first obtains the Euler-Lagrange equation in the form

$$\boldsymbol{\nabla}\left(\frac{\partial f}{\partial \boldsymbol{\nabla}\phi}\right) + \frac{\partial f}{\partial \phi} = 0 \qquad \text{wherefrom} \qquad \epsilon\epsilon_0\, \boldsymbol{\nabla}\mathbf{E}(\mathbf{r}) = \sum_i z_i e_i \rho_i(\mathbf{r}). \qquad (4)$$

This is obviously nothing but the Poisson equation. By further minimising the free energy (Equation 3) with respect to ionic densities n_i one obtains the Euler-Lagrange equation in the form

$$\frac{\partial f}{\partial n_i} = 0 \qquad \text{wherefrom} \qquad z_i e_i \phi(\mathbf{r}) + k_B T \log\left(\frac{n_i(\mathbf{r})}{n_i^0}\right) = 0. \qquad (5)$$

Taking into account the Poisson equation (Equation 4), one can rewrite the preceding equation as the Poisson-Boltzmann equation,

$$\boldsymbol{\nabla}^2 \phi(\mathbf{r}) = -\frac{1}{\epsilon\epsilon_0} \sum_i \rho_i(\phi(\mathbf{r})) = -\frac{e_0}{\epsilon\epsilon_0} \sum_i z_i n_i^0 e^{-\beta z_i e_0 \phi(\mathbf{r})}, \qquad (6)$$

that gives the equilibrium (mean-field) profile of the electric field in the system. If one integrates the Poisson equation over the whole volume available to the mobile ionic species and takes into account the theorem of Gauss-Ostrogradsky, one gets

$$\epsilon \epsilon_0 \oint_{\partial V} (\mathbf{E} \cdot \mathbf{n}) \, d^2\mathbf{r} = \int_V \sum_i e_i \rho_i(\mathbf{r}) d^3\mathbf{r} = \oint_{\partial V} \sigma d^2\mathbf{r}. \tag{7}$$

Here, a further assumption is that the system is overall electroneutral, and thus the total volume charge should be matched by the neutralising surface charge of surface charge density σ on the boundaries of the volume. Further details on the derivation of the Poisson-Boltzmann (PB) equation can be found in Andelman (1995).

If the magnitude of the surface charges is not too large, one has $\beta e_0 \phi(\mathbf{r}) \ll 1$ in the whole accessible volume. Assume further that we have only two mobile charged species, $e_1 = e_0$ and $e_2 = -e_0$, together with $n_1 = n_+$ and $n_2 = n_-$, in equilibrium with a bulk reservoir with $n_i^0 = n_0$. In this case the Poisson-Boltzmann equation can be linearised and reduced to the Debye-Hückel equation:

$$\nabla^2 \phi(\mathbf{r}) = -\frac{e_0 n_0}{\epsilon \epsilon_0} \sum_i z_i e^{-\beta z_i e_0 \phi(\mathbf{r})} \approx -\frac{e_0 n_0}{\epsilon \epsilon_0} \sum_i (z_i - \beta(z_i)^2 e_0 \phi(\mathbf{r}) + \dots) = \kappa_D^2 \phi(\mathbf{r}). \tag{8}$$

By assumption $\sum_i z_i = 0$ and $\sum_i (z_i)^2 = 2$. Here we have introduced

$$\kappa_D^2 = 8\pi \ell_B n_0 \qquad \text{where} \qquad \ell_B = e_0^2 / 4\pi \epsilon \epsilon_0 k_B T \tag{9}$$

is the Bjerrum length, defined to be the separation between two elementary charges at which their electrostatic interaction energy equals the thermal energy. At room temperature in water, $\ell_B = 7.1$ Å. The constant $\kappa_D = \lambda_D^{-1}$ is interpreted as the inverse Debye screening length. For monovalent salts at room temperature with concentration expressed in moles [M] per litre, $\lambda_D = 3.05/\sqrt{[M]}$ in Å. The salient feature of electrostatic interaction on the linearised PB equation level is the screening quantified by the Debye length λ_D, leading to an approximately exponential decay of the interactions as a function of the separation between the charges. This picture of screened interactions has to be drastically modified in the case of polyvalent counter-ions (Kjellander 2001).

For further details on the PB equation, see Andelman's chapter in this volume and Andelman (1995). Here we note that the linearisation of the PB equation is usually justified at asymptotic conditions, meaning usually a small surface charge and/or a large separation between charged macroions, as well as by the fact that it is easily amenable to analytic solutions. The full non-linear PB equation represents a much tougher mathematical problem analytically solvable only for special geometries.

2.2   The cell model

In many colloidal systems, most notably in the case of ordered DNA phases, one seldom deals with isolated molecules in ionic solutions. Quite often one has a phase of densely packed macroions that complicates the problem of evaluating the electrostatic interactions even further. A simple way around this problem is the polyelectrolyte cell model (Fuoss *et al.* 1951, Lifson and Katchalsky 1954), which is a variant of the Wigner-Seitz model of electrons in the crystalline lattice and substitutes the complicated colloidal geometry with a cell containing a single colloid. The effect of the rest is assumed to be mimicked by the cell wall where the electrostatic potential should have a zero derivative by symmetry.

For a long cylindrical molecule such as DNA in a dense phase where molecules are on the average oriented in one direction, a cylindrical cell model should capture the main features of the molecular environment. There are, nevertheless, important caveats that one has to be aware of in the context of the cell model (Tamashiro and Schiessel 2003). The linearised PB equation in cylindrical coordinates (r, θ, z) reads

$$\frac{1}{r}\frac{d}{dr}\left(r\frac{d\phi}{dr}\right) = \kappa_D^2\phi,$$ (10)

with the boundary condition at the inner wall, (i.e., at the surface of the central molecule with a radius a) being

$$\frac{d\phi}{dr}(r = a) = -\frac{\sigma}{\epsilon\epsilon_0} = -\frac{e_0}{2\pi\epsilon\epsilon_0 ab},$$ (11)

where b is the length of the molecule per one elementary charge on the surface. For DNA, its structural charge would correspond to one charge per 1.7 Å. The boundary condition at the outer surface of the cell, assumed to be located at $r = R$, is by symmetry

$$\frac{d\phi}{dr}(r = R) = 0.$$ (12)

The radius of the cell R is obtained from the macromolecular density of the system. If the density of the macromolecules is n_M, then $n_M^{-1} = \pi(R^2 - a^2)b$. The macromolecular density thus determines the radius of the cell.

Obtaining a solution of the linearised Poisson-Boltzmann equation in the cylindrical cell model is quite straightforward and leads to the electrostatic potential that can be expressed via the cylindrical Bessel functions $K_0(x)$ and $I_0(x)$ as

$$\phi(r) = \frac{e_0}{2\pi\epsilon\epsilon_0 b\,(\kappa_D a)}\frac{K_0(\kappa_D r)I_1(\kappa_D R) + I_0(\kappa_D r)K_1(\kappa_D R)}{K_1(\kappa_D a)I_1(\kappa_D R) - I_1(\kappa_D a)K_1(\kappa_D R)}.$$ (13)

From here we get for the radial component of the electrostatic field the expression

$$E_r(r) = -\frac{d\phi}{dr} = \frac{e_0}{2\pi\epsilon\epsilon_0 ba}\frac{K_1(\kappa_D r)I_1(\kappa_D R) - I_1(\kappa_D r)K_1(\kappa_D R)}{K_1(\kappa_D a)I_1(\kappa_D R) - I_1(\kappa_D a)K_1(\kappa_D R)}.$$ (14)

One should note here that for a single cylinder, i.e., in the limit of $R \to \infty$, the electrostatic potential and the electrostatic field reduce to

$$\lim_{R\to\infty} \phi(r) = \frac{e_0}{2\pi\epsilon\epsilon_0 b\,(\kappa_D a)}\frac{K_0(\kappa_D r)}{K_1(\kappa_D a)} \quad \text{and}$$

$$\lim_{R\to\infty} E_r(r) = \frac{e_0}{2\pi\epsilon\epsilon_0 ba}\frac{K_1(\kappa_D r)}{K_1(\kappa_D a)}.$$ (15)

Additional terms in the cell model solution, Equations 13 and 14, are thus due to the finite concentration of DNA and obviously depend on its density via the radius of the outer cell wall R. One should note here that the linearised solutions of the PB equation are quite accurate in the case that the salt concentration is not too small (Shkel *et al.* 2000, Shkel *et al.* 2002). The non-linear PB equation has an analytical solution (Tracy and Widom 1997) only for a single cylinder in infinite ionic solution.

2.3 Osmotic pressure

The forces between macromolecules mediated by the equilibrium distribution of counter-ions and salt ions between them can be obtained via the stress tensor at the outer surface of the cell, which gives the force acting on this surface. This force per unit area of the cell is obviously nothing but the osmotic pressure. The stress tensor contains the Maxwell electrostatic part and the osmotic part (Gelbart *et al.* 2000),

$$\sigma_{ij} = \epsilon\epsilon_0 \left(E_i E_j - \tfrac{1}{2}E^2 \delta_{ij} \right) - k_B T \sum_i n_i(\phi(\mathbf{r}))\delta_{ij}. \tag{16}$$

The last term is simply the van 't Hoff ideal gas pressure corresponding to the ideal gas entropy in the free energy *ansatz*. The negative sign comes from the general continuum mechanics argument that positive pressure should lead to a decrease in volume.

We can evaluate the stress tensor on any plane of cylindrical symmetry within the cell. In many cases it is simplest to take the stress tensor at the surface of the central macromolecule, in which case the forces are obtained via a contact theorem (Gelbart *et al.* 2000). Equivalently we can take the stress tensor at the outer wall of the cell where by symmetry the electric field is zero, and thus the stress tensor contains only the van 't Hoff part. Since the latter case is simpler, we thus find for the radial components of the stress tensor (all the other ones vanish at the outer wall)

$$\sigma_{rr} = -k_B T \sum_i n_i(r = R). \tag{17}$$

The osmotic pressure $\Pi(R)$ in the cell is, because of mechanical equilibrium, negative and equal to the force per surface area on the outer wall; thus

$$\Pi(R) = -\sigma_{rr} = k_B T \sum_i n_i(r = R). \tag{18}$$

We can use this expression to derive the form of the osmotic pressure in the case of the linearised PB equation in a cylindrical cell. Remember that we have assumed that we have only two mobile charged species, $e_1 = e_0$ and $e_2 = -e_0$, together with $n_1 = n_+$ and $n_2 = n_-$, in equilibrium with a bulk reservoir where $n_1^0 = n_2^0 = n_0$. The osmotic pressure difference between the cell and the bulk reservoir, which alone is measurable, can be evaluated *via* Equation 18:

$$\begin{aligned} \Pi &= k_B T(n_+(r = R) + n_-(r = R)) - 2k_B T n_0 \\ &= 2k_B T n_0 \left(\cosh \beta e_0 \phi(r = R) - 1 \right). \end{aligned} \tag{19}$$

This is the complete expression for osmotic pressure difference between the inside of the cell and the bulk. Obviously, the only molecular species contributing here to the osmotic pressure are the mobile ions. The macromolecule itself does not contribute to the osmotic pressure. We will see later that this point of view is not entirely correct.

Another approach to the evaluation of the osmotic pressure, independent of the cell model, would be to calculate the complete equilibrium free energy $\mathcal{F}$ from Equation 3 *via* a volume integral over all the space available to the mobile charges,

$$\mathcal{F}[n_M] = \int d^3\mathbf{r}\, f(\mathbf{E}(\mathbf{r}), n_i(\mathbf{r})) \tag{20}$$

where the electrostatic field and the densities of the mobile charges are obtained from the solution of the full or linearised PB equation (Equation 6). This free energy is of course a function of the concentration of the macromolecules n_M. The osmotic pressure in the system would then be obtained from the standard thermodynamic relation

$$\Pi = -\left(\frac{\partial \mathcal{F}(n_M)}{\partial V_M}\right)_{T,\mu}. \tag{21}$$

Constant temperature and chemical potential of the mobile charged species are of course assumed in the preceding expression. In the case that the macromolecular solution is ordered and exhibits certain symmetries this expression can be simplified even further. If we again assume that the effects of the packing symmetry of the molecules can be captured by a cylindrical cell model of radius R and that the length of the molecules is L, we can write for the osmotic pressure

$$\Pi = -\left(\frac{\partial \mathcal{F}(n_M)}{\partial V_M}\right)_{T,\mu} = -\left(\frac{\partial (\mathcal{F}(R)/L)}{2\pi\, R \partial R}\right)_{T,\mu}. \tag{22}$$

Equation 22 connects the expression for the osmotic pressure in the cell model with the one obtained from the complete partition function. Discrepancies between the values for osmotic pressure obtained from the two expressions are due to the approximate nature of the cell model.

2.4 Interaction between cylindrical macromolecules

Let us now investigate the osmotic pressure in the cylindrical cell model on the level of the linearised PB equation. In this case Equation 19 can be written as

$$\begin{aligned}
\Pi &= 2k_B T n_0 \left(\cosh\beta e_0\phi(r=R) - 1\right) \\
&= 2k_B T n_0 (\beta e_0)^2 \phi^2(r=R) + \cdots = \kappa_D^2 \epsilon\epsilon_0 \phi^2(n=R) + \ldots ,
\end{aligned} \tag{23}$$

and represents the osmotic pressure difference between the wall of the cell and the bulk reservoir. Using now the solution of the linearised PB equation, Equation 13, we end up with the following expression for Π:

$$\Pi = \frac{\sigma^2}{\epsilon\epsilon_0 K_1^2(\kappa_D a)}\, K_0^2(\kappa_D R)\, p^2(\kappa_D R, \kappa_D a), \tag{24}$$

where we have introduced the correction factor $p(x,y)$ given by

$$p(\kappa_D R, \kappa_D a) = \frac{1 + \frac{K_1(\kappa_D R) I_0(\kappa_D R)}{I_1(\kappa_D R) K_0(\kappa_D R)}}{1 - \frac{K_1(\kappa_D R) I_1(\kappa_D a)}{I_1(\kappa_D R) K_1(\kappa_D a)}}. \tag{25}$$

The factor $p^2(\kappa_D R, \kappa_D a)$ obviously represents the effect of the finite concentration of the macromolecules, i.e., DNA, or in other words the effect of the walls of the cell. For small DNA densities this correction factor goes to unity, since in that limit only the first neighbours of the central molecule are important. This would be equivalent to taking only the first neighbours into account in the evaluation of the free energy.

Equation 24 for osmotic pressure is relatively complicated. In order to avoid the details that will later prove to be irrelevant, we investigate only its asymptotic form, valid for large values of $\kappa_D R$ or small values of the macromolecular concentration. In this limit we have $p(\kappa_D R, \kappa_D a) \longrightarrow 1$ and the approximate form of the osmotic pressure can be derived as

$$\Pi \sim \frac{\sigma^2}{\epsilon\epsilon_0 K_1^2(\kappa_D a)} K_0^2(\kappa_D R) = \frac{\sigma^2}{\epsilon\epsilon_0 K_1^2(\kappa_D a)} \frac{\pi}{2} \frac{e^{-2\kappa_D R}}{\kappa_D R}, \tag{26}$$

where we have taken into account the asymptotic form of the Bessel function $K_0(x) \longrightarrow \sqrt{\frac{\pi}{2}} exp(-x)x^{-1/2}$. From the expression for the osmotic pressure (Equation 22), we can now derive the interaction free energy per unit length between the central molecule and its neighbours. We obtain

$$\mathcal{F}(R)/L \sim \frac{\pi^2}{2} \frac{\sigma^2}{\epsilon\epsilon_0 \kappa_D^2 K_1^2(\kappa_D a)} e^{-2\kappa_D R} = k_B T \frac{\ell_B}{b^2} e^{-2\kappa_D R}. \tag{27}$$

Here we introduced an effective separation b between charges along the cylinder $1/b = \pi\sigma\lambda_D/\sqrt{2}e_0 K_1(a/\lambda_D)$ (Brenner and Parsegian 1974) (for DNA $b \sim l_{PO_4}$, where $l_{PO_4} \sim$ 1.7 Å is the separation between the phosphate charges along DNA (Bloomfield *et al.* 2000)). This expression derived via the cell model with cylindrical symmetry is very close to the expression derived via a pair-interaction energy evaluated on the linearised PB level (Brenner and Parsegian 1974). This is not surprising since, apart from geometric factors, the cell model and the pairwise free energy should give the same result at vanishing macroion concentrations.

In the case of very large surface charges the non-linearities of the PB equation effectively change the surface charge density entering Equation 27 via $b \longrightarrow \ell_B$ while leaving the separation dependence largely unaltered (Andelman 1995). Sometimes these non-linearity effects, which become pronounced depending on whether the self-consistent Manning parameter $\zeta = \ell_B/b$ is larger or smaller than one, are referred to as Manning condensation (Schmitz 1993).

3   Equation of state: No thermal fluctuations

An equation of state in general means a connection between the osmotic pressure Π and the macromolecular density n_M of an assembly,

$$\Pi = \Pi(n_M), \tag{28}$$

and can in principle be obtained exactly via a complete statistical mechanical treatment of a solution composed of cylindrical macromolecules and the bathing medium. Since this is usually not feasible, a helpful shortcut is to evaluate the osmotic pressure in the cell model that mimics the finite macromolecular density as explained earlier.

The system that we are studying is composed of macromolecules that have inter-molecular as well as intramolecular degrees of freedom since they are usually not in-finitely rigid. A most naive approach to the equation of state would be to simply forget the intramolecular degrees of freedom, assume that the macromolecules are ideally rigid and that they assemble into a crystal of hexagonal symmetry with perfect positional or-der. In this case the osmotic pressure of such a system can be obtained via Ewald force summation (in the case of short-range interactions even this is dispensable) involving the intermolecular potentials leading directly to the equation of state or again via the cell model; the latter approach being certainly simpler then the Ewald summation since it is effectively a single-particle model. The osmotic pressure, i.e., the equation of state, in the cell model would be given by

$$\Pi(n_M) = - \left(\frac{\partial (\mathcal{F}(R)/L)}{2\pi R \partial R} \right)_T \quad \text{where} \quad \mathcal{F}(R) = k_B T \frac{\ell_B}{b^2} e^{-2\kappa_D R}, \qquad (29)$$

where L is the length of the molecules in the array. As already stated, in writing this free energy we have assumed that the van der Waals interactions as well as the short-range non-electrostatic structural interactions do not make an essential contribution to the equation of state. For the latter this is true only at not very high ionic concentrations and not too high concentration of the macromolecules (for details see Strey *et al.* (1999)).

We can now take this equation of state and compare it with experiments performed on DNA (Strey *et al.* 1999) at various solution conditions. After performing this type of exercise, one is immediately convinced that something crucial is missing as there is practically no correspondence (except at very high densities) between the experiment and this type of simple-minded theory (see Figure 2). In the case of DNA the magni-tude of the surface charge is taken at the Manning value (Bloomfield *et al.* 2000), thus $b = \ell_B$. Obviously this equation of state underestimates the energetics of the system. In the following section we will try to amend it by taking into consideration also the intramolecular degrees of freedom of the macromolecular ions.

4 Equation of state: The effect of thermal fluctuations (1)

While the intermolecular degrees of freedom are taken into account through interaction potentials described in the previous section, the intramolecular degrees of freedom are usually treated within a mesoscopic elastic model that substitutes macroscopic elasticity for the complicated short-range intramolecular potentials acting between different seg-ments of the macromolecules (Chaikin and Lubensky 2000). Elastic models for cylindri-cal macromolecules have been well worked out (Petrov 1999). The general idea is that the trace over all the microscopic degrees of freedom is assumed to result exactly in the mesoscopic Hamiltonian itself, which one then uses in the partition function to evaluate the trace over mesoscopic degrees of freedom.

Because in the regime of relevant densities the hexagonal array is either in the line hexatic phases or the cholesteric phases (Strey *et al.* 1999), the mesoscopic elastic Hamil-tonians pertaining to them can be used. The goal here will be to combine the effects

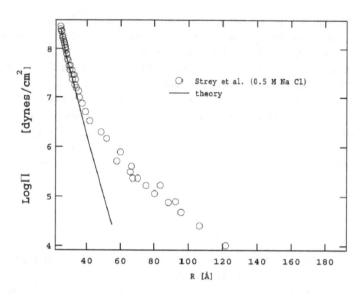

Figure 2. *A comparison of the DNA equation of state, Equation 29, with experimental data (Strey et al. 1999). This form of the equation of state assumes that the molecules are infinitely rigid and crystalline. The magnitude of the electrostatic part of the interaction is obtained by assuming that the charge density of the DNA is correctly given by the Manning value (Bloomfield et al. 2000). Obviously there are large discrepancies on this level between the theoretical predictions and actual data.*

of thermally driven elastic fluctuations of the macromolecules with the interactions between them and calculate their combined effect on the equation of state. This approach is based on ideas first introduced by Helfrich in the 1970s and later worked out in detail by Lipowsky, Leibler and others in the 1980s (Lipowsky 1995), who showed in the context of membranes that thermal conformational fluctuations can have a profound effect on interactions between flexible macromolecules.

4.1 A macroscopic theory of the equation of state in an ordered macromolecular array

Let us first consider the elastic free energy of a nematic: a 3D liquid with long-range orientational order with an average director **n** along the z axis. Such phases are typically formed by solutions of rod-like or disc-like objects. There are three kinds of deformations in quadratic order of **n** with symmetry $C_{\infty h}$: splay, twist and bending. The corresponding elastic constants for these deformations are the Frank constants K_1, K_2 and K_3 (Chaikin and Lubensky 2000).

$$\mathcal{F}_N = \frac{1}{2} \int d^2\mathbf{r}_\perp dr_z \left[K_1 \left(\nabla \cdot \mathbf{n} \right)^2 + K_2 \left(\mathbf{n} \cdot \left(\nabla \times \mathbf{n} \right) \right)^2 + K_3 \left(\mathbf{n} \times \left(\nabla \times \mathbf{n} \right) \right)^2 \right].$$

$$(30)$$

For small deviations of the director field $\mathbf{n(r)}$ around its average orientation along the z axis $\mathbf{n}(\mathbf{r}) \approx (\delta n_x(\mathbf{r}), \delta n_y(\mathbf{r}), 1)$, the free energy assumes the form

$$\mathcal{F}_N = \frac{1}{2} \int d^2\mathbf{r}_\perp dz \left[K_1 \left(\nabla_\perp \cdot \delta\mathbf{n} \right)^2 + K_2 \left(\nabla_\perp \times \delta\mathbf{n} \right)^2 + K_3 \left(\partial_z \delta\mathbf{n} \right)^2 \right], \tag{31}$$

where $\nabla = (\nabla_\perp, \partial_z)$. For polymer nematics we now have to consider that the director field and the density of polymers in the (x, y) plane, $\rho = \rho_0 + \delta\rho$, are coupled (de Gennes and Prost 1993, Meyer 1982). If the polymers were infinitely long and stiff the coupling would be given by the continuity equation:

$$\partial_z \delta\rho + \rho_0 \nabla_\perp \cdot \delta\mathbf{n} = 0. \tag{32}$$

This constraint, however, is softened if the polymer has a finite length L or a finite persistence length $\mathcal{L}_p$. The persistence length is defined through the bending modulus of a single chain K_C as $K_C = k_B T \, \mathcal{L}_p$. For DNA, the persistence length is $\mathcal{L}_p \sim 50$ nm. The total bending elastic constant then has the form $K_3 = \rho_M K_C$, where ρ_M is the 2D density of the polymers perpendicular to their long axes (see also the chapters by Warren and MacKintosh in this volume).

On length scales larger than $\mathcal{L}_p$, the polymer can either fill the voids with its own ends or fold back on itself (Semenov and Khokhlov 1988). On these length scales the polymer nematic can splay without density change. Following Nelson (2002), this can be expressed by introducing G, a measure of how effectively the constraint is enforced. Density changes are expanded to second order in density deviations $\delta\rho(\mathbf{r}_\perp, z) = \rho(\mathbf{r}_\perp, z) - \rho_0$. B is the bulk modulus for compressions and dilations normal to the chains. The total free energy can now be written as

$$\mathcal{F} = \mathcal{F}_0(\rho_0) + \frac{1}{2} \int d^2\mathbf{r}_\perp dz \left[B \left(\frac{\delta\rho}{\rho_0} \right)^2 + G \left(\partial_z \delta\rho + \rho_0 \nabla_\perp \cdot \delta\mathbf{n} \right)^2 \right] + \mathcal{F}_N \,,$$

where $G = \frac{k_B T L}{2\rho_0}$, where L can not exceed $\mathcal{L}_p$ (Nelson 2002). In the limit of finite polymer length, G is also finite and can be obtained from the observation that $\partial_z \delta\rho + \rho_0 \nabla_\perp \cdot \delta\mathbf{n}$ equals the difference between the number of polymer heads and tails (Meyer 1982). From here one derives that G is the concentration susceptibility for an ideal mixture of heads and tails; thus $G = k_B T / (\rho_H + \rho_T)$, where ρ_H and ρ_T are the average concentrations of heads and tails, with $\rho_H, \rho_T = \rho_M$. The macromolecular density on the other hand equals $\rho_M = \rho_0/L$, wherefrom $G = k_B T \ell / 2\rho_0$. The corresponding structure factor can be written as

$$\mathcal{S}(q_\perp, q_z) = <|\delta\rho(q_\perp, q_z)|^2> = k_B T \, \frac{\rho_0^2 q_\perp^2 + k_B T \frac{\mathcal{K}(\mathbf{q})}{G}}{B q_\perp^2 + k_B T \left(\frac{B}{G\rho_0^2} + q_z^2 \right) \mathcal{K}(\mathbf{q})}, \tag{33}$$

where we defined

$$\mathcal{K}(\mathbf{q}) = \frac{K_1 q_\perp^2 + K_3 q_z^2}{k_B T}. \tag{34}$$

For long-fragment DNA the limit $L \longrightarrow \infty$ is appropriate, leading to the structure factor proposed by Selinger and Bruinsma (1991):

$$\mathcal{S}(q_\perp, q_z) = k_B T \, \frac{\rho_0^2 q_\perp^2}{K_1 q_\perp^2 q_z^2 + K_3 q_z^4 + B q_\perp^2}. \tag{35}$$

In order to calculate the contribution to the free energy due to fluctuations in nematic order we have to sum over all the density modes, obtaining

$$\mathcal{F} = \frac{1}{2} k_B T \iint \frac{d^2 q_\perp dq_z}{(2\pi)^3} \log \left(K_1 q_\perp^2 q_z^2 + K_3 q_z^4 + \mathcal{B} q_\perp^2 \right). \tag{36}$$

The problem here is that the above integral requires a cutoff and that the higher order terms in $q_\perp$ are more important than the ones we have kept here. For a moment let us assume that this free energy is valid and we may calculate

$$\frac{\partial \mathcal{F}}{\partial \mathcal{B}} = \frac{1}{2} k_B T V \iint \frac{q_\perp dq_\perp dq_z}{(2\pi)^2} \frac{q_\perp^2}{K_1 q_\perp^2 q_z^2 + K_3 q_z^4 + \mathcal{B} q_\perp^2}. \tag{37}$$

The q_z integral can be done straightforwardly and we are left with

$$\frac{\partial \mathcal{F}}{\partial \mathcal{B}} = \frac{1}{2} k_B T \frac{V}{(2\pi)^2} \frac{\pi}{2} \int \frac{q_\perp^3 dq_\perp}{\sqrt{\mathcal{B} q_\perp^2} \sqrt{K_1 q_\perp^2 + 2\sqrt{\mathcal{B} K_3 q_\perp^2}}}. \tag{38}$$

This integral depends essentially on the upper cutoff for $q_\perp = q_{\perp max}$ and we obtain

$$\frac{\partial \mathcal{F}}{\partial \mathcal{B}} = k_B T \frac{V}{4\pi} \frac{\mathcal{B} K_3}{K_1^2 \sqrt{\mathcal{B} K_1}} F \left(\frac{q_{\perp max}}{2\sqrt{\frac{\mathcal{B} K_3}{K_1^2}}} \right), \tag{39}$$

where the function $F(x)$ is defined as

$$F(x) = \int_0^x \frac{u^{3/2} du}{\sqrt{1+u}} = \frac{1}{4} \left(\sqrt{x}\sqrt{1+x}(2x-3) + 3 \operatorname{areasinh}\sqrt{x} \right)$$

$$= \begin{cases} \frac{2}{5} x^{5/2} & ; x \ll 1 \\ \frac{1}{2} x^2 & ; x \gg 1 \end{cases}. \tag{40}$$

From here we obtain the two limiting forms of the free energy as

$$\mathcal{F} \simeq \frac{k_B T}{5 \times 2^{3/2} \pi} \sqrt[4]{\frac{\mathcal{B}}{K_3}} q_{\perp max}^{5/2} + \cdots \quad ; q_{\perp max} \ll 2\sqrt{\frac{\mathcal{B} K_3}{K_1^2}}, \tag{41}$$

$$\mathcal{F} \simeq \frac{k_B T}{16\pi} \sqrt{\frac{\mathcal{B}}{K_1}} q_{\perp max}^2 + \cdots \quad ; q_{\perp max} \gg 2\sqrt{\frac{\mathcal{B} K_3}{K_1^2}}. \tag{42}$$

Obviously the long-wavelength physics is very complicated and depends crucially on the values of typical polymer length and the ratios of elastic constants. However, it is also dependent on the $q_\perp$ cutoff. We have to either eliminate the cutoff by including higher-order terms in the original Hamiltonian or choose a meaningful cutoff. Higher-order terms will capture the short-wavelength physics and remove the divergence (Strey *et al.* 1999).

One can show (Strey *et al.* 1999) that a consistent value of the cutoff has to be proportional to the Brillouin zone radius $q_{\perp max} \simeq \frac{\pi}{D}$, where D is the effective separation between the polymers in the nematic phase. This is a physically meaningful and appropriate cutoff because the underlying macroscopic elastic model has, by definition, to break down at wavelengths comparable to the distance between molecules.

Putting in the numbers valid for DNA arrays, one realises that in the regime of densities considered here we are always in the Equation 41 limit. We would now have to derive the mesoscopic elastic moduli from the microscopic interactions described via a pair potential $\mathcal{F}(R)$ (the interaction free energy) between the segments of the macromolecules. At present this program is too ambitious and we simply exploit the standard *ansatz* for the different elastic moduli (de Gennes and Prost 1993) expressed via the cell model free energy

$$
\begin{aligned}
K_1 &= K_2 \simeq \mathcal{F}(R)/R \\
K_3 &\simeq \rho_0 K_C + \mathcal{F}(R)/R \\
\mathcal{B} &\simeq V\frac{\partial^2 \mathcal{F}(V)}{\partial V^2} = \frac{1}{4\pi}\left(\frac{\partial^2(\mathcal{F}/L)}{\partial^2 R} - \frac{1}{R}\frac{\partial(\mathcal{F}/L)}{\partial R}\right),
\end{aligned}
\tag{43}
$$

where ρ_0 is the 2D density of the macromolecules perpendicular to their long axes, K_C is the elastic rigidity modulus of a single polymer molecule and we assume that the polymers have an average separation R between first neighbours (Strey *et al.* 1999).

The macroscopic free energy (Equation 41) together with the values of elastic constants (Equation 43) already point to the salient features of the fluctuation modified equation of state. Obviously the thermal fluctuations make the free energy much longer ranged than the underlying microscopic interaction potential. If the interaction potential decays exponentially with characteristic length λ, then the free-energy equation (Equation 41) decays with four times the characteristic length! The factor of four is a simple consequence of mesoscopic elasticity.

We now use the form of the bare interaction free energy $\mathcal{F}(R)$ appropriate for a DNA array (Equation 29). One can see that at all relevant densities $K_3 \simeq \rho_0 K_C$. We are now able to fit the calculated equation of state obtained from Equation 41 to the experimental equation of state (Strey *et al.* 1999, Figure 3). The values for the DNA bending rigidity and the Debye length obtained from such a fit are comfortably within the expected range (Strey *et al.* 1999). It should be mentioned here that the effective charge evaluated from the fit is about half the amount expected on the basis of the Manning condensation theory.

A fundamental drawback of this formulation for the equation of state in an assembly of flexible molecules is most clearly seen in the *ansatz* equation, Equation 43. The elastic moduli are not really calculated on the same level as the free energy but are assumed to have a form that, at least for the compressibility modulus, would be strictly valid only for rigid molecules. The preceding formulation is thus not completely self-consistent, and we will make an attempt to improve it in the next section. The failure of this attempt will make us aware of some fundamental properties of the nature of the positional order in DNA arrays.

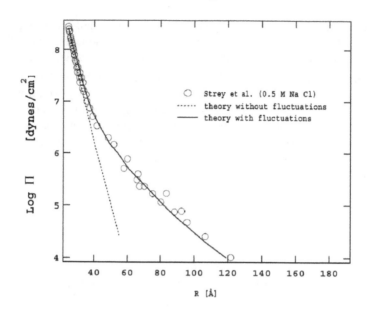

Figure 3. *A fit of the lowest-order fluctuation equations of state (Equation 41) to DNA data (Strey et al. 1999). Dashed line — theory without fluctuations as in Figure 2. Full line — theory with conformational fluctuations of the molecules taken into account on a harmonic level (Equation 41). The value of the effective charge on the DNA surface is obtained from the fit to experiment and is found to be about half the Manning condensation value.*

5 Equation of state: The effect of thermal fluctuations (2)

At this point, one can use any of the advanced theories that take into account the thermal fluctuations at a deeper level than the macroscopic theory of the previous section and that allow for the thermal fluctuation effects that can modify the compressibility modulus itself (Lipowsky 1995). Such theories are well worked out in other areas of physics also, such as magnetic vortex arrays in type II superconductors. There are different approaches that one can follow. One could either formulate the problem in the language of the functional renormalisation group (Lipowsky 1995) or on the level of a variational calculation of the compressibility modulus in the Feynman-Kleinert style (Kleinert 1995). To remain as close as possible to the approach outlined before, we chose the second, i.e., variational, approach. It usually fares quite well even when compared to the more powerful renormalisation group approach.

5.1 Variational calculation of the osmotic pressure in a hexagonal array

For a system with a hexagonal local symmetry, I follow closely the calculation of Volmer and Schwartz (1999) derived for the system of magnetic vortex lines in type II superconductors. Apart from the difference between elastic energies of a vortex line vs. a flexible polymer, the two cases are analogous. The interaction Hamiltonian of an oriented DNA polymer array with hexagonal local symmetry can be written in the form

$$\mathcal{H} = \tfrac{1}{2} K_C \sum_{n,m=1}^{\infty} \int dz \left(\frac{\partial^2 \mathbf{r}_\perp{}^{(n,m)}(z)}{\partial z^2} \right)^2 + \tfrac{1}{2} \sum_{n,m \neq n'm'}^{\infty} \int dz \, V \left(\mathbf{r}_\perp{}^{(n,m)}(z) - \mathbf{r}_\perp{}^{(n',m')}(z) \right),$$

(44)

where $\mathbf{r}_\perp{}^{(n,m)}(z)$ is the local displacement of a polymer chain at the (n,m) lattice position perpendicular to the long axis, z, while $V \left(\mathbf{r}_\perp{}^{(n,m)}(z) - \mathbf{r}_\perp{}^{(n',m')}(z) \right)$ is the interaction potential between different macromolecules at the same value of z. In principle, the indices n, m would run through all the positions of the polymers at a certain planar cross section through the nematic. Because of the short-range nature of the interaction and computational convenience, I restrict them to nearest neighbours (Volmer and Schwartz 1999). K_C is, of course, the elastic modulus of DNA given by $K_C = k_B T \mathcal{L}_P$, where $\mathcal{L}_P$ is the persistence length.

Instead of using this nonharmonic Hamiltonian, I will take a simpler reference Hamiltonian of a general harmonic form. Let us start with the following parametrisation

$$\begin{aligned} \mathbf{r}_\perp{}^{(n,m)}(z) &= \mathbf{R}_{nm} + \mathbf{u}^{(n,m)}(z), \\ \mathbf{u}^{(n,m)}(z) &= \sum_{\mathbf{Q}_\perp, q_z} \mathbf{u}(\mathbf{Q}_\perp, q_z) e^{iq_z z + i\mathbf{Q}_\perp \mathbf{R}_{nm}}, \end{aligned}$$

(45)

where $\mathbf{R}_{nm} = n\mathbf{a}_1 + m\mathbf{a}_2$, with $\mathbf{a}_1$ and $\mathbf{a}_2$ as the two basis vectors of the macromolecular lattice perpendicular to the long axis, z, of the molecules and $\mathbf{Q}_\perp$ the appropriate reciprocal lattice vectors. I take the reference Hamiltonian in the general harmonic form in the reciprocal Fourier space,

$$\mathcal{H}_0 = \tfrac{1}{2} \sum_{\mathbf{Q}_\perp, q_z} \left(K_C q_z^4 \delta_{ik} + \mathcal{B}_{ik}(\mathbf{Q}_\perp) \right) u_i(\mathbf{Q}_\perp, q_z) u_k(-\mathbf{Q}_\perp, -q_z) + V_0(\mathbf{a}),$$

(46)

where $\mathcal{B}_{ik}(\mathbf{Q}_\perp) = \mathcal{B}_{ik} \sum_{\mathbf{a}} 4 \sin^2 \frac{\mathbf{Q}_\perp \mathbf{a}}{2}$ and the sum over $\mathbf{a}$ refers to summation over the positions of nearest neighbours.

The idea of the Feynman-Kleinert variational principle is to now use the Hamiltonian (Equation 46) as an harmonic *ansatz* whose effective parameters, such as $\mathcal{B}_{ik}$ and $V_0(\mathbf{a})$, are determined variationally by minimising the upper bound for the free energy (Kleinert 1995). This approach has already been used in the context of multilamellar systems (Podgornik and Parsegian 1992).

Let us start with what is usually referred to as the Gibbs-Bogolyubov inequality. Taking the exact free energy corresponding to the Hamiltonian $\mathcal{H}$ and an approximate one corresponding to $\mathcal{H}_0$, it is straightforward to derive

$$\mathcal{F} \leq \mathcal{F}_0 + \langle \mathcal{H} - \mathcal{H}_0 \rangle_{\mathcal{H}_0}.$$

(47)

The average $\langle \dots \rangle_{\mathcal{H}_0}$ is performed with respect to the Hamiltonian of Equation 46. Obviously the above inequality defines an upper bound for the free energy. By evaluating explicitly the terms in Equation 47 with the reference Hamiltonian given by Equation 46, one is left with

$$\mathcal{F}_0 = \tfrac{1}{2} \sum_{\mathbf{Q}_\perp, q_z} \left(K_C q_z^4 \delta_{ik} + \mathcal{B}_{ik}(\mathbf{Q}_\perp) \right) \langle u_i(\mathbf{Q}_\perp, q_z) u_k(-\mathbf{Q}_\perp, -q_z) \rangle_{\mathcal{H}_0} + V_0(\mathbf{a}), \quad (48)$$

where the positional correlation function is given by

$$\langle u_i(\mathbf{Q}_\perp, q_z) u_k(-\mathbf{Q}_\perp, -q_z) \rangle_{\mathcal{H}_0} = \sigma_{ik}(\mathbf{Q}_\perp, q_z) = \frac{k_B T}{K_C q_z^4 \delta_{ik} + \mathcal{B}_{ik}(\mathbf{Q}_\perp)} \quad (49)$$

via the equipartition theorem since we have a general quadratic form of the reference Hamiltonian. On the other hand, the second term of Equation 47 can be derived as

$$\langle \mathcal{H} - \mathcal{H}_0 \rangle_{\mathcal{H}_0} = \tfrac{1}{2} V_{\sigma_{ij}(\mathbf{Q}_\perp, q_z)}(\mathbf{a}) - 2 \mathcal{B}_{ij} \sum_{\mathbf{Q}_\perp, q_z} \sum_{\mathbf{a}} \sin^2 \tfrac{\mathbf{Q}_\perp \mathbf{a}}{2} \sigma_{ik}(\mathbf{Q}_\perp, q_z) - V_0(\mathbf{a}), \quad (50)$$

where in the first term of the preceding expression we have introduced the fluctuation modified form of the interaction potential given by

$$V_{\sigma_{ij}(\mathbf{Q}_\perp, q_z)}(\mathbf{a}) = \int d^2 \mathbf{r}_\perp \, V(\mathbf{r}_\perp) \sum_{\mathbf{a}} \sum_{\mathbf{k}} e^{\left(i\mathbf{k}(\mathbf{a} - \mathbf{r}_\perp) - 2\beta k_i k_j \sum_{\mathbf{Q}_\perp, q_z} \sin^2 \frac{\mathbf{Q}_\perp \mathbf{a}}{2} \sigma_{ij}(\mathbf{Q}_\perp, q_z) \right)}.$$

$$(51)$$

This simply follows from the fact that $\mathcal{H}_0$ is a quadratic function for which one has $\langle e^{iA_i u_i} \rangle_{\mathcal{H}_0} = e^{-\frac{1}{4} A_i A_k \langle u_i u_k \rangle_{\mathcal{H}_0}}$. All this is very closely related the analysis of Volmer and Schwartz (1999) for the system of magnetic vortex lines in type II superconductors. The only difference is in the conformational energy of a polymer (elastic energy) and a vortex line (tension energy). The above variational formulation is obviously based on two parameters: $\sigma_{ik}(\mathbf{Q}_\perp, q_z)$ and $V_0(\mathbf{a})$.

What the Feynman-Kleinert variational principle is aiming at is to minimise the second term of Equation 47. Let us first consider the minimal value of $V_0(\mathbf{a})$. Clearly, the second term of Equation 47 is non-negative. It is in fact minimal if

$$V_0(\mathbf{a}) = \tfrac{1}{2} V_{\sigma_{ij}(\mathbf{Q}_\perp, q_z)}(\mathbf{a}) - \tfrac{1}{2} \sum_{\mathbf{Q}_\perp, q_z} \mathcal{B}_{ik}(\mathbf{Q}_\perp) \, \sigma_{ij}(\mathbf{Q}_\perp, q_z), \quad (52)$$

thus making $\mathcal{F} = \mathcal{F}_0$ or *in extenso*

$$\mathcal{F}_0 = -k_B T \log \langle e^{-\beta \mathcal{H}_0} \rangle = V_0(\mathbf{a}) + \frac{k_B T}{2} \mathrm{Tr} \sum_{\mathbf{Q}_\perp, q_z} \log \left(\delta_{ik} + \frac{\mathcal{B}_{ik}(\mathbf{Q}_\perp)}{K_C q_z^4} \right). \quad (53)$$

It is obvious from the above expression what the fluctuations do to the free energy on this level of approximations. Without the thermal noise the second term of Equation 53 would be zero, and we would be back to the equation of state without any intramolecular degrees of freedom. The fluctuations again, just as in the previous section, effectively boost the bare intermolecular interactions this time quantified by $V_0(\mathbf{a})$.

As for the second parameter that needs to be minimised, one gets simply

$$2 \mathcal{B}_{ij} \sum_{\mathbf{a}} \sin^2 \frac{\mathbf{Q}_\perp \mathbf{a}}{2} = \tfrac{1}{2} \frac{\partial V_{\sigma_{ij}(\mathbf{Q}_\perp, q_z)}(\mathbf{a})}{\partial \sigma_{ij}(\mathbf{Q}_\perp, q_z)}. \quad (54)$$

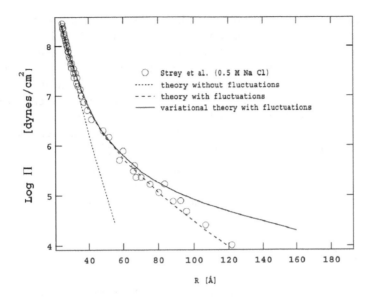

Figure 4. *A fit of the variational equation of state to the experimental data. Here the values of the elastic moduli are obtained self consistently and also include a fluctuation contribution. Surprisingly this fit with a variational equation of state for DNA obtained from Equation 53 fares much worse than the macroscopic fluctuation theory, see Figure 3. The reason lies in the nature and range of the positional order in an ordered assembly of DNA, see main text.*

Equations 52 and 54 represent the solution to the minimisation problem. Very similar equations have already been derived in the case of magnetic vortex arrays (Volmer and Schwartz 1999).

The most important quantity of the preceding formulation of the fluctuational renormalisation of the interactions in a macromolecular array is the pair interaction potential per unit length between two polymer segments $V\left(\mathbf{r}_\perp{}^{(n,m)}(z) - \mathbf{r}_\perp{}^{(n',m')}(z)\right)$, given again by Equation 29, that enters Equation 53. One should appreciate the main difference between the calculation described in this section and the previous section: before the elastic moduli were given by different expressions, involving only the *bare* interaction, Equation 43; now the elastic modulus is given as a function of fluctuation renormalised interaction (Equation 54). The last calculation should thus in principle be more accurate in describing the effect of thermal fluctuations.

However, when we compare the equation of state obtained from Equation 53 with the macroscopic fluctuation equation of state from Equation 41, we see that it fares much worse when compared with experiments (see Figure 4). This fact is surprising, since the whole idea was to get an even better estimate of the fluctuation effects, and demands an explanation.

The reason for this discrepancy turns out to be quite simple. The variational *ansatz* based on positional correlation function $\sigma_{ik}(\mathbf{Q}_\perp, q_z)$ only makes sense if this quantity itself is well defined, i.e., if $\sigma_{ik}(\mathbf{Q}_\perp, q_z) < \infty$. This constraint fits best the description of a solid, with long-range positional order, and finite correlations between any two macromolecular positions. DNA, however, at the relevant densities, is not a solid. It is a hexatic liquid crystal with only short-range positional order and is thus more akin to a fluid. For a fluid, of course, with only short-range positional correlations, the positional correlations (as opposed to density correlations, which of course remain finite) diverge in the thermodynamic limit and cannot be described with a finite correlation function. Therefore, the variational theory, strictly applicable only to a hexagonal crystal, fares much worse than the mesoscopic fluctuation theory of the previous section if applied to a DNA array at densities between the crystalline and isotropic phases.

A variational theory similar in spirit to what I developed above was also put forth by de Vries (1994) basing his analysis on previous work by Odijk (1992, 1993). Numerical results of this approach are indistinguishable from those presented here.

6 Conclusion

I have given a broad overview of the work on the equation of state of DNA in aqueous ionic solutions. Most of what I described applies only to monovalent salt solutions. Higher valency salts have a very different effect on the properties of DNA (Bloomfield *et al.* 2000) falling outside of my immediate interests. The picture that I developed here shows that, depending on the monovalent salt concentration, conformational fluctuations mask the direct electrostatic interactions at all but very low salt levels. It might thus come as a surprise that for DNA, which is a very highly charged polyelectrolyte, the effect of electrostatic interactions on the equation of state is modified in an essential way by the conformational fluctuations of DNAs. Direct electrostatic effects for this highly charged polyelectrolyte are thus counter-intuitively discernible only in a very limited range of salt concentrations.

Acknowledgements

I would like to thank Adrian Parsegian, Per Lyngs Hansen, Stephanie Tristram-Nagle and John Nagle for numerous discussions on the various aspects of the work described here.

References

Andelman D, 1995, in *Structure and Dynamics of Membranes* 603, eds Lipowsky R and Sackmann E (North Holland, New York).

Bloomfield V A, Crothers D M and Tinoco I, 2000, *Nucleic Acids: structures, properties and functions* (University Science Book, Sausalito).

Brenner S L and Parsegian V A, 1974, *Biophys J* **14** 327–334.

Chaikin P M and Lubensky T C, 2000, *Principles of Condensed Matter Physics* (Cambridge University Press, Cambridge).

de Gennes P and Prost J, 1993 *The Physics of Liquid Crystals*, second edition (Oxford University Press, Oxford).

de Vries R, 1994, *J Phys II France* **4** 1541.

Fuoss R, Katchalsky A, and Lifson S, 1951, *Proc Natl Acad Sci USA* **37** 579.

Gelbart W M, Bruinsma R F, Pincus P A and Parsegian V A, 2000, *Physics Today* **53** 38.

Israelachvili J, 1998, *Intermolecular and Surface Forces*, second edition (Academic Press, London).

Kjellander R, 2001, in *Electrostatic effects in soft matter and biophysics*, eds Holm C, Kekicheff P and Podgornik R, NATO Science Series II - Mathematics, Physics and Chemistry **46**.

Kleinert H, 1995, *Path Integrals in Quantum Mechanics, Statistics, and Polymer Physics* (World Scientific Pub Co, Singapore).

Lifson S and Katchalsky A, 1954, *J Polymer Sci* **XIII** 43.

Lipowsky R, 1995, in *Structure and Dynamics of Membranes*, eds Lipowsky R and Sackmann E (North Holland, New York).

Mahanty J and Ninham B W, 1976, *Dispersion Forces* (Academic Press, London).

Meyer R, 1982, in *Polymer Liquid Crystals* 133, eds Ciferri A, Krigbaum W and Meyer R, (Academic, New York).

Nelson D R, 2002, *Defects and Geometry in Condensed Matter Physics* (Cambridge University Press, Cambridge).

Odijk T, 1992, *Langmuir* **8** 1690.

Odijk T, 1993, *Europhys Lett* **24** 177.

Petrov A G, 1999, *The Lyotropic State of Matter: Molecular Physics and Living Matter Physics* (Gordon and Breach Science Publishers).

Podgornik R, Strey H H and Parsegian V A, 1998, *Curr Op in Colloid and Interf Sci* **3** 534.

Podgornik R and Parsegian V A, 1992, *Langmuir* **8** 557.

Schmitz K S, 1993, *Macroions in Solution and Colloidal Suspension* (VCH Publishers, New York).

Selinger J and Bruinsma R, 1991, *Phys Rev A* **43**, 2910.

Semenov A and Khokhlov A, 1988, *Sov Phys Usp* **31**, 988.

Shkel I A, Tsodikov O V and Record Jr. M T, 2000, *J Phys Che.* **104** 5161–5170.

Shkel I A, Tsodikov O V and Record Jr. M T, 2002, *Proc Natl Acad Sci* **99** 2597–2602.

Strey H H, Podgornik R, Rau D C and Parsegian V A, 1998, *Curr Opin Struc Biol* **8** 309–313.

Strey H H, Parsegian V A and Podgornik R, 1999, *Phys Rev E* **59** 999–1008.

Tamashiro M N and Schiessel H, 2003, *Phys Rev E* **68** 066106.

Tracy CA and Widom H, 1997, *Physica A* **244** 402–413.

Verwey E G and Overbeek J T G, 1948, *The Theory of the Stability of Lyophilic Colloids* (Elsevier, Amsterdam).

Volmer A and Schwartz M, 1999, *Eur Phys J B* **7** 211.

Sequence-structure relationships in proteins

Ron Elber, Jian Qiu, Leonid Meyerguz and Jon Kleinberg

Department of Computer Science, Cornell University
4130 Upson Hall, Ithaca NY 14853-7501, US

ron@cs.cornell.edu

1 Introduction

In this chapter, we provide a brief description of the protein-folding and inverse-folding problems, and then move on to discuss the design of energy functions for protein recognition based on machine learning approaches. The energy functions are applied to estimate the sequence capacity of all known protein folds and to compute the evolutionary temperature(s) of embedding sequences in known protein structures.

The chapter is divided into three sections. We start with a brief introduction to proteins, continue with the design of energies for recognition of protein folds, and conclude with an application to protein evolution, studying the sequence capacity of different structures.

Proteins are linear polymers that are sometimes cross-linked (via sulfur bonds) but are never branched. (See Warren's chapter in this volume for an introduction to polymer physics.) They serve diverse and numerous functions in the cell as facilitators of many biochemical reactions, signalling processes and providers of essential skeletal structures. These linear polymers consist of 20 different types of monomers that share the same backbone atoms (exceptions are proline and glycine) and have different (short) side chains. The chemical composition of a protein molecule is determined by the linear sequence of amino acids called the primary structure. The sequence starts at the so-called N terminal and ends at the C terminal; the two ends are not equivalent, i.e., running the sequence backward does not produce the same protein. Typical lengths of protein sequences are a few hundred amino acids. The extremes are a few tens to a few thousands of amino acids.

One of the remarkable features of proteins is their ability to fold into a well-defined three-dimensional structure in aqueous solutions, the so-called protein-folding problem. This chapter is concerned with a few indirect aspects of this problem. The working hypothesis is that the three-dimensional shape is determined uniquely by the sequence of the amino acids and is the thermodynamically stable state. Anfinsen (1973) put forward this extraordinary hypothesis that protein molecules are stable in isolation with no support from other components of the living system. From a computational and theoretical viewpoint, the Anfinsen hypothesis makes it possible to define and use in predictions an (free) energy function of an isolated protein molecule (in an aqueous solution). This function leads to significant simplification and saving of computational resources compared to studying a complete cellular environment. The free energy has a global minimum that coincides with the three-dimensional structure observed experimentally.

Structures of proteins are classified in terms of secondary structure elements, domains and individual chains (tertiary structure), and packing of tertiary structural elements (quaternary structure). Secondary structure is determined according to the hydrogen bond patterns of the backbone atoms that form small structural elements (ten to twenty amino acids). These elements are assembled to form the stable three-dimensional compact structure of the protein. Typical elements of secondary structure are the helices and sheets, where helices provide local chain structure and beta sheets connect pieces of the chain that can be far apart along the sequence (but close in space). The formation of local structure restricts the number of allowed conformations of the peptide chain and facilitates more accurate and rapid folding compared to a comprehensive search through all self-avoiding 'walks' of the polymer chain.

Domains are fragments of a protein chain. Each domain includes several secondary structure elements and is 'self-sustained'. It is expected that the average number of contacts between amino acids that belong to the same domain (the total number of contacts in the domains divided by the number of amino acids) is much larger than the average number of contacts between amino acids situated at different domains. Domains have evolutionary implications. Empirically, domains were shown to swap between genes and proteins, suggesting an evolutionary mechanism in which a significant segment of one protein (a domain) is inserted in, or exchanged with, another protein. This is in contrast to the alternative evolutionary mechanism of a point mutation, a process that modifies one amino acid at a time. Identifying relevant domains is likely to assist us in the characterisation of basic building blocks of evolutionary processes and the mechanisms that guide them.

A complete (single) protein chain defines the tertiary structure. The quaternary structure is an aggregate of a few protein chains that work cooperatively on a biological task. The discussion in the present chapter considers only isolated chains, and we therefore stop at the tertiary structure. This is clearly an approximation since some of the relevant interactions arise from nearby chains. Nevertheless, as in the domain picture, we anticipate that at least some of the individual chains are stable and can be studied in isolation. (For a more detailed introduction to proteins, see Poon's chapter in this volume.)

In statistical mechanics the folded conformation of a protein can be found by minimising the potential of mean force that we loosely call the free energy, F. The free energy is defined by the following integration:

$$F(X) = -k_{\mathrm{B}}T log \left[\int \exp \left[-\frac{U(X, R)}{k_{\mathrm{B}}T} \right] dR \right]. \tag{1}$$

The probability of finding the system in equilibrium specified by a temperature T at X is proportional to $\exp[-F(X)/k_{\mathrm{B}}T]$. The microscopic potential is $U(X, R)$, k_{B} is the Boltzmann constant and T is the absolute temperature. In Equation 1 the free energy is a function of a subset of the total number of coordinates, X, which includes (for example) bond rotations. The vector R includes the remaining coordinates that we eliminate by the integration on the right-hand-side equation. Examples of typical coordinates of the R vector are the positions of the solvent (water) molecules and bond vibrations within a protein. The free energy is defined in terms of protein coordinates (e.g., torsions) that remain quite large in number. The number of torsions more than doubles the number of amino acids in the protein and is therefore between a few hundreds to a few thousands for a single protein chain. Since each of the torsions has about three rotamer states, a significant entropic contribution to the reduced free energy remains and a minimum alone cannot be the whole story. However, in the discussion we do not consider the question of stability or chain entropy (i.e., if the minimum of the potential of mean force is sufficiently deep to overcome the entropy of the misfolded state). At present we are happy to identify the minimum with the correct structure, even if the stability energy is not available.

It is clear from Equation 1 that for a reasonable microscopic potential $U(X, R)$ (so that the integral is well defined) the free energy $F(X)$ is computable. However, we cannot determine for the general case a simple and transferable functional form for $F(X)$, even if the microscopic potential is known (and this is not guaranteed either). By transferable potentials we mean a single set of parameters for a given type of an amino acid regardless of the position of the amino acid along the sequence or the specific protein chain the amino acid is embedded in. The transferable formulation is similar in spirit to that of the microscopic potential and leads to more general and simpler parameterisation.

Therefore, to ensure transferable potentials and ease of computations many applications to protein folding assume the functional form of $F(X)$ and do not compute it as outlined in Equation 1. A set of potential parameters is optimised within the preset functional form. Assuming an empirical functional form for $F(X)$ is a natural extension of the approach used for the atomically detailed potential, $U(X, R)$. The last is also set empirically in most applications to proteins since the full exact calculations (including explicitly the electrons in the system) are just too expensive. Only a limited number of calculations (that are severely restricted in time) employ the full electronic structure model. For example, the following free-energy functional is assumed to be a sum of pair interactions between all amino acids, a convenient but ad hoc proposition:

$$F = \sum_{i>j} F_{ij} \left(\alpha_i, \beta_j, r_{ij} \right) \qquad \forall i, j \quad (i, j \quad amino \quad acids). \tag{2}$$

The distance between the geometric centres of the amino acid side chains is $\mathbf{r}_{ij}$ (Meller and Elber 2002). The free energy of each interacting pair depends on the distance between the pair, $\mathbf{r}_{ij}$, and their type, α_i, β_j (but not their position along the sequence). The impact of the averaging formulated in Equation 1 is subtle. For example, averaging of the solvent interactions yields a repulsive potential of mean force between charged amino acids even if they have opposite electric charges that attract in vacuum. The preferred state of charged amino acids is to be surrounded by water molecules, far from the low dielectric medium typical of the interior of proteins. The tendency to be well solvated is observed only indirectly (since the solvent is not present explicitly in the model), and results in effective repulsion between well solvated (charged) amino acids. The hydrophobic (apolar) residues 'attract' each other since they disrupt the hydrogen bond structure of the water molecules and their aggregation minimises this effect. These interactions are weak, require the cancellation of many large terms, and are difficult to reproduce by direct averaging for proteins. The solvent-induced interactions are small in magnitude, and the integrals in Equation 1, which are performed stochastically, may not be accurate to the level required to fold proteins. An alternative approach that avoids the integration in Equation 1, and which we consider in Section 2, is to 'learn' the free-energy surface from a set of experimentally determined protein structures.

We conclude our discussion of energy with another comment from the school of sceptics. The hypothesis that the native structures of proteins are global (free) energy minima is not always true. Some post-folding modifications (for example, cutting a leading peptide or the start of the protein chain) make the global (free) energy minimum of the original chain different from the native (modified) structure. A classic example is of the protein insulin (Lehninger 1982). Other examples are proteins that do not fold spontaneously and require external help of other macromolecules (chaperones) to adopt their correct three-dimensional shape. Nevertheless, despite the considerable complexity of the biological machinery that folds proteins (which suggests that some proteins cannot be studied in isolation), we do find numerous proteins that follow the Anfinsen hypothesis. Therefore, the following discussion, seeking a functional form for the free energy and its global minimum, is a valid approach to determine structures of many proteins.

While the path from sequence to structure is considered to be *the* protein-folding problem, this chapter focuses on another intriguing question: the inverse folding problem. Anfinsen's hypothesis argues that every sequence corresponds to one unique structure of the protein. Is the reverse true, i.e., can any structure of a protein be linked to a unique sequence of amino acids? This question, the reverse of the protein-folding problem (from structure to sequence), is answered by a definite 'no'. There are many sequences that are known (experimentally) to fold to the same or similar shapes. Consider the Protein Data Bank (Berman *et al.* 2000) *(http://www.rcsb.org)*, which is the digital repository of protein shapes. The PDB includes 25,960 protein structures as of June 15, 2004. These structures include many redundant shapes and can be reduced to a few hundred distinct protein families. The structural families are defined by shape similarities regardless of the amino acid sequences of the compared proteins. Hence, on the average, there are hundreds of sequences in the PDB that fold into the same protein shape. The 'seed' shape defines a fold family.

Consider another important database of proteins, PIR Non-Redundant Reference Protein Database (PIR-NREF, *http://pir.georgetown.edu/pirwww/search/pirnref.shtml*), which includes sequences only. A significant fraction of the millions of sequences in the PIR-NREF Database can be associated with a known fold family. The observed redundancy in mapping from sequences to structures in PIR-NREF is even larger than the redundancy implied by the PDB. It is a mapping from the many (sequences) to the relatively few (structures). Evolutionary processes that modify and generate new protein sequences by changing one amino acid at a time are 'stuck' in the neighbourhood of individual structures (islands in sequence space) and produce new proteins that have essentially the same shape (note that the evolutionary process we consider here is *not* the domain swap mentioned earlier). The seed shape is used over and over again for alternative sequences. The variations in sequences in the neighbourhood of a given fold may adjust the function of the protein while maintaining the same overall structure. For example, a small change in activity would create a modified enzyme with enhanced (or reduced) affinity to the same ligand. A large change will use the same structural template for enzymes with different ligands or chemical reactions. Since there are numerous examples of the second kind (large change in function), it is difficult to predict protein function based on structure similarity only.

An intriguing follow-up research direction is the sequence capacity of a structure. Given a shape of a protein X and energy E (which is a function of the sequence and the structure), what is the number of sequences $N(E, X)$ that fit this shape with energy lower than E? We will demonstrate that the number of sequences is so large that statistical mechanical analysis is suggestive. Following the usual notion of entropy in statistical thermodynamics, we define a 'selection temperature' for the ensemble of sequences that fit a particular structural family. We finally speculate on evolutionary implications of our work.

2   Energy functions for fold recognition

For meaningful calculations of macromolecular properties, we must have a free-energy function that weighs the importance of different structures. The lower the free energy, the more probable the structure. The design, choice of functional form, and optimisation of parameters for the free-energy function are the focus of the present section. Traditionally, energy functions for simulations of condensed phases and macromolecules (and proteins *are* macromolecules) were built according to chemical principles, starting with small molecular models and interpolating to the large macromolecules such as proteins. A typical atomically detailed energy function is of the following form

$$U(X, R) = U_c(X, R) + U_n(X, R). \tag{3}$$

The energy $U_c(X, R)$ includes the covalent terms: bonds, angles and torsions. These are two, three and four body terms, respectively (see Figure 1). Most of the time, the bonds and angles of the protein chain remain near their equilibrium values. It therefore makes sense to model the bond and the angles with (stiff) harmonic energy terms. An alternative is to use holonomic constraints and to fix them at their ideal values.

In the present chapter we do not directly consider the contribution of the covalent part U_c but focus instead on $U_n(X, R)$. The covalent term is considered (and enforced) indirectly by using a subset of structures that satisfy the holonomic constraints on bonds and angles. The energy of the above covalent coordinates at their ideal value is zero, which makes it unnecessary to add the contributions of the bond and angle energies.

The case of torsions is different from the bonds and the angles. Torsions are allowed to change significantly and are not constrained. However, the torsion energy contribution

Figure 1. *A schematic of a polymer chain with covalent degrees of freedom denoted by arrows. A 'stick' connects two atoms at its edges. A bond term describes the distance between the two atoms. An angle (term) is between two connected bonds, and a torsion measures the angle between the planes defined by three sequential bonds (the first and the second bond define the first plane, the second and the third bonds define the second plane).*

is small and is set to zero in some potential functions, which is the approach we take here. An exception to the 'rule' of torsions with small energy contribution is rotations around double bonds (e.g., amide planes). The rotations around double bonds are fixed at their ideal values similarly to the bonds and the angles.

As we argued in the introduction, the function $F(X)$ may be computed from the detailed potential $U(X, R)$ or $U_n(X, R)$ following the integration outlined in Equation 1. However, this is computationally intractable, and so far no one has done it comprehensively and accurately for proteins (even if we are willing to forget about the transferability issue). A more pragmatic approach is to accept that a function $F(X)$ exists, and to search for a functional form and parameters that make general physical sense and are successful in identifying the correct folds of proteins.

2.1   Statistical potentials

The idea of statistical potentials was pioneered by Scheraga (Tanaka and Scheraga 1976) and popularised by Miyazawa and Jernigan (Miyazawa and Jernigan 1984). It is probably the most widely used functional form of an energy function in the protein-folding field. At this point it is useful to introduce some probabilistic arguments to motivate the following computational approach. As argued above, the free energy is directly related to the probability of finding the system at a particular state. Consider the following question: what is the probability that in the set of known protein folds we will find an amino acid of type α_k in a (given) structural site with exposed surface area A_i and secondary structure s_j? Here is the formula:

$$P\left(\alpha_k \mid A_i, s_j\right) = \frac{P\left(A_i, s_j, \alpha_k\right)}{P\left(A_i, s_j\right)}. \tag{4}$$

The probability of the event z is $P(z)$. The amino acid type is α_k. We will define the secondary structure by a discrete variable s_j ($0 = \alpha$ helix, $1 = \beta$ sheet structure, $2 = 3/10$ helix, $3 =$ bend and turn, $4 = \pi$ helix, and the rest), and the exposed surface area A_i is binned into eight discrete states. Note that the variables A_i and s_j take the role of the X coordinate vector that we discussed abstractly in the previous section. Identifying the relevant reduced variables is of crucial importance, and here we are using our intuition on protein structure and energy. A surface term motivates polar residues to be on the surface of the protein, and such events should be observed with high frequency. Similarly, hydrophobic residues are buried in the protein matrix, a frequent observation.

The preceding conditional probability is related to the inverse protein-folding problem that was mentioned in the introduction. Alternatively, and more related to the protein-folding problem, we may consider the probability that an amino acid α_k will be found in a structural site characterised by (A_i, s_j). We will use the exposed surface area and secondary structure as non-bonded variables to describe the state of the protein:

$$P(A_i, s_j \mid \alpha_k) = \frac{P(A_i, s_j, \alpha_k)}{P(\alpha_k)}. \tag{5}$$

For a protein chain with a sequence $\alpha_1 \alpha_2 ... \alpha_L$, we write the probability of having a sequence of structural sites characterised by $(A_1, s_1)(A_2, s_2)...(A_L, s_L)$ as a product. This is clearly an approximation in which we assume no correlation between the sites. Nevertheless, let us push a little further in that direction:

$$P((A_1, s_1)...(A_L, s_L)|\alpha_1...\alpha_L) = \prod_{l=1} \frac{P(A_l, s_l, \alpha_l)}{P(\alpha_l)}. \tag{6}$$

Since the free energy F of a state X is related to the probability of observing that state, $P(X) \propto \exp[-F(X)/k_\mathrm{B}T]$, we can use reverse engineering and write the free energy of folding as

$$
\begin{aligned}
F((A_1, s_1)...(A_L, s_L)|\alpha_1...\alpha_L) = {} & -k_\mathrm{B}T \sum_l \log[P(A_l, s_l, \alpha_l)] \\
& + k_\mathrm{B}T \sum_l \log[P(\alpha_l)].
\end{aligned}
\tag{7}
$$

Note that approximating the probability as a product results in a free energy that is a sum. The free-energy components depend on the properties of site i only. Of the two functions at the right-hand side of Equation 7, $P(\alpha_i)$ is trivial to estimate (and probably irrelevant if our focus is on a fixed sequence with only the protein coordinates as variables). The more challenging function to estimate is $-k_\mathrm{B}T \log[P(A_l, s_l, \alpha_l)]$. It is based on averaging over all possible conformations of the protein chain and solvent coordinates that are consistent with the values of the predetermined secondary structure and surface area (Equation 1, here we go again).

Besides the technical difficulties, it is important to note that the absolute free energy as given in Equation 7 is not necessarily what we need. It is more useful to consider the free-energy difference between the folded and unfolded states, since a protein is always in one of these states and we are attempting to estimate which of the two states is more probable:

$$\Delta F_{FU} = F_F - F_U = -kT \sum_l \log\left[\frac{P\left(A_{lk}^F, s_{lk}^F, \alpha_{lk}\right)}{\sum P\left(A_{lk}^U, s_{lk}^U, \alpha_{lk}\right)}\right]. \tag{8}$$

The index l runs over the sequence, and k is used to denote the type of the site characterised by surface area, secondary structure, and the amino acid embedded in it. The summation in the denominator includes all structures that we assigned to the unfolded state.

The expression in Equation 8 is very general, so more details on computability are required. To make the formula meatier we need to come up with a feasible computational scheme of the free energy per structural site. The first step is to construct a model for the unfolded state, since direct summation over all possible unfolded coordinates is impossible in practice. In the unfolded state we expect the structural characteristics (surface area and secondary structure) to be weakly dependent on the amino acid types. We also expect it to be independent of the specific *misfolded* structure under consideration. Hence, rapidly exchanging misfolded structures are expected to be similar on average. Note that we differentiate between misfolded and unfolded structures. Unfolded structures make a larger set than misfolded structures. The set of unfolded structures includes non-compact shapes that do not resemble true protein conformations. Misfolded structures are protein-like shapes that (nevertheless) are incorrect. This assumption makes it possible to estimate the direct sum in the denominator (right-hand side of Equation 8) using a statistical argument:

$$\Delta F_{FU} = F_F - F_U = -kT \sum_l \log \left[\frac{P\left(A_{lk}^F, s_{lk}^F, \alpha_{lk}\right)}{(N-1) \cdot \overline{P}\left(A_{lk}^U, s_{lk}^U\right) \overline{P}(\alpha_{lk})} \right]. \tag{9}$$

The total number of structures at hand is N. One of the structures is correct and the rest of the $(N-1)$ structures represent a misfolded state. The symbol $\overline{P}$ denotes probability of a structural site averaged over the set of misfolded structures. Since $(N-1)$ is fixed, it adds a constant value to the free-energy difference. This constant affects the absolute stability of the current model but not the ranking of the structures according to their probability. Accurate estimation of the free energy of stability is important but difficult to obtain computationally since it requires comprehensive summation of all possible (unfolded) structures. The good news is that absolute stability is not required to detect which of the candidate structures in our set is more likely to be the correct fold. We approach the more moderate goal by considering two fold candidates i and j, and compare their free-energy differences ΔF_i and ΔF_j. This is a good point to define and use the statistical potential $V(A_{lk}, s_{lk}, \alpha_{lk})$:

$$V(A_{lk}, s_{lk}, \alpha_{lk}) = \left[\frac{P\left(A_{lk}, s_{lk}, \alpha_{lk}\right)}{P\left(A_{lk}^U, s_{lk}^U\right), P(\alpha_{lk})} \right]. \tag{10}$$

The statistical potential can be used to estimate which fold is preferred. We have

$$\Delta F_i - \Delta F_j = \sum_l V(A_{lk}^i, s_{lk}^i, \alpha_{lk}) - \sum_l V(A_{lk}^j, s_{lk}^j, \alpha_{lk}). \tag{11}$$

All that remains is to estimate the numerical value of the entries to the table $V(A_p, s_q, \alpha_r)$ (the triplet of indices (p,q,r) identifies the type of the structural site and the amino acid, and replaces the single index k used in Equation 11). Perhaps the most remarkable feature of the statistical potential approach to fold recognition (identifying the correct fold) is the way in which the table is generated. The probabilities in Equation 10 are estimated directly from the PDB. Having a set of non-redundant protein structures defines the N candidate structures that we are using to generate the tables for $P(A_p, s_q, \alpha_r)$ and $P(A_p^U, s_q^U)$ (computing $P(\alpha_r)$ is trivial). We first consider all correctly folded proteins. For each protein we have binned the number of occurrences of the

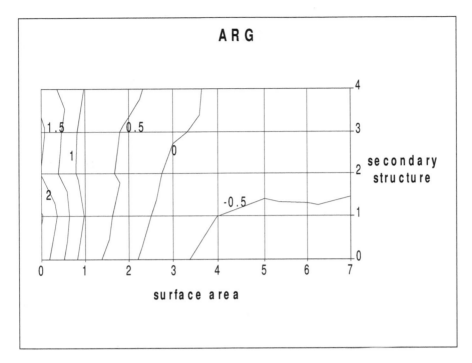

Figure 2. *The statistical potential of an arginine is plotted as a function of the secondary structure and of the fraction of solvent-exposed surface area. The secondary structure is parameterised as follows (0=α helix, 1=β sheet structure, 2=3/10 helix, 3=bend and turn, 4=π helix, and the rest). The exposed surface area is normalised with respect to a maximum found in a tri-peptide Gly-X-Gly or in the proteins. Note that the two variables are sometimes correlated.*

triplet A_p, s_q, α_r. We have a non-redundant sample of about 6000 proteins with lengths between a few tens to thousands of amino acids. The number of bins is 800, which is significantly smaller than (roughly) 1,000,000 data points, allowing for sufficient sampling. The next task of estimating the probability of misfolded sites, $P(A_p^U, s_q^U)$, is done in a similar way by collecting the same *structural* data in 40 bins. By ignoring the correlation between structural sites and sequences, we assume that the distribution of the structural sites represents misfolded (but compact) structures. As argued earlier, our prime interest is in ranking proposed plausible folds. We avoid the more difficult calculation of stability, which must take into account truly unfolded non-compact structures in order to estimate the free energy of stability. The set representing the misfolded structures should include $(N-1)$ shapes that are incorrect and exclude the correct fold. However, removing the native shape from the set of 6000 structures will have a small effect on the statistics and will make it necessary to generate a separate $P(A_p^U, s_q^U)$ for every fold. It is much simpler to generate this function including all the structures only once. The difference in the probabilities is expected to be negligible anyway.

Note also that the set of structures that we considered has nothing to do with the normal thermal energy (after all, these are folded structures picked from the PDB and

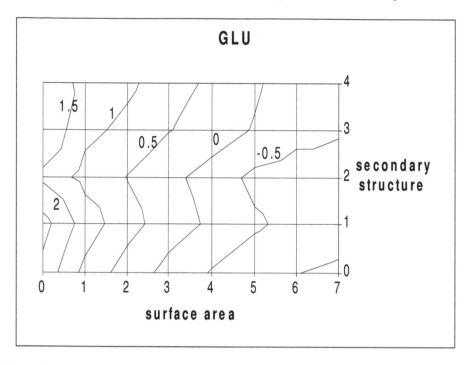

Figure 3. *Statistical potential for glutamic acid. See legend of Figure 2 for more details.*

not from thermal denaturation experiments). The multiplying factor $k_B T$ in Equation 10 determines the energy scale and not the relative ordering of different structures. It can be chosen arbitrarily, and in the calculations that follow we set it to 1.

In Figures 2, 3 and 4 we show statistical potentials parameterised by exposed surface area, secondary structure, and type of amino acid. We show three cases (different amino acids) of two-dimensional cross sections of the computed statistical potential.

The first example (Figure 2) is of arginine, a charged residue. It follows the usual expectation from polar residues. As with other charged residues it has a significant tendency to form a helix, although from the plot the weight of a beta sheet structure is similar. The second example (Figure 3) is of another charged residue (glutamic acid). Glutamic acid also strongly prefers maximal exposure. It has a tendency to an alpha helical structure, with a 3/10 helix the second best. Our third and last example (Figure 4) of this kind is of a hydrophobic residue (valine).

The potentials in Figures 2, 3 and 4 look reasonable and coincide with our physical and chemical intuition about proteins. However, they do not reflect the usual physical principles of free energies and their definition by statistical mechanics; the set of structures we consider is not thermal. These potentials were proven useful for 'fishing' templates for structural modelling. However, they should not be used to estimate thermodynamic properties.

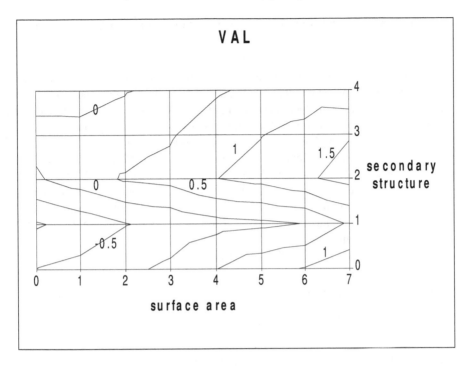

Figure 4. *Statistical potential for valine. Note the strong tendency of valine to be buried and to adopt a beta sheet conformation. See legend of Figure 2 for more details.*

2.2 Potentials from mathematical programming

The attractive idea of statistical potentials is the use of experimentally determined protein structures to learn folding potentials, a radically different approach from the chemical physics bottom-up approach, where parameters are derived from small molecules and the potential is scaled to large molecules (such as proteins). The difficulty in the chemical approach for proteins is that they are only marginally stable, and slight inaccuracies in building up parameters for small molecules will be enhanced when applied to macromolecules (as proteins are). The statistical potentials are easy to construct and to use, and were successful in identifying the correct folds in numerous cases. These advantages kept the statistical potentials in wide use. On the other hand, the derivation and design of the statistical potentials require numerous assumptions, putting into question our ability to use these entities in the calculation of physical and chemical properties (besides ranking the candidate structures for modelling). In this section we propose an alternative learning scheme that is considerably more flexible (in the choice of the functional form, and the parameter set) and makes it possible to pick a potential that is not inconsistent with known chemical and physical properties. The method we have in mind is that of mathematical programming. Just as in statistical potentials, we learn the potential from protein structures and not from data or calculations on small molecules. However, we learn it in a way that is consistent with the chemical physics principles of the system.

Consider the free energy, $F(X)$, which is a function of the reduced set of coordinates,

X. A minimal requirement from this free energy, either from recognition or physical perspectives, is

$$F(X_i) - F(X_n) > 0 \qquad \forall \ i. \tag{12}$$

Related inequalities were written and solved by Maiorov and Crippen (1992), and Vendruscolo *et al.* (2000). We denote the coordinates of the correct (native) structure by X_n and the coordinates of a wrong (decoy) structure by X_i. The preceding condition, that the free energy of the correct structure is lower than the free energy of any alternative structure, is expected from the true energy function as well as from a successful measure of fold templates. How to use the flexible information in Equation 12 to estimate functional form and parameters is a problem that can be addressed efficiently with mathematical programming tools. We first note that $F(X)$ (as with any function) can be expanded by a (complete) basis set with linear coefficients. In the following 'learning formulation' the decoy and the correct structures are known and the linear coefficients are the unknowns that we wish to determine:

$$F(X) = \sum_k \alpha_k \phi_k(X). \tag{13}$$

Substituting the linear expansion in Equation 12, we have

$$\sum_k \alpha_k \left[\phi_k(X_i) - \phi_k(X_n) \right] = \sum_k \alpha_k \Delta\phi_k \left(X_i, X_n \right) > 0 \qquad \forall i, n. \tag{14}$$

Equation 14 defines a set of linear inequalities in α_k (for all decoy and native structures) that we wish to determine. We call $\Delta\phi$ the structural difference function. Linear inequalities can be solved efficiently using mathematical programming techniques. We may write Equation 14 as a condition on a scalar product of two vectors $\hat{a}$ and $\widehat{\Delta\phi}$. The two vectors must be parallel to satisfy the constraint (positive scalar product). Every inequality divides the space of parameters (linear coefficients) into two, a half that satisfies the inequality and another half that does not. Gathering the constraints of all the inequalities can result in one of two outcomes: (a) there is a feasible volume of parameters, every point that belongs to that volume satisfies all the inequalities or (b) there is no set of parameters for which all the inequalities are satisfied, i.e., the problem is infeasible.

A schematic of the determination of two parameters with two inequalities is shown in Figure 5. Note that the actual number of inequalities that we solve in practice is much larger than the number of potential parameters that we wish to determine (the linear expansion coefficients). Typically, millions of constraints are solved with a few hundred parameters. In fact, we can use the number of constraints that have been solved as a test of the quality of the model. The more inequalities we are able to solve with the same number of parameters, the better is the functional form that we have chosen for the energy function. Hence, in some cases increases in potential complexity (and number of parameters) are not justified since the number of inequalities that we solve after adding more parameters does not increase in a substantial way. In the field of 'machine learning' in computer science, such an ineffective way of increasing model complexity and over fitting parameters is a major concern and is called 'overlearning'.

For the set of proteins that follow the Anfinsen's hypothesis, we expect that the free-energy function exists and the set of inequalities is feasible (after all, nature has already solved that problem). However, since in practice our base functions are always incomplete, infeasibilities are at least as likely to indicate the failure of the current model as

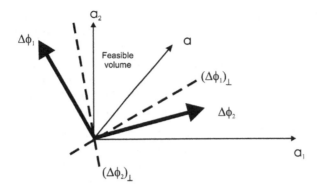

Figure 5. *A schematic of the parameter space and the inequalities we used to solve for the potential parameters. The diagram is for two parameters and two constraints. The dotted lines are hyperplanes (lines in two dimensions) perpendicular to the corresponding $\Delta\phi_k$, the structural difference vectors, and are called $(\Delta\phi_k)_\perp$. A solution, the vector $\hat{a}$, must be between the enclosing hyperplanes, and a thin arrow denotes a sample solution. Note that the norm of the solution, $|\hat{a}|$, is arbitrary.*

a failure of the Anfinsen's hypothesis. We therefore use the onset of infeasibility as a sign that the model is not good enough and seek a better basis set. This observation is in contrast to the statistical potential approach that does not provide such a self-test. A statistical potential that fails to recognise the correct fold of a protein does not offer an alternative path to further improve the potential.

A limitation of the mathematical programming approach, which is similar to the statistical potential calculations, is the energy scale. It is not possible using the inequality in Equation 14 to determine the absolute value of the coefficients α_k. For any solution $\hat{\alpha}$, the product $\lambda\alpha_k$, where λ is a positive constant, is also a solution. The norm of the parameter vector that solves the inequalities of Equation 14 is unbounded. The scale can be determined using experimental information (when available). For example, measurements of the stability energy, or the free-energy gap between the folded and unfolded states, can be used to determine the absolute scale. The connection is, however, not trivial since the stability energy is a free-energy difference that requires a model with all the uncertainties associated with it. This is the typical case, and most experimental observations that determine an energy scale require expensive computational averages to be accounted for by simulations.

An important advantage of the mathematical programming approach compared to the statistical potentials is the ability to learn from incorrect shapes. Statistical potentials 'learn' only from correct structures. Misfolded structures are learnt in an average way, and unfolded shapes are not considered at all. The inequalities make it possible to consider all alternatives shapes, misfolded and unfolded structures alike (provided that they are available). The limitation is technical and not conceptual (how many decoys we can effectively solve), and our collaborators are working to develop codes for parallel study of exceptionally large sets of inequalities (Wagner *et al.* 2004).

Another limitation of the mathematical programming approach is the ability of this approach to optimise (efficiently) only convex quadratic functions. Clearly, general thermodynamic properties will have more complex dependence on the potential parameters, and this restriction affects our ability to make the best choice of a parameter set from a feasible volume. Nevertheless, there are a few guidelines that help us make an educated guess. These guidelines are statistical in nature and are based only on the information we have at hand (a limited set of protein structures). We therefore do not use thermodynamic information in the procedure described in the following text.

Note also that the mathematical programming approach learns from the tail of the distribution of free-energy difference (and not the average as is done in statistical potentials). The tail is of prime importance since we wish to put the native shape at the extreme left of the distribution. For the mathematical programming algorithm, every new inequality can have a significant impact if it cuts through the so-far feasible parameter space. The statistical potentials learn only average misfolded structures. Adding new information in the form of one or a few new conditions (after considerable data were already put in) is unlikely to change significantly the statistical potentials. In contrast, one or a few new data points can have a profound effect on potentials computed with mathematical programming. We emphasise that our data is without noise, and we do expect to find a true potential that solves all the data exactly.

The training procedure that we described here will always be limited by the availability of data. The space of alternate protein conformations is tremendous in size and is unlikely to be explored in full for proteins of average size. For an average number of conformations per amino acid, Z, and protein of length L, a rough estimate to the number of possible states of the chain is Z^L (for $Z = 3$ and $L = 100$ we have $5 \cdot 10^{47}$). The largest set of constraints that we solved is of the order 10^7, much smaller than 10^{45}. Given the sparsity of the data, the feasible volume of parameters will never be determined exactly, and significant deviations, especially near the boundaries of the feasible volume, are expected. We therefore do not wish to select a parameter set near the boundaries, since the uncertainties may result in false predictions. We anticipate, however, that deep in the feasible volume (assuming that some depth is available) interpolations to new datasets are more likely to be accurate. In other words, we expect that the centre of the feasible volume will be sufficiently far from the boundaries, which are not well determined and are more prone to errors. If our learning is sound, most new data points will fall in the neighbourhood of the borders, the centre of the feasible volume is expected to remain feasible.

How do we define and find the centre of the feasible volume? We are working with two different approaches. In the first approach, we exploit the properties of the interior point algorithm (Meller *et al.* 2002), an optimisation procedure to solve constrained problems of the type of Equation 14. In the interior point algorithm, continuous logarithmic barriers replace the inequalities. A continuous minimisation problem is solved that is guaranteed to converge in a polynomial number of steps. If the system is bound, the minimum of the target function will be the analytical centre, a position in which all the forces from the logarithmic barriers balance each other. The analytical centre is the sum of the forces of all inequalities that were used; some of the inequalities are redundant. An example of a trivial redundancy are the two constraints $\alpha_1 > 3$ and $\alpha_1 > 5$. Clearly, the inequality $\alpha_1 > 5$ is sufficient. However, the interior point algorithm uses both inequalities to generate forces towards the centre. In that sense a direction with many redundant inequalities is more repulsive than a direction that was sampled sparsely. The result is therefore not the geometric centre of the feasible volume that is defined by a minimal set of inequalities. Instead, the centre of the interior point algorithm (with no function to minimise) is a weighted average of forces from all the constraints. In practice, the analytical centring procedure, which means using the interior point algorithm without a function to optimise, provided the best potentials measured by a maximal number of proteins recognised with a minimal number of parameters.

The second option is an intriguing subfield of machine learning in computer science, namely the SVM approach (support vector machine, Cristianini and Shawe-Taylor 2001). In the language of the problem at hand, it is possible to use statistical learning theory to come up with a mathematical programming formulation similar to Inequality 14 and to obtain meaningful results even if the set is not feasible. Here we consider only the feasible case (i.e., there are parameters such that all the inequalities are satisfied). The discussion about the infeasible set is beyond the scope of this chapter. Returning to the task at hand, we cosmetically adjust the inequalities in Equation 14 to read

$$\sum_k \alpha_k \left[\phi_k(X_i) - \phi_k(X_n) \right] = \sum_k \alpha_k \Delta \phi_k (X_i, X_n) > \delta \qquad \forall i, n. \tag{15}$$

The new variable δ defines an energy gap (the difference in energy between the folded and the misfolded/unfolded shapes). We wish to maximise this distance to increase the stability of predictions made by the energy function. This is only a cosmetic change since maximising the gap directly is unproductive. The energy gap according to Equation 15 and the norm of the vector of coefficients, $\hat{\alpha}$, are not bound. The undetermined energy gap is a result of the missing energy scale that we mentioned earlier. To get around this problem we redefine the coefficient vector $\hat{\alpha}$ to be $\hat{\alpha}/\delta$, which sets the energy (and the energy gap) to be dimensionless. Minimisation of the newly defined vector of parameters will maximise the dimensionless energy gap. The problem we solve is

$$\sum_k \alpha_k \Delta \phi_k (X_i, X_n) > 1 \qquad min[\hat{\alpha}^t \hat{\alpha}] \qquad \forall i, n. \tag{16}$$

If an energy scale is determined by other sources, we can always enforce the scale by replacing the '1' on the right-hand side by the appropriate constant. The parameter vector so determined maximises the distance from the planes that are closing the feasible volume. This procedure is much closer in spirit to a geometric interpretation of the centre of the feasible volume than the analytical centre of the interior point algorithm mentioned earlier. Nevertheless, emphasising the importance of constraints that are sampled very frequently, even if they are redundant (as is done in the interior point algorithm), does have merit. In practical applications, the potentials we derived from analytical centres tend to perform better than potentials derived from the SVM procedure. In the following text we describe a specific potential (THOM2, Meller and Elber 2001) that was calculated with the centring of the interior point algorithm.

THOM2 is a specific realisation of the structural function, $\phi_k(X_i)$, based on bio-chemical intuition, which was motivated by the lukewarm success of another potential, THOM1 (see the following text). THOM1 and THOM2 are exploiting (in different ways) properties similar to the solvent-exposed surface area that we discussed earlier. Instead of surface area we count the number of contacts to a site as another measure of solvent accessibility. A contact is defined between the geometric centres of two side chains that are separated by no more than 6.5 Å. A site with a large number of contacts (to other protein residues) is less likely to be exposed to the solvent. This type of site is likely to host apolar amino acids such as phenylalanine or isoleucine. On the other hand, sites with a small number of contacts are appropriate for charged residues such as lysine that strongly prefer a water environment. THOM1 is an energy function that builds on the aforementioned intuition. We construct a table $T1(\alpha, n)$ that assigns an energy value to a site along the protein chain, according to the type of the amino acid (α) embedded in the site, and the number of contacts with the site (n). The total (free) energy of a protein is given by the sum of contributions from different sites:

$$F(X) = \sum_l T1(\alpha_l, n_l). \tag{17}$$

The summation index l is over the protein sequence (and structural sites). We have assumed separability of the free-energy function to decouple contributions from individual sites. This separation is similar to what we have done with the statistical potentials. It is convenient to write Equation 17 with a sum over all the types of sites K (K is the product of the number of amino acid types and the number of neighbours a site may have):

$$F(X) = \sum_{k=1}^{K} m_k T1(\alpha_k, n_k). \tag{18}$$

The integer m_k is the number of times a site of a given type was sampled in a structure (for example, we may have in a specific protein five alanine residues embedded in sites with exactly four neighbours, in which case the corresponding m will be five). Using the last formulation, inequalities for THOM1 parameter training are written as

$$\Delta F = \sum_{k=1}^{K} \left(m_k^i - m_k^{(n)} \right) T1(\alpha_k, n_k) > 0 \qquad \forall i, (n). \tag{19}$$

The table entries are the unknown coefficients to be determined (in this case with the interior point algorithm). The indices of the inequalities are for misfolded structures i, or native shape (n). The number of parameters for THOM1 is 200 (twenty amino acids and contact numbers vary from 0 to 9), which was determined using a few million inequalities (Meller and Elber 2002). It turns out that the THOM1 capacity to recognise native shapes is limited. We were therefore looking for a more elaborate model with a better recognition capacity, hence THOM2.

The THOM2 scoring scheme is also about contacts. In contrast to THOM1, which scores sites, THOM2 scores individual contacts. Different contacts score differently according to the number of contacts to that site and the amino acid embedded in the prime site. Consider a site with n_1 neighbours that we call the primary site. One of the contacts of the prime site is with a secondary site that has n_2 neighbours. THOM2 is an energy table that scores a contact between the two sites according to the type of amino acid in the primary site, and the number of contacts n_1 and $n_2 - T2(\alpha_1, n_1, n_2)$. The free energy of a protein in the THOM2 framework is therefore written as

$$\Delta F = \sum_{k=1}^{K} \left(m_k^i - m_k^{(n)} \right) T2(\alpha_k, n_1 k, n_2 k) > 0 \qquad \forall i, (n). \tag{20}$$

	ALA	ARG	ASN	ASP	CYS	GLN	GLU	GLY	HIS	ILE
(1,1)	0.225	−0.029	−0.033	−0.082	−0.822	−0.259	0.091	0.286	0.072	−0.117
(1,5)	−0.207	−0.257	−0.103	0.196	−1.109	−0.005	−0.075	0.002	0.029	−0.306
(1,9)	−6.011	−4.086	−5.419	−6.137	−7.266	−5.878	−5.801	−5.808	−4.753	−5.455
(3,1)	−0.006	−0.096	−0.172	0.023	−0.496	−0.091	0.108	0.307	0.043	−0.104
(3,5)	−0.078	0.177	0.153	0.129	−0.693	0.115	0.236	0.037	−0.029	−0.287
(3,9)	−0.295	0.056	−0.327	0.082	−0.780	0.182	0.018	−0.128	−0.469	−0.597
(5,1)	0.134	−0.206	0.045	0.222	−0.147	−0.113	0.076	0.480	0.191	−0.148
(5,3)	0.064	0.165	0.202	0.169	−0.596	0.040	0.127	0.183	−0.038	−0.245
(5,5)	−0.654	0.681	−0.264	−0.195	−0.821	−0.092	0.427	−0.365	−0.194	−0.469
(7,1)	6.291	5.499	5.558	6.020	5.090	5.547	5.681	6.102	5.697	5.591
(7,5)	0.172	0.289	0.363	0.386	−0.276	0.285	0.450	0.327	0.277	−0.080
(7,9)	0.082	0.409	−0.003	−0.154	−0.297	0.038	−0.275	0.052	0.039	
(9,1)	10.000	4.497	6.050	5.215	3.999	5.936	10.000	10.000	10.000	10.000
(9,5)	0.259	0.305	0.261	0.712	0.412	−0.017	0.323	0.828	−0.091	1.256
(9,9)	0.195	0.042	−0.367	−1.340	−1.186	0.469	1.374	−1.358	1.055	−1.991
(0,0)										

	LEU	LYS	MET	PHE	PRO	SER	THR	TRP	TYR	VAL	GAP
(1,1)	−0.159	−0.016	0.213	−0.204	0.029	0.047	−0.065	−0.502	−0.637	−0.280	8.900
(1,5)	−0.230	−0.132	−0.147	−0.292	−0.231	0.067	−0.093	−0.605	−0.398	−0.358	5.700
(1,9)	−5.855	−4.905	−4.967	−5.826	−6.169	−5.887	−5.886	−5.254	−6.791	−6.989	10.000
(3,1)	−0.099	0.106	−0.196	−0.170	−0.015	0.405	0.061	−0.311	−0.295	−0.053	10.000
(3,5)	−0.213	0.141	0.080	−0.315	−0.054	0.058	0.079	−0.364	−0.278	−0.168	10.000
(3,9)	−0.487	0.086	−0.851	−0.065	0.195	0.234	0.150	−0.151	0.034	−0.272	10.000
(5,1)	−0.319	−0.056	−0.152	−0.271	0.169	0.190	0.342	−0.068	0.016	0.190	10.000
(5,3)	−0.187	0.258	−0.259	−0.283	0.089	0.114	0.017	−0.365	−0.297	−0.270	10.000
(5,5)	−0.423	0.336	0.319	0.074	0.549	0.218	0.005	0.038	−0.459	−0.584	10.000
(7,1)	5.262	6.082	5.642	5.797	5.819	5.226	5.477	6.419	5.170	5.530	10.000
(7,5)	−0.008	0.497	0.243	−0.158	0.421	0.126	0.337	0.042	−0.083	−0.029	10.000
(7,9)	−0.175	0.668	0.061	0.032	−0.706	0.825	0.242	−0.362	0.142	−0.246	10.000
(9,1)	6.222	5.593	4.915	6.021	9.614	10.000	10.000	5.885	10.000	10.000	10.000
(9,5)	−0.150	0.525	−0.194	0.431	3.066	0.426	0.524	−0.080	0.081	0.206	10.000
(9,9)	−0.248	−0.293	1.411	−1.330	6.939	3.223	−0.538	0.815	−0.533	−0.515	10.000
(0,0)											1.000

Table 1. *The table of the THOM2 energy as a function of the contact type and the amino acid type (i is the primary site, j the secondary site). Note that the number of neighbours of a site is 'coarse grained' and means the following actual number of neighbours $1 \rightarrow 1, 2\,3 \rightarrow 3, 4\,5 \rightarrow 5, 6\,7 \rightarrow 7, 8\,9 \rightarrow\, \geq 9$.*

The sum in Inequality 20 is over contact types (not sites). The counters for the un-folded structure m_k^i and the native shape $m_k^{(n)}$ are characteristics of the structure that are scored according to table $T2(\alpha, n_1, n_2)$ to be determined. The index k is equivalent to the triplet (α, n_1, n_2) and is used in Equation 20 in addition to the triplet for clarity. The THOM2 energy was designed subject to about 30 million constraints (Meller and Elber 2001). The set that was found feasible with the 300 parameters comprises the entries to the $T2$ table of THOM2. It is remarkable that only 300 parameters capture the informa-tion contained in 30 million constraints, suggesting that this functional form is indeed useful.

It is also interesting that some of the entries to the table are undetermined (the entries with values of 10.00). Hence, the number of parameters that we actually required to satisfy all the inequality constraints was even smaller than 300 (291 parameters). The combination of a site with the maximal number of neighbours, interacting with a site with the smallest number of neighbours, was exceptionally rare in our data and left many of these parameters (for different types of amino acids) undetermined.

The THOM2 potential derived as discussed in the preceding text will be used in the study of evolutionary capacity of structures in the next section.

3   The evolutionary capacity of proteins

One of the remarkable properties of proteins is the redundancy of sequence space with respect to structure space. There are numerous sequences that fold into the same shape. An obvious question is how large 'numerous' is, and in this section we attempt to address this problem (Meyerguz *et al.* 2004). More concretely, we compute the entity we name 'structure capacity' — the number of sequences that a particular protein can accommodate up to an energy E. We consider protein sequences that improve on the stability of the native structure, i.e., sequences that are more stable than the native sequence of a particular (experimentally determined) protein structure. We find an exponentially large number of 'better' (more stable) sequences. The observation that one may improve stability (in a considerable way) compared to the natural sequence is perhaps not that surprising, since protein sequences are not optimised for structural stability only. True biological sequences are subject to constraints that are related to their function. Proteins need to be flexible, to have recognition sites, and other biological features that are at variance with the single criterion (stability) we use here. Despite the limitations of studying stability only, there is still considerable interest in it, providing insight into the space in which further design and evolutionary refinement of sequences can be made. The stability constraint is an obvious one. It is always part of the equation, and therefore studying it in isolation is likely to provide meaningful information, even if it is highly permissive (as we indeed find to be the case).

So much for philosophy. To be specific we compute the function $N(\Delta F)$. It is the number of sequences with free-energy gaps larger than ΔF. The problem now resembles the calculation of a microcanonical partition function, with a small (but important) difference. The microcanonical partition function is the number density — the number of sequences in the neighbourhood of ΔF:

$$\Omega(\Delta F) = \frac{dN(\Delta F)}{d\Delta F}. \tag{21}$$

It is useful to reiterate the definition of the problem. The space in which we count events is of sequences and not of Cartesian coordinates. The sequence space is discrete, and the maximum number of sequences that may fit to a protein of length L is 20^L (twenty types of amino acids). During the counting the structure is kept fixed while we generate sequences that may fit into this particular structure with a present stability criterion ΔF. The total number of sequences with free-energy gap below ΔF is given by $N(\Delta F)$. The number density is a useful entity to build on a 'thermodynamic' description of sequence space. The entropy, S, of sequence space (constrained to the neighbourhood of one structure) is given by

$$S = \log\left[\Omega(\Delta F)\right]. \tag{22}$$

To obtain a comprehensive view (as much as possible) on the structural templates of sequence evolution we repeat the calculations of sequence capacities for all distinct folds in the protein databank. To determine the distinct folds we employed a similarity measure of our design and compared all the structures in the protein databank against each other. Starting from a seed structure, new structures were added to the non-redundant set only if they were sufficiently different from all the structures already included in the set. This procedure gave 3660 non-redundant shapes (Meller and Elber 2002). Repeating the procedure with a different similarity measure (a measure produced by the CE structural alignment program (Shindyalov and Bourne 1998)) provided comparable results.

The sequence space of each of these folds was counted separately. This counting is approximate since we ignore potential overlap of sequence space between different protein shapes. For example, the same sequence A may be found to have a low energy in two proteins P_1 and P_2. Obviously, the sequence A can match with one structure only. Computing the sequence space for one structure at a time ignores this possibility, and overcounting of sequences is a possibility. The extent of the overcounting is unclear and is a topic of future work.

A restricted counting is made in which no deletions or insertions of amino acids are allowed during our model of the evolutionary process. That is, the lengths of the template structure and the sequence are the same and are fixed. Related counting and evolutionary studies that did not probe the complete protein data bank were pursued by other groups (Shakhnovich and Gutin 1993, Saven and Wolynes 1997, Betancourt and Thirumalai 2002, Lau and Dill 1990, Koehl and Levitt 2002, Larson *et al.* 2002, Huynen *et al.* 1996, Lipman and Wilbur 1991).

3.1   The counting algorithm

We emphasise that the following algorithm is not Metropolis, though it is still a randomised algorithm. The procedure is based on the umbrella sampling of Torrie and Valleau (Torrie and Valleau 1997) and the knapsack algorithm of Morris and Sinclair (Morris and Sinclair 1999). We consider a sequence A_0 embedded in a structure X with a free-energy difference $\Delta F_0 \equiv \delta F(A_0, X)$. We wish to determine the ratio $N\left(\Delta F^{(1)}\right) / N\left(\Delta F^{(2)}\right)$, where $N\left(\Delta F^{(1)}\right)$ is the number of sequences with energies up to $\Delta F^{(1)}$. The energy of the starting sequence ΔF_0 is set below $\Delta F^{(2)}$. The algorithm goes as follows:

1. Pick at random one of the amino acids, α_{ij}, in the current sequence A_i and change it at random to one of the twenty amino acids to generate a new intermediate sequence $\overline{A}_i$.

2. Check the energy of the intermediate sequence $\Delta F\left(\overline{A}_i, X\right)$. If it is larger than $\Delta F^{(2)}$ reject the step, change the sequence back to the original sequence A_i, and return to 1. Otherwise, accept the step (set A_{i+1} to be equal to $\overline{A}_i$), and continue to step 3.

3. Compare $\Delta F\left(\overline{A}_i, X\right)$ to $\Delta F^{(1)}$ and $\Delta F^{(2)}$. Updates the counters l_1 and l_2 (l_i is the number of sampled sequences with energy smaller than $\Delta F^{(i)}$).

4. Check stopping criteria (number of steps, convergence of the ratio $l_1/l_2 \cong$ $N\left(\Delta F^{(1)}\right) / N\left(\Delta F^{(2)}\right)$ that approximate the function we are after). Go to 1 if criteria were not satisfied.

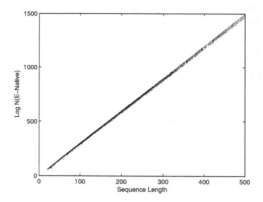

Figure 6. *Counting the number of sequences that can be embedded in different folds (3660 in total) with energies better than the energy of the native sequence. The log of the number of sequences is plotted as a function of protein length to emphasise the exponential dependence. While the most obvious feature is the linear dependence, we should note that the line has significant thickness, which is significant since a log function was used. The plots include counting results from two potentials. One set is from THOM2, the potential that is discussed in this chapter. The second potential (TE13, Tobi and Elber 2000) is discussed elsewhere. The results of the two potentials are practically identical.*

It is necessary for the energies $\Delta F^{(1)}$ and $\Delta F^{(2)}$ to be sufficiently close to each other, so the ratio will be close to one and converging rapidly. Calculation of ratios could be aggregated together (a collection of rapidly converging randomised counting). From the preceding equation it is clear that we can get a sequence of ratios similar to

$$\prod_{i=1,..,n} \frac{l_i}{l_i + 1} = \frac{l_1}{l_N} \equiv \frac{N\left(\Delta F_{(1)}\right)}{N\left(\Delta F_{(N)}\right)}. \tag{23}$$

From Equation 23 we can estimate the number of sequences for any energy $\Delta F_{(N)}$ provided that we know the density at anchor energy $\Delta F_{(1)}$. Anchors are not hard to get if we know the sequence with the lowest possible energy, since the number of alternative sequences in the neighbourhood of that sequence is small and directly countable. If the minimum energy is not known, one could use an energy that can be sampled easily. For example, it is not difficult to estimate the median energy and the number of sequences below the median (exactly half of the total number of sequences, 20^L).

3.2   Computing temperatures for all protein folds

We have computed the number of sequences for all relevant free-energy differences $N(\Delta F)$. This function has a strong (exponential) dependence on the protein length, which is easy to rationalise. The total number of possible sequences is exponential in length (20^L). The actual number of accepted sequences is expected to grow as M^L ($M < 20$) (still grows exponentially with the length). Every length extension of the

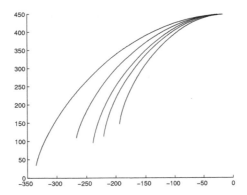

Figure 7. *Computing sequence capacity for five proteins of the same length (150 amino acids), from the set of 3660 proteins that we analyse are shown in detail. The proteins are (from left to right): 1f3g, 1nul, 1ash, 1br1, 1bbr.*

protein molecule and the addition of a new structural site will allow a few more amino acids (per site) to be accommodated, increasing exponentially the number of accessible sequences. Counting for the complete set of 3660 proteins that differ significantly in length was performed.

In Figure 6 we show $\log\left(N\left(\Delta F\right)\right)$ as a function of ΔF. The obvious linearity of the plot strongly supports the preceding assertion of exponential growth in $N\left(\Delta F\right)$ as a function of the protein length L. In Figure 7 we show a sample of a few complete curves of $\log\left[N\left(\Delta F\right)\right]$ vs. the free-energy difference ΔF.

The energy that we used for the counting was THOM2, for which the determination of the lowest energy sequence is trivial, making it possible to identify the lowest energy sequence and its corresponding degeneracy. We finally compute the temperature associated with the energy of the native sequence using

$$\frac{1}{T} = \frac{d}{dE_n}\log\left(\Omega(E)\right). \tag{24}$$

In Figure 8 we show the distribution of temperatures computed for THOM2 energy and for the set representing the protein databank. The distribution of temperatures is highly peaked but still quite broad.

The calculations of the temperature employ a standard thermodynamic formula (Landau and Lifshitz 1986). However, the meaning of this temperature is not obvious. What are the implications of the temperatures? Can we propose a mechanism that might lead to this set? Here is a possible model that can help us think about the data. We consider the selection of a particular native sequence with energy E_n and write the probability that it will be observed biologically, $P(S_n)$, as a product of two terms: the number of sequences at $E_n - \Omega(E_n)$ and a selection function $G(E_n)(P\left(S_n(E_n)\right) \equiv P(E_n) = \Omega(E_n)G(E_n)$. The selection function of nature depends on more than the energy. For example: flexibility, binding site and electric field are important to protein function and exert evolutionary pressures. The number of sequences at a particular energy E_n is a rapidly increasing

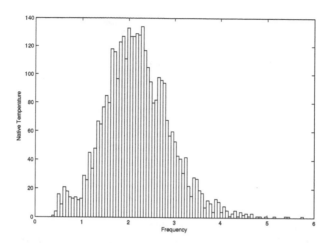

Figure 8. *The distribution of selection temperatures computed for all 3660 folds at their native energies.*

function of the energy. To find an optimal (probable) sequence with a low energy, it is necessary that the selection function will be rapidly decreasing, leading to a maximum in $P(E_n)$. More conveniently we seek a (equivalent) maximum in $\log[P(E_n)]$. We have

$$\left.\frac{d\log[P]}{dE}\right|_{E=E_n} = \left.\frac{d\log(\Omega(E))}{dE}\right|_{E=E_n} + \left.\frac{d\log((E))}{dE}\right|_{E=E_n} = 0$$

$$\frac{1}{T_n} = \left.-\frac{d\log(\Omega(E))}{dE}\right|_{E=E_n}. \tag{25}$$

Hence, the selection temperature is telling us something about the selection function. Equation 25 makes it possible to compute a relationship between the number of sequences (that we can compute) and sequence selection. The selection functions computed for different protein shapes at their native energies are therefore quite similar (as the temperatures are).

How can a universal selection mechanism be realised? The simplest answer is the universality of the genetic code and *mutation mechanisms* (e.g., UV radiation on DNA base pairs). All genes coded on the DNA are likely to be mutated in roughly the same way, providing the same level of 'sequence-thermal-excitation' (temperature) to all genes (proteins). The other option to explain the data, which is more intriguing (but not necessarily more correct), is to have *all the folds connected via paths in sequence space*, i.e., a sequence that belongs to one structural family can be mutated to a different structural family. In this case (regardless of the underlying mutation mechanism) the temperature should be the same. A way to prove or to disprove the preceding proposition is to try to identify plausible paths connecting sequence islands that are associated with a given structure. Simple models have been studied (Kleinberg 1999). However, studies of known protein folds and the interactions of their corresponding sequence spaces are required.

References

Anfinsen C B, 1973, Principles that govern folding of protein chains, *Science* **181**, 223.

Berman H M, Westbrook J, Feng Z, Gilliland G, Bhat T N, Weissig H, Shindyalov I N and Bourne P E, 2000, The protein data bank, *Nucleic acids research* **28** 235.

Betancourt M R and Thirumalai D, 2002, Protein sequence design by energy landscaping, *Journal of Physical Chemistry* **106** 599.

Cristianini N and Shawe-Taylor J, 2001, *An introduction to support vector machines and other kernel-based learning methods* 189, first edition (Cambridge University Press, Cambridge).

Huynen M A, Stadler P F and Fontana W, 1996, Smoothness within ruggedness: The role of neutrality in adaptation, *Proceeding of the National Academy of Science USA* **93** 397.

Kleinberg J, 1999, Efficient algorithms for protein sequence design and the analysis of certain evolutionary fitness landscapes, *Proc 3rd ACM RECOMB Intl Conference on Computational Molecular Biology*, 226-237.

Koehl P and Levitt M, 2002, Protein topology and stability define the space of allowed sequences, *Proceeding of the Natural Academyof Sciences USA* **99** 1280.

Landau L D and Lifshitz E M, 1986, *Statistical Physics I* 34–36 (Pergamon Press, Oxford).

Larson S M, England J L, Desjarlais J R and Pande V S, 2002, Thoroughly sampling sequence space: Large-scale protein design of structural ensembles, *Protein science* **11** 2804.

Lau K F and Dill K, 1990, Theory for proteinmutability and biogenesis, *Proceeding of the National Academy of Science USA* **87** 638.

Lehninger A L, 1982, *Principles of Biochemistry* 723 (Worth Publishers Inc., New York).

Lipman D and Wilbur W, 1991, Modeling the neutral and selective evolution of protein folding, *Proceeding of the Royal Society London B* **245** 7.

Maiorov V N and Crippen G M, 1992, Contact potential that recognizes the correct folding of globular proteins, *Journal of Molecular Biology* **227**, 876.

Meller J and Elber R, 2001, Linear Optimization and a double statistical filter for protein threading protocols, *Proteins-Structure, Function and Genetics* **45**, 241.

Meller J and Elber R, 2002, Protein recognition by sequence-to-structure fitness: Bridging efficiency and capacity of threading models, in *Advances in chemical physics*, ed F Richard (John Wiley & Sons).

Meller J, Wagner M and Elber R, 2002, Maximum Feasibility Guidline in the Design and Analysis of Protein Folding Potentials, *J Comput Chem* **23**, 111.

Meyerguz L, Grasso C, Kleinberg J and Elber R, 2004, Computational analysis of sequence selection mechanisms, *Structure* **12** 547.

Miyazawa S and Jernigan R L, 1984, Estimation of effective interresidue contact energies from protein crystal structures: Quasi chemical approximation, *Macromolecules* **18** 534.

Morris B and Sinclair A, 1999, Random walks on truncated cubes and sampling 0-1 knapsack solutions, *Proc. IEEE Foundations of Computer Science* 130-240.

Saven J G and Wolynes P G, 1997, Statistical Mechanics of the Combinatorial Synthesis and Analysis of Folding Macromolecules, *Journal of Physical Chemistry B* **101** 8375.

Shakhnovich E I and Gutin A M, 1993, A new approach to the design of stable proteins, *Protein Engineering* **6** 793.

Shindyalov I N and Bourne P E, 1998, Protein structure alignment by incremental combinatorial extension (CE) of the optimal path, *Protein Engineering* **11** 739.

Tanaka S and Scheraga H A, 1976, Statistical Mechanical Treatment of Protein Conformation. 1. Conformational properties of amino acids in proteins, *Macromolecules* **9** 142.

Tobi D and Elber R, 2000, Distance dependent, pair potential for protein folding: Results from linear optimization, *Proteins-Structure, Function and Genetics* **41** 40.

Torrie G M and Valleau J P, 1997, Non-physical sampling distributions in Monte-Carlo free energy estimation - umbrella sampling, *Journal of Computational Physics* **23** 187.

Vendruscolo M, Mirny L A, Shakhnovich E I and Domany E, 2000, Comparison of two opti-
    mization methods to derive energy parameters for protein folding: Percepton and Z score,
    Proteins-Structure, Function and Genetics **41**, 192.
Wagner M, Meller J and Elber R, 2004, Large-scale linear programming techniques for the design
    of protein folding potentials, *Mathematical Programming Series B* **101**, 301.

Physical and functional aspects of protein dynamics

J C Smith, T Becker, S Fischer, F Noe, A L Tournier, G M Ullmann & V Kurkal

Interdisciplinary Center for Scientific Computing, Ruprecht Karls University
D-69120 Heidelberg, Germany

jeremy.smith@iwr.uni-heidelberg.de

1 Introduction

Whereas the second half of the 20th century was notable for having seen the development and application of techniques for solving the three-dimensional structures of biological macromolecules, the 21st century may well be that in which the internal dynamics required for function are finally elucidated. Motions in proteins are inherently difficult to characterise in detail, due to their wide range of forms and timescales and their inherent anharmonicity due to the irregularity of protein energy surfaces. Therefore, computational methods, such as molecular dynamics simulation, must play a predominant role in sorting out which motions occur and which are required for function.

Here we broadly review some aspects of this field that are of particular physical interest, ranging from the glass transition in proteins through to large-scale conformational change. In this way the uninitiated reader will gain insight into the complexity of the protein dynamical landscape and the various ways in which dynamics can influence function.

2 Hydration effects and the dynamical transition

2.1 The dynamical transition in proteins

Various experimental techniques such as neutron scattering, Mössbauer spectroscopy and x-ray scattering have shown the presence of a temperature-dependent transition in protein

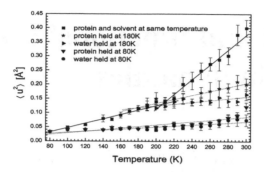

Figure 1. *Mean-square fluctuations, $\langle u^2 \rangle$, of the protein non-hydrogen atoms for different sets of simulations.*

dynamics at around 180-220 K (Doster *et al.* 1989, Dunn *et al.* 2000, Parak *et al.* 1981, Tilton *et al.* 1992). In this temperature range the dynamics of proteins, as represented by the mean-square displacement, $\langle u^2 \rangle$, of the protein atoms, change from harmonic behaviour below the transition temperature to anharmonic behaviour above. This dynamical transition has also been seen using molecular dynamics (MD) simulation techniques (Smith *et al.* 1990, Hayward and Smith 2002, Bizzarri *et al.* 2000). Figure 1 shows such a simulation, in which a typical transition in the protein $\langle u^2 \rangle$ can be seen at 220 K. Experiments have shown that in several proteins biological function ceases below the dynamical transition (Ferrand *et al.* 1993, Rasmussen *et al.* 1992, Parak *et al.* 1980).

The protein dynamics transition has features in common with the glass transition (Green *et al.* 1994, Angell 1995). Much debate is still going on to determine whether a protein can be considered a glass. Proteins share the stretched exponential behaviour seen in glasses; however, they do not have a precisely defined transition temperature, T_g, and characteristic sharp jump in heat capacity at T_g as seen in glasses.

2.2   The role of solvent in the dynamical transition

A number of experiments have indicated that when a protein is solvated, the dynamical transition is strongly coupled to the surrounding solvent (Ferrand *et al.* 1993, Reat *et al.* 2000, Paciaroni *et al.* 2002, Teeter *et al.* 2001, Cordone *et al.* 1999). The observed dependence of the dynamical transition behaviour on the solvent composition leads to the question of the role of solvent in the dynamical transition (Reat *et al.* 2000, Hayward and Smith 2002). However, whether solvent drives the dynamical transition in a hydrated protein is still open to question. In recent MD simulations solvent effects were probed by using dual heatbath methods that set the protein and its solvent at different temperatures. This approach enables features inherent to the protein energy landscape to be distinguished from features due to properties of the solvent.

To perform the dual heatbath simulations, the Nosé-Hoover-Chain (NHC) algorithm was implemented and added in the CHARMM package (Brooks *et al.* 1983, Tuckerman and Martyna 2000). The model system consisted of myoglobin surrounded by one shell of solvent (492 water molecules). In the NHC method the different parts of the system are each regulated not by one but by two heatbaths, the first one regulating the system and the second regulating the first heatbath. NHC has the advantage over the Nosé-Hoover algorithm in that it reproduces exact canonical behaviour and is more stable.

In Figure 1 are shown the results of simulations in which the protein and solvent are set at the same temperature. These reproduce the experimentally observed dynamical transition in $\langle u^2 \rangle$ in myoglobin at ~220 K. $\langle u^2 \rangle$ is seen to increase relatively slowly up to ~220 K, whereas beyond this temperature it increases more sharply with temperature, giving rise to the characteristic dynamical transition feature. The data from this initial set of runs was analysed to investigate which parts of the proteins are subject to the dynamical transition. The $\langle u^2 \rangle$ of sidechain atoms was found to be 6 times greater than the backbone $\langle u^2 \rangle$ at 80 K, and twice as large at 300 K (data not shown). The innermost atoms were seen not to show any dynamical transition feature, their $\langle u^2 \rangle$ increasing linearly with temperature. In contrast, the outer shells of atoms exhibit a marked transition at ~220 K, the outermost solvent-exposed atoms being the most affected (data not shown). Thus, the atoms found to be most influenced by the dynamical transition are the side-chain atoms on the outer layers of the protein, i.e., the protein atoms interacting with the solvent shell.

2.3   Solvent caging of protein dynamics

Figure 1 also presents the protein fluctuations calculated from the dual heatbath simulations, performed fixing the temperature of one component below the dynamical transition while varying the temperature of the other component. Fixing the solvent temperature at 80 K or 180 K suppresses the dynamical transition, the protein $\langle u^2 \rangle$ increasing linearly with temperature up to 300 K. Therefore, low temperature solvent cages the protein dynamics. Figure 1 also shows that holding the protein temperature constant at 80 K or 180 K and varying the solvent temperature also abolishes the dynamical transition behaviour in the protein.

In summary then, holding either component at a low temperature suppresses the protein dynamical transition. Cold (80 K and 180 K) solvent is seen to effectively cage protein dynamics over the whole range of protein temperatures examined (from 80 K up to 180 K). This indicates the important role of solvent in influencing protein dynamics.

2.4   Dynamical transition and protein function

Protein function is dependent on protein flexibility. As during the dynamical transition there is a significant increase in flexibility, a loss of function might be expected below the dynamical transition temperature. A number of studies have indeed shown that at least some proteins cease to function below dynamical transition (Daniel *et al.* 2003). However, the temperature dependence of motions in a cryosolution of the enzyme glutamate dehydrogenase when examined and compared with enzyme activity (Daniel *et al.* 1999) showed that the enzyme activity remains below the measured picosecond-timescale dynamical transition at ≈ 220 K with no significant deviation of activity from Arrhenius behaviour down to 190 K. These results suggest that there exists a range of temperatures

(190-220 K) at which the enzyme rate limiting step does not require, and is not affected by the anharmonic motions taking place on the picosecond timescale. Another important aspect of enzyme activity is the hydration threshold, i.e., the minimum hydration required for an enzyme to be active. It is widely accepted that dry enzymes are non-functional and a commonly discussed threshold value is 0.2g of water/g of protein required for activity. However, experiments using gas phase substrates, in which the critical limitations imposed by diffusion were removed, showed that activity is possible at very low hydration levels (3%), well below the proposed threshold. This activity at hydration as low as 3% may be related to the role of water in inducing anharmonic motions (Kurkal *et al.* 2005). The presence of water reduces the barrier energy for anharmonic jumps causing the activity at lower temperature compared to the thermal energy required for the barrier crossing in the absence of water. Hence, an enzyme will show activity at 3% hydration provided the water molecules present can lower the barriers to an extent that the functional anharmonic motion could take place at room temperature. However, it still is unclear as to what extent motions activated in the dynamical transition describe functional protein dynamics. The understanding of physical characteristics of proteins such as the dynamical transition should allow further understanding of protein equilibrium fluctuations and their relation to function.

3   Neutron scattering from proteins

Neutron scattering is widely used to probe picosecond-nanosecond timescale dynamics of condensed-phase molecular systems (Bee 1988, Lovesey 1987). In hydrogen-rich molecules like proteins the scattering will be dominated, due to the high incoherent scattering length of hydrogen atoms, by incoherent scattering. The basic quantity measured is the incoherent dynamic structure factor $S(\mathbf{Q}, \omega)$, where $E = \hbar\omega$ is the energy transfer and $\mathbf{Q} = \mathbf{k}_f - \mathbf{k}_i$, with $\mathbf{k}_i$ and $\mathbf{k}_f$ being the incident and final wave vector of the scattered neutrons, respectively. The scattering function contains information on both the timescales and the spatial characteristics of the dynamical relaxation processes involved.

The scattering function can be written as space and time Fourier transformation of the van Hove autocorrelation function $G(\mathbf{r}, t)$:

$$S(\mathbf{Q}, \omega) = \frac{1}{(2\pi)^2} \int dt \, e^{-i\omega t} \int d\mathbf{r} \, e^{-i\mathbf{Q}\cdot\mathbf{r}} G(\mathbf{r}, t) \tag{1}$$

$$G(\mathbf{r}, t) = \frac{1}{N} \sum_i \int d\mathbf{r}' \langle \delta(\mathbf{r} - \mathbf{r}' + \mathbf{R}_i(0)) \delta(\mathbf{r}' - \mathbf{R}_i(t)) \rangle. \tag{2}$$

In Equations 1 & 2 $\mathbf{R}_i(t)$ is the position vector of atom i $(i = 1 \ldots N)$ at time t and the brackets $\langle \cdots \rangle$ indicate an ensemble average. $G(\mathbf{r}, t)$ is the probability that a certain particle can be found at position $\mathbf{r}$ at time t, given that it was at the origin at $t = 0$ (here and throughout the chapter only the classical limit of these functions is considered). For the dynamical transition in proteins, the elastic and quasielastic parts of the scattering function are of primary interest. In systems of spatially confined atoms, such as proteins, the elastic incoherent neutron scattering is of relatively high intensity and is thus used to obtain an estimate of the dynamics present. Quasielastic scattering gives access to typical timescales and geometries of diffusive dynamics involved. (See further Egelhaaf's chapter in this volume for other aspects of dynamic scattering.)

Equation 2 shows that the average over all hydrogen atoms determines the scattering function. Given that hydrogens are equally distributed over the protein, we see that neutron scattering gives an average over all motions present in the protein. The guiding picture in interpreting dynamic neutron scattering is that of a potential energy surface or 'energy landscape'. The shape of this energy landscape determines the associated microscopic dynamics. For proteins, however, the energy landscape can be complex and rugged with many local minima. This leads to the presence of a wide range of vibrational and diffusive dynamical processes. The simplicity of looking at only one atom type is therefore somewhat overshadowed by the difficulty that a wide range of dynamical processes may have to be considered to adequately describe the scattering function.

Several simplified models have been used to describe the dynamics activated at the dynamical transition, including continuous diffusion (Kneller and Smith 1994), jumping between minima (Frauenfelder *et al.* 1979, Elber and Karplus 1987, Doster *et al.* 1989, Lamy *et al.* 1996, Frauenfelder *et al.* 1991), mode-coupling theory (Doster and Settles 1999, Perico and Pratolongo 1997, La Penna *et al.* 1999), stretched-exponential behaviour (Dellerue *et al.* 2001) and 'effective force constants' (Zaccai 2000). Although these models are sometimes qualitatively different, all can reproduce available experimental data well. Recently, a method has been presented for extracting useful information from experimental elastic incoherent neutron scattering data without assuming a specific dynamical model (Becker and Smith 2003). The method proceeds in two stages: fitting to the Q-dependence of the elastic scattering, followed by decomposition of the resulting $\langle \Delta r^2 \rangle$.

3.1 Analysing elastic scattering at low Q

To analyse elastic incoherent scattering from proteins without restricting the interpretation to a specific model the use is proposed of a heuristic function of the form:

$$S(Q,0) = e^{-\frac{1}{6}Q^2 \langle \Delta r^2 \rangle} \left(1 + \sum_{m=2}^{\infty} b_m \cdot (-Q^2)^m \right) \tag{3}$$

$$\approx e^{-\frac{1}{6}Q^2 \langle \Delta r^2 \rangle} \left(1 + b \cdot Q^4 \right). \tag{4}$$

Here, b_m are parameters fitted to reproduce the experimentally-observed elastic incoherent scattering. This expansion of the elastic scattering function reflects the idea that for low Q, the scattering is a Gaussian function in Q with increasing deviations at higher Q-values. Two different aspects contribute to the parameters b_m. The dynamics of single atoms can lead to deviation from a Gaussian scattering function for higher Q-values as well as a distribution of mean-square displacements. Looking at these two possible causes of non-Gaussian behaviour, it was shown (in Becker and Smith 2003) that both aspects lead to a function of the form of Equation 3. It was also shown that in systems in which heterogeneity makes the dominant contribution to the heuristic parameters b_m, b_1 is the variance of the distribution of $\langle \Delta r^2 \rangle$. However, to what extent each of these two

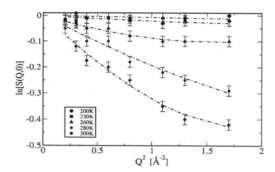

Figure 2. *Equation 4 (dot-dashed line) fitted to experimental data from Daniel et al. (1998). See Table 1 for the resulting parameters.*

T [K]	$\langle \Delta r^2 \rangle$ [Å^2]	b [Å^4]
200	0.02 ± 0.02	0.009 ± 0.009
230	0.08 ± 0.05	0.09 ± 0.09
260	0.37 ± 0.06	0.45 ± 0.1
280	0.6 ± 0.2	0.3 ± 0.2
300	1.1 ± 0.1	0.8 ± 0.1

Table 1. *Mean-square displacement and variance from fitting Equation 4 to the experimental data obtained in Daniel et al. (1998).*

effects contributes will vary from system to system and is not known *a priori*. Therefore, using Equation 3 and treating b_m as heuristic parameters is equivalent to making minimal assumptions about the system. In the low Q-range, as long as deviations from Gaussian behaviour are small, i.e., $b_m \cdot (Q^2)^m \ll 1$, we can neglect higher-order terms and can derive two parameters from the elastic scattering, $\langle \Delta r^2 \rangle$ and b (see Figure 2 and Table 1).

3.2   Measured mean square displacement

Figure 3 shows $\langle \Delta r^2 \rangle_{Expt}$ as a function of temperature. The data exhibit a dynamical transition at ~220 K involving a sharp increase in $\langle \Delta r^2 \rangle_{Expt}$. The next step in neutron data analysis is to interpret $\langle \Delta r^2 \rangle$ as obtained from Equation 4.

In Becker and Smith (2003) the following form for the measured mean-square displacement was derived:

$$\langle \Delta r^2 \rangle_{Expt} = \langle \Delta r^2 \rangle_{Conv} - a \frac{2}{\pi} \arctan \left(\frac{\Delta \omega}{\lambda} \right). \tag{5}$$

Here, $\langle \Delta r^2 \rangle_{Conv}$ is the time-converged mean-square displacement consisting of a vibra-

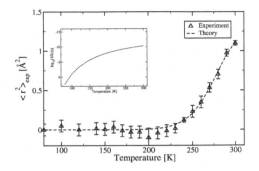

Figure 3. $\langle \Delta r^2 \rangle_{Expt}$ *determined on a protein solution (glutamate dehydrogenase in 70%CD$_3$OD/30%D$_2$O) using the instrument IN6 at the ILL (Daniel et al. 1998), and fitted using Equation 5. Insert: Characteristic relaxation time, $\lambda^{-1}(T)$ as a function of temperature.*

tional part and contributions from the elastic incoherent scattering function of the protein. The second term on the right-hand side represents the finite energy resolution of the instrument. Here $\Delta\omega$ is the full-width half-maximum resolution of the instrument, λ is a typical relaxation frequency of the protein and a is the maximal amount that the relaxation process can contribute to the time-converged $\langle \Delta r^2 \rangle$.

Equation 5 shows that two different processes can lead to a temperature-dependent transition in $\langle \Delta r^2 \rangle_{Expt}$: a non-linear change in $\langle \Delta r^2 \rangle_{Conv}$ with T or equally well a temperature dependence of the relaxation frequency λ. A discussion of both possibilities is now given.

3.3 Temperature-dependent $\langle \Delta r^2 \rangle_{Conv}$

Models involving a non-linear temperature dependence of $\langle \Delta r^2 \rangle_{Conv}$ have been frequently invoked to explain dynamical transition behaviour. In these models the dynamical transition results from a change with T of the equilibrium, converged, long-time atomic probability distribution *i.e.*, $\langle \Delta r^2 \rangle_{Conv}$. One example of such models is in Doster *et al.* (1989), which consists of a two-state potential with a free-energy difference between the states of ΔU, separated by a distance, d. The increased population of the higher energy state with increasing temperature leads to a transition in $\langle \Delta r^2 \rangle_{Conv}$ and thus $\langle \Delta r^2 \rangle_{Expt}$. Another model is in Zaccai (2001) and Bicout and Zaccai (2001) in which the energy landscape is approximated by two harmonic potentials with different force constants. Here, the probability of atoms occupying the lower force-constant potential increases with temperature, thus also leading to an increase of $\langle \Delta r^2 \rangle_{Conv}$.

The characteristic the above models have in common is that they lead to $\langle \Delta r^2 \rangle_{Expt}$ being independent of the instrumental resolution provided that resolution is sufficiently high such that all the relaxation processes in the system are accessed.

3.4   Temperature-dependent $\langle \Delta r^2 \rangle_{Res}$

An alternative mechanism for nonlinear behaviour of $\langle \Delta r^2 \rangle_{Expt}$ involves a nonlinear increase with T of $\langle \Delta r^2 \rangle_{Res}$ due to motions becoming fast enough to be detected. In principle, this effect can lead to apparent dynamical transition behaviour in the absence of any change in $\langle \Delta r^2 \rangle_{Expt}$. Figure 3 shows a fit of Equation 5 to the experimentally-determined $\langle \Delta r^2 \rangle$ from Daniel *et al.* (1998). The insert to Figure 3 shows the associated relaxation time, $\tau(T) = \frac{1}{\lambda(T)}$. τ changes from the nanosecond to the picosecond timescale with increasing temperature, passing into the instrumental time resolution window of ~ 100 ps. This figure demonstrates that dynamical transition behaviour can appear in a dynamic neutron scattering experiment without any change with T in the long-time, converged dynamics.

3.5   Dynamical transition and neutron frequency windows

Detecting the dynamical transition on a neutron scattering instrument depends on the relation between the timescale of the relaxation processes responsible for the increased $\langle u^2 \rangle$ and the energy resolution of the neutron scattering spectrometer. As seen in the case of glutamate dehydrogenase solutions (Vurkal *et al.* 1999), the dynamical transition temperature in the same solution depends strongly on the timescale of the motions observed and the transition temperature shifts from ≈ 220 K to ≈ 150 K with improvement in the energy resolution. This can qualitatively be explained in terms of temperature-dependent timescales of internal protein motions. An increase in temperature results in faster extra fluctuations that move into the accessible timescale window of the spectrometer used.

Two possible scenarios occur for the existence of the dynamical transition in proteins (Becker *et al.* 2004). A change with temperature of the long-time probability distribution of the single-atom displacements forms the first 'equilibrium' scenario. Here, the characteristic relaxation frequencies of the dynamics are all within the energy resolution of the instrument used, and the dynamical transition, in principle, can then lead to a characterisation of energy levels occupied by different conformational substates. In the second scenario, there is no change in the time-converged atomic probability distribution with temperature. This is called the 'frequency-window' scenario. This would correspond, for example, to the systems possessing multiple minima with the minima having the same energy. In this case, temperature-dependent apparent dynamical transition behaviour will be observed when the relaxation frequencies of the dynamics responsible for the mean-square displacement are too slow to be detected by the finite energy resolution instrument. If the frequencies increase with temperature such that they fall into the frequency window of the instrument, then the dynamical transition is observed.

For systems where the frequency window scenario dominates, the dynamical transition provides information about the timescales of the motion that fall into the energy window of the instrument. For the dynamics involving barrier crossing, the dynamical transition analysis provides information regarding the barriers on the timescales accessed.

4 Protonation reactions in proteins

Electrostatic interactions are important for understanding biochemical systems. Acid-base reactions create or destruct unit charges in biomolecules and can thus be fundamental for their function. Together with association reactions and chemical modifications such as phosphorylations, acid-base reactions are the main cause of changes in protein properties. Protonation or deprotonation of titratable groups can cause changes in binding affinities, enzymatic activities, and structural properties. Moreover, very often protonations or deprotonations are the key events in enzymatic reactions. The reduction or oxidation of redox-active groups has a similar importance. In particular, the reduction of disulfide bonds can cause unfolding or functionally important conformational transitions. Consequently, the function of most proteins depends crucially on the pH and on the redox potential of the solution. For example, acidic denaturation of proteins in the stomach is a prerequisite for protein degradation during digestion. Besides this rather unspecific effect, the environment can tune the physiological properties of proteins in a specific manner. Different values of pH or redox potential in different organs, tissues, cells, or cell compartments steer protein function. Physiological redox and pH buffers such as glutathione and phosphates control these environmental parameters in living systems strictly. A few examples emphasise the physiological importance of pH and redox potential. The pH gradient in mitochondria or chloroplasts drives ATP synthesis. This pH gradient is in both systems generated by several proton transfer steps that couple to a sequence of redox reactions. In hemoglobin, pH influences O_2 binding and thus regulates O_2 release during blood circulation, a behaviour known as the Bohr effect. Pepsinogen cleaves itself in an acidic environment to the highly active peptidase pepsin. Finally, membrane fusion during influenza virus infection involves large pH-induced structural changes of the protein hemagglutinin.

5 Coupling between conformational and protonation state changes in membrane proteins

Many membrane proteins transport electrons and protons across a membrane (Ullmann 2001). Protonatable groups play a prominent role in these reactions, because they can either function as proton acceptors or donors in proton transfer reactions or they can influence the redox potential of adjacent redox-active groups. The titration behaviour of protonatable groups in proteins can often considerably deviate from the behaviour of isolated compounds in aqueous solution. This deviation is caused by interactions of the protonatable group with other charges in the protein and also by changes in the dielectric environment of the titratable group when the group is transferred from aqueous solution into the protein (Ullmann and Knapp 1999, Beroza and Case 1998, Briggs and Antosiewicz 1999, Sham *et al.* 1997). The situation can be even more complicated because owing to the fact that the charge of protonable residues depends on pH, their interaction is pH-dependent. This can lead to titration curves that can not be described by usual sigmoidal functions (Onufriev *et al.* 2001, Ullmann 2003).

The photosynthetic reaction centre (RC) is the membrane protein complex that performs the initial steps of conversion of light energy into electro-chemical energy (Oka-

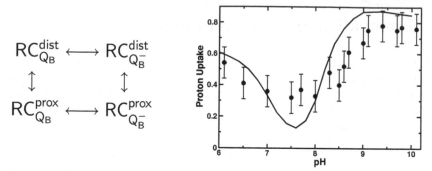

$$RC^{dist}_{Q_B} \longleftrightarrow RC^{dist}_{Q_B^-}$$

$$\updownarrow \qquad\qquad \updownarrow$$

$$RC^{prox}_{Q_B} \longleftrightarrow RC^{prox}_{Q_B^-}$$

Figure 4. *pH dependence of the proton uptake upon Q_B reduction. The symbols in the diagram show the experimentally-determined proton uptake. The line shows a proton uptake calculated by electrostatic calculations using a model that takes conformational transition between the two different reaction centre positions RC^{dist} and RC^{prox} into account.*

mura *et al.* 2000, Sebban *et al.* 1995a) by coupling electron transfer reactions to proton transfer. The bacterial RC of *Rhodobacter sphaeroides* is composed of three subunits: L, M and H. The L and M subunits have pseudo-two-fold symmetry. Both the L and M subunits consist of five transmembrane helices. The H subunit caps the RC on the cytoplasmic side and possesses a single N-terminal transmembrane helix. The RC binds several cofactors: a bacteriochlorophyll dimer, two monomeric bacteriochlorophylls, two bacteriopheophytins, two quinones, a non-heme iron and a carotenoid. The non-heme iron lies between the two quinone molecules. The primary electron donor, a bacteri-ochlorophyll dimer called the special pair, is located near the periplasmic surface of the complex, and the terminal electron acceptor, a quinone called Q_B, is located near the cytoplasmic side. While Q_A is a one-electron acceptor and does not protonate directly, Q_B accepts two electrons and two protons to form the reduced Q_BH_2 molecule. The first reductions of Q_A and of Q_B are accompanied by pK_a shifts of residues that interact with the semiquinone species (Wraight 1979). The reductions induce substochiometric proton uptake by the protein (Rabenstein *et al.* 1998a, 1998b, Rabenstein *et al.* 2000). The number of protons taken up by the protein upon reduction of the quinones is an observable that is directly dependent on the energetics of the system and intimately coupled to the thermodynamics of the electron transfer process between the states $Q_A^-Q_B$ and $Q_AQ_B^-$. The pH-dependence of the proton uptake associated with the formation of Q_A^- and Q_B^- in wild-type RCs have been determined for *Rb. sphaeroides* and *Rb. capsulatus* (Maroti and Wraight 1988, McPherson *et al.* 1988, Tandori *et al.* 2002, Sebban *et al.* 1995b). Using x-ray structural analysis, it has been shown that a major conformational difference exists between the RC handled in the dark (the ground state) or under illumination (the charge-separated state) (Stowell *et al.* 1997). The main difference between the two structures concerns Q_B itself, which was found in two different positions about 4.5 Å apart. In the dark-adapted state in which Q_B is oxidised, Q_B is found mainly in the distal position and only a small percentage in the proximal position. Under illumination, i.e., when Q_B is reduced, Q_B is seen only in the proximal position. The crystal was grown at pH=8 (Allen 1994). The reaction centre structures with proximal or distal Q_B are called RC^{prox} and RC^{dist}, respectively (Lancaster and Michel 1997). The proton uptake upon the first Q_B

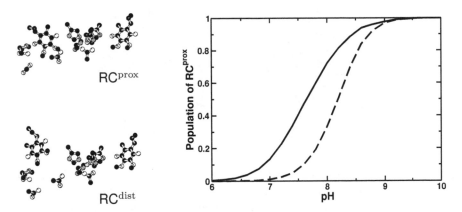

Figure 5. *Conformational equilibrium between RC^{prox} and RC^{dist} structures. The population of RC^{prox} shown for oxidised (dashed line) and reduced (solid line) Q_B depends on pH. In the neutral pH range, both conformations are populated.*

reduction and the pH-dependent conformational equilibrium between RC^{prox} and RC^{dist} are shown in Figures 4 and 5, respectively.

Using continuum electrostatic calculations, we investigated the pH-dependence of the proton uptake associated with the reduction of Q_B (Taly *et al.* 2003). Two experimentally observed conformations of the RC were considered: with Q_B bound in the proximal or the distal binding site. Comparing the calculated and experimental pH-dependence of the proton uptake revealed that a pH-dependent conformational transition is required to reproduce the experimental proton uptake curve (Figure 5). Neither the individual conformations nor a static mixture of the two conformations with a pH-independent population are capable of reproducing the experimental proton uptake profile. The result is a new picture of RC function in which the position of Q_B depends not only on the redox state of Q_B, but also on pH (Taly *et al.* 2003). This model will now be tested experimentally by performing X-ray crystallography of the RC system at different pH values.

6 Analysis of conformational changes in proteins

Proteins often have multiple stable macrostates. The conformations associated with one macrostate correspond to a certain biological function. Understanding the transition between these macrostates is important to comprehend the interplay between the protein in question and its environment and can even help to understand malfunctions that lead to diseases like cancer. While these conformational transitions are usually too fast to be measured experimentally, they also occur too rarely to observe them by running standard molecular dynamics simulations. They thus pose a difficult challenge to theoretical molecular biophysicists. In this final section, we will briefly summarise computational methods which have been proposed to analyse conformational changes in macromolecules and identify possible reaction pathways. In particular we will be interested

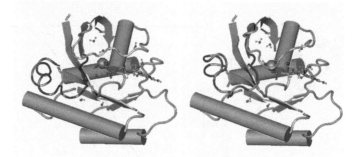

Figure 6. *The GTP-bound (left) and GDP-bound (right) state for the conformational switch of ras p21. The transition involves a complex reconfiguration of the backbone fold, providing a considerable challenge for computational pathfinding methods.*

in the analysis of complex transitions in proteins such as the conformational switch in Ras p21 (Figure 6).

Computationally, the problem of finding one possible reaction path between two macrostates corresponds to identifying a low-energy path on the potential energy surface of the protein between two representative end-conformations. A continuous path can be represented by a series of points in conformational space $P = [\mathbf{r}_0, \mathbf{r}_1, ..., \mathbf{r}_M - 1, \mathbf{r}_M]$ and some predefined way of interpolation between adjacent points, e.g., linear or spline interpolation. Here, $\mathbf{r}_0$ and $\mathbf{r}_M$ are the given reactant and product endstates whereas the intermediate points have to be found. Starting from an initial guessed path, the intermediate points can be optimised locally using different methods to obtain a low-energy reaction pathway.

6.1   Penalty function methods

One broad class of methods defines a path cost (or penalty) function, C:

$$C = \sum_{k=0}^{M-1} c(\mathbf{r}_k)|\mathbf{r}_{k+1} - \mathbf{r}_k|, \qquad (6)$$

where $c(\mathbf{r}_k)|\mathbf{r}_{k+1} - \mathbf{r}_k|$ assigns a cost for moving along the path from position $\mathbf{r}_k$ to $\mathbf{r}_{k+1}$. The initial guess path can be improved straightforwardly by minimising its cost function in the space of all possible paths using standard techniques such as steepest descent, conjugate gradient or simulated annealing. A rather straightforward definition for a cost function (Elber and Karplus 1987) is given by:

$$c(\mathbf{r}) = E(\mathbf{r})/L, \qquad (7)$$

where $E(\mathbf{r})$ is the potential energy of conformation $\mathbf{r}$ and L is the total path length, defined as $\sum_{k=0}^{M-1} |\mathbf{r}_{k+1} - \mathbf{r}_k|$. The best path is thereby defined as the path of minimum mean potential energy. With some additional constraints, as described below, this functional is called a self-penalty walk. It is, in its spirit and results, very similar to the nudged elastic band method (Mills and Jonsson 1994).

While having the merit of simplicity, the above definition has the following difficulty: through averaging, many small barriers can produce a cost comparative to a single very high barrier. This is not very realistic as the probability of a state decays exponentially with energy height. In this respect, the MaxFlux algorithm (Huo and Straub 1997) proposes a physically better-motivated cost function:

$$c(\mathbf{r}) = e^{E(\mathbf{r})/k_B T}, \tag{8}$$

where $1/k_B T$ is the Boltzmann coefficient. This temperature dependence allows shorter paths with higher energy barriers to become preferred at higher temperatures, an effect that also reflects physical reality.

The penalty function methods require a number of constraint terms to be added to the cost function: (1) constraints to remove the relative rigid-body translations and rotations of the structures along the path (to yield meaningful values for $|\mathbf{r}_{k+1} - \mathbf{r}_k|$), (2) equidistance constraints to avoid an increase of point density along the path in its low-energy segments and a correspondingly low resolution in segments crossing the saddle region, (3) self-avoidance terms to prevent the path folding back upon itself.

6.2   Heuristic methods

Rather than defining an objective cost function, heuristic methods improve the initial path by following a specific set of rules. Arguably, this approach is mathematically not as elegant as the penalty function approach. A clear disadvantage is that, because of the absence of an objective function, standard optimisation procedures cannot be simply combined with these methods. On the other hand, these methods can be made more efficient than penalty function methods by tailoring the optimisation of the path intermediates according to the position of these intermediates (e.g., to spend more optimisation effort on the points in the saddle region than on other points).

As an example, we summarise one heuristic method that has proven to be robust and efficient in very large systems: conjugate peak refinement (CPR) (Fischer and Karplus 1992). It is summarised by the pseudocode below. The basic approach of the algorithm is that path points are added and/or optimised by performing a one-dimensional maximisation along the path and minimisations in the space conjugate to the path direction. This yields path points which follow the streambed of the energy surface and along which the energy increases monotonously from the minima to the saddle points. The path points corresponding to the energy maxima along the path have converged to saddle points of first order because they are in a local maximum along the path direction and in a local minimum along all other directions of the set of conjugate vectors. CPR therefore returns an approximation to a steepest descent path whose only energy maxima correspond to first-order saddle points, i.e., a minimum-energy path (MEP). The algorithm does not evaluate the Hessian explicitly. Furthermore, most effort is concentrated on the location of saddle points. For these reasons, the algorithm is fast and can be used to compute reaction paths in proteins with thousands of atoms and for complicated transitions involving hundreds of saddle points (Olsen *et al.* 2000, Sopkova *et al.* 2000, Dutzler *et al.* 2002).

CPR$(P = [r_0, r_1, ..., r_M])$

(1) Let r_{max} be a local maximum with the highest energy value along the path P not yet flagged as a saddle point. If there is no such point, exit.
   Let s_0 be the tangent vector to the path at r_{max}

(2) If r_{max} lies between two existing path-points r_i and r_{i+1}, move it closer to the streambed by calling $r_{new} :=$ approachMEP(r_{max}, s_0).
   Add new point: $P := [r_0, r_1, ..., r_i, r_{new}, r_{i+1}, ..., r_M]$. Go to (1)

(3) If r_{max} lies on an existing path-point r_k, check whether the energy has a nearby maximum along s_0.
   If no, then r_k is an unwanted deviation from the path. Remove it and go to (1).
   If yes, then replace r_k by a point closer to the streambed by calling $r_k :=$ approachMEP(r_k, s_0). Go to (1)

approachMEP(x_0, s_0)

(1) For j from 0 to $D - 1$ repeat
   (1.1) Build a new conjugate vector s_{j+1} with respect to the Hessian at x_0.
   (1.2) If s_{j+1} is no longer conjugate to s_0, RETURN(x_j)
   (1.3) Obtain x_{j+1} by performing a line minimisation along s_{j+1}, starting from x_j.

(2) If the energy gradient at x_D is vanishing, flag x_D as saddle point. RETURN(x_D)

6.3   The initial pathway problem

Complex conformational transitions that include a reconfiguration of the backbone fold (as is the case in the conformational switch of Ras p21, Figure 6) impose a major challenge on reaction-path-finding methods. The first difficulty here is to find any MEP at all that has energy barriers low enough to be consistent with the experimental reaction rate. The results of both penalty function methods and heuristic methods rely on an initially guessed path prior to the optimisation. Unfortunately, it is very difficult for complex backbone rearrangements to find an initial guess that will optimise to a path with low barriers.

When the initial guess is generated by standard interpolation techniques (e.g., linear interpolation in cartesian or internal coordinates), the optimised path often has unphysical saddle points, such as the crossing of bonds (interpenetration of two bonded atom pairs) or the passage of water molecules through aromatic side chains. We have developed interpolation techniques that are specifically tailored for polymers such as proteins or nucleic acids to construct appropriate intermediate structures for the initial path. With these, we have been able to apply CPR in a fully automated manner to very complicated conformational transitions while systematically avoiding unphysical saddle points (to be published elsewhere).

6.4   Multiple pathways

When a single low-energy path has been found, it is in general not known whether it is just one example of a number of different paths. It has been shown that the MaxFlux method above can be used to identify slightly different pathways in small peptides (Huo and Straub 1999) and even short pathways in moderately sized polypeptides (Straub *et al.* 2002) by varying the initial guess. However, as yet no general method exists for systematically generating many different minimum energy pathways in large molecules. Such a method is highly desirable, since it would not only help to identify different possibilities for a conformational transition but also benefit the search for globally optimal pathways.

In summary, then, a significant advance in methodology is still required to properly study the many complex reaction pathways of biological interest.

7   Conclusions

It is hoped that the above cornucopia of dynamical phenomena and associated methodology gives the reader some appreciation for the remaining challenges in biomolecular dynamics research. Deepening our understanding requires methodological advances ranging from instrumentation improvement (e.g., in neutron spectroscopy), and methods of analysing experimental data through theoretical and computational advances for tackling these complex systems. It is to be expected that the burgeoning field of biophysics will devote considerable skill and resources in the next decades to understanding this aspect of the working of the many wonderful tiny molecular machines in the living cell.

References

Allen J, 1994, *Proteins* **20** 283.

Angell C A, 1995, *Science* **267** 1924.

Becker T and Smith J C, 2003, *Physical Review E* **67** 21904 .

Becker T, Hayward J, Finney J L, Daniel R M and Smith J C, 2004, *Biophys J* **87** 1436.

Bee M, 1988, *Quasielastic Neutron Scattering: Principles and Applications in Solid-state Chemistry, Biology and Materials Science* (Hilger, Bristol).

Beroza P and Case D A, 1998, *Methods Enzymol* **295** 170.

Bicout D J and Zaccai G, 2001, *Biophys J* **80** 1115.

Bizzarri A R, Paciaroni A and Cannistraro S, 2000, *Physical Review E: Statistical Physics, Plasmas, Fluids, and Related Interdisciplinary Topics* **62** (**3B**) 3991.

Briggs J M and Antosiewicz J, 1999, *Rev Comp Chem* **13** 249.

Brooks B R, Bruccoleri R E, Olafson B D, States D J, Swaminathan S and Karplus M, 1983, *Journal of Computational Biology* **4** (**2**) 187. original Charmm paper.

Cordone L, Ferrand M, Vitrano E and Zaccai G, 1999, *Biophysical Journal* **76** (**2**) 1043.

Daniel R M, Smith J C, Ferrand M, Hery S, Dunn R and Finney J L, 1998, *Biophysical Journal* **75** (**5**) 2504 .

Daniel R M, Finney J L, Reat V, Dunn R and Smith J C, 1999, *Biophys J* **77** 2184.

Daniel R M, Dunn R V, Fineey J L and Smith J C, 2003, *Ann Rev of Biophys and Biomolecular structure* **32** 69.

Dellerue S, Petrescu A-J, Smith J C and Belissent-Funel M-C, 2001, *Biophys J* **81** (**3**) 1666.

Doster W, Cusack S and Petry W, 1989, *Nature*, **337**(6209) 754.

Doster W and Settles M, 1999, in *Hydration Processes in Biology*, eds Bellissent-Funel M-C and Teixera J, volume **305** of *NATO Science Series: Life Sciences* (IOS Press).

Dunn R V, Reat V, Finney J, Ferrand M, Smith J C and Daniel R M, 2000, *Biochemical Journal* **346** (**2**) 355.

Dutzler R, Schirmer T, Karplus M and Fischer S, 2002, *Structure* **10** 1273.

Elber R and Karplus M, 1987, *Science* **235** 318.

Frauenfelder H, Petsko G A and Tsernoglou D , 1979, *Nature* **280** 558.

Frauenfelder H, Sligar S and Wolynes P, 1991, *Science* **254** 1598.

Ferrand M, Dianoux A J, Petry W and Zaccai G, 1993, *Proceedings of the National Academy of Sciences of the United States of America* **90 (20)**.

Fischer S and Karplus M, 1992, *Chemical Physics Letters* **194 (3)** 252.

Green J L, Fan J and Angell C A, 1994, *Journal of Physical Chemistry* **98 (51)** 13780.

Hayward J A and Smith J C, 2002, *Biophysical Journal* **82 (3)** 1216.

Huo S and Straub J E, 1997, *Journal of Chemical Physics* **107 (13)** 5000.

Huo S and Straub J E, 1999, *Proteins* **36** 249.

Kneller G R and Smith J C, 1994, *J Mol Biol* **242** 181.

Kurkal V, Daniel R M, Finney J L, Tehei M, Dunn R V and Smith J C, 2005, *Biophys J* **89** (in press).

Lamy A, Souaille M and Smith J C, 1996, *Biopolymers* **39** 471.

Lancaster C R D and Michel H, 1997, *Structure* **5** 1339.

La Penna G, Pratolongo R and Perico A, 1999, *Macromolecules* **32** 506.

Lovesey S, 1987, *Theory of Neutron Scattering from Condensed Matter*, volume 1 (Oxford University Press, Oxford).

McPherson P H, Okamura M Y and Feher G, 1988, *Biochim Biophys Acta* **934** 348.

Maróti P and Wraight C A, 1988, *Biochim Biophys Acta* **934** 329.

Mills G and Jonsson H, 1994, *Physical Review Letters* **72** 1124.

Olsen K, Fischer S and Karplus M, 2000, *Biophysical Journal* **78** 394A.

Onufriev A, Case D A and Ullmann G M, 2001, *Biochemistry* **40** 3413.

Paciaroni A, Cinelli S and Onori G, 2002, *Biophysical Journal* **83 (2)** 1157.

Parak F, Frolov E N, Kononenko A A, Mossbauer R L, Goldanskii V I and Rubin A B, 1980, *FEBS Letters* **117 (1)** 368.

Parak F, Frolov E N, Mossbauer R L and Goldanskii V I, 1981, *Journal of Molecular Biology* **145** 825.

Perico A and Pratolongo R, 1997, *Macromolecules* **30** 5958.

Rabenstein B, Ullmann G M and Knapp E W, 1998a, *Biochemistry* **37** 2488.

Rabenstein B, Ullmann G M and Knapp E W, 1998b, *Eur Biophys J* **27** 628.

Rabenstein B, Ullmann G M and Knapp E W, 2000, *Biochemistry* **39** 10487.

Rasmussen B F, Stock A M, Ringe D and Petsko G A, 1992, *Nature* **357 (6377)** 423.

Reat V, Dunn R, Ferrand M, Finney J L, Daniel R M and Smith J C, 2000, *Proceedings of the National Academy of Sciences of the United States of America* **97 (18)** 9961.

Sebban P, Maróti P and Hanson D K, 1995a, *Biochimie* **77** 677.

Sebban P, Maróti P, Schiffer M and Hanson D, 1995b, *Biochemistry* **34** 8390.

Sham Y Y, Chu Z T and Warshel A, 1997, *J Phys Chem B* **101** 4458.

Smith J, Kuczera K and Karplus M, 1990, *Proceedings of the National Academy of Sciences of the United States of America* **87 (4)** 1601.

Sopkova J, Fischer S, Guilbert C, Lewit-Bentley A and Smith J, 2000, *Biochemistry* **39** 14065.

Stowell M H B, McPhillips T M, Rees D C Soltis S M, Abresch E and Feher G, 1997, *Science* **276** 812.

Straub J E, Guevara J, Huo S and Lee J P, 2002, *Accounts of Chemical Research* **35** 473.

Taly A, Sebban P, Smith J C and Ullmann G M, 2003, *Biophys J* **84** 2090.

Tandori J, Valerio-Lepiniec J M M, Schiffer M, Maroti P, Hanson D and Sebban P, 2002, *Photochem Photobiol* **75** 126.

Teeter M M, Yamano A, Stec B and Mohanty U, 2001, *Proceedings of the National Academy of Sciences of the United States of America* **98 (20)** 11242.

Tilton R F, Dewan J C and Petsko G A, 1992 *Biochemistry* **31 (9)** 2469.

Tuckerman M E and Martyna G J, 2000, *Journal of Physical Chemistry* **194** 159.

Ullmann G M and Knapp E W, 1999, *Eur Biophys J* **28** 533.

Ullmann G M, 2001, Charge Transfer Properties of Photosynthetic and Respiratory Proteins, in *Supramolecular Photosensitive and Electroactive Matrials* 525–584, ed Nalwa H S (Academic Press, New York).

Ullmann G M, 2003, *J Phys Chem B* **107** 1263.

Wraight C A, 1979, *Biochim Biophys Acta* **548** 309.

Zaccai G, 2000, *Science* **288** 1604.

Models of cell motility

Jean-François Joanny

Physico-Chimie Curie, Institut Curie Section Recherche
26 rue d'Ulm, 75248 Paris Cedex 05, France

jean-francois.joanny@curie.fr

1 Summary

Motility is a fundamental property of cells and bacteria. Unicellular organisms constantly move in search for food; eggs would not be fertilised in the absence of sperm cell motion; macrophages move to infection sites; fibroblast cells motion allows the remodelling of connective tissues and the rebuilding of damaged structures. Cell motion also plays an important role in cancer with the formation of metastases. Cells and bacteria can swim when they are propelled by the beating of cilia or flagellae or by the polymerisation of an actin gel. They can also crawl inside the extracellular matrix or on surfaces. The crawling motion of cells requires a deformation of the cytoskeleton and occurs in several steps: protrusion in which new cytoskeleton polymerises in front of the cell leading edge, adhesion of the cell on the substrate that allows momentum transfer and depolymerisation and contraction of the rear of the cell in which the adhesion sites are broken and the rear parts of the cytoplasm are dragged forward.

The cell cytoskeleton has the mechanical properties of a soft elastic gel made of three types of filaments; actin, microtubules and intermediate filaments. The most important component as far as the mechanical properties of the cytoskeleton are concerned is actin. The elastic modulus of this gel is of the order of $10^3 - 10^4$ Pa. Actin filaments are polar and have a well-defined barbed end (or plus end) and a well-defined pointed end (or minus end). The actin filaments are constantly polymerising at the plus end at the front of the moving cell and depolymerising at their minus end towards the back of the cell. This phenomenon is known as treadmilling; it requires ATP and energy is thus consumed in the polymerisation process. For most cells, the actin filaments of the cytoskeleton interact with myosin molecular motors. Myosin molecular motors aggregate on the cytoskeleton gel and move towards the plus ends of the filaments. Their motion generates forces and also consumes energy in the form of ATP. If a motor aggregate is bound to two or more filaments, the action of the motors induces internal stresses in the cytoskeleton gel. We call such a gel where energy is constantly injected and generates internal stresses an active gel. An active gel is by essence a structure that is not at thermal equilibrium.

My lectures at the Edinburgh school did not give a general view of cell motility but rather presented some of our recent results on very specific examples of cell motility that emphasise one of the general processes involved. I give here only a very short summary of the problems discussed. References are given to our original work where the important relevant references to other work can be found.

Listeria bacteria swim inside host cells. Their motion is due to the formation of an actin comet gel at the rear of the bacteria by polymerisation of the actin of the surrounding cell. There are no myosin motors involved in this motion. The deformation of the gel creates elastic stresses that drive the motion. The polymerisation kinetics is also coupled to the stress distribution; the polymerisation velocity is large at places where the gel pulls on the bacteria and small at places where the gel pushes the bacteria (to create the motion). A general discussion of the motility of the bacteria is given in the work of Prost (2002). Our recent work considers a bio-mimetic system 'Soft Listeria', where the bacteria are replaced by liquid oil drops that show a Listeria-like motion by formation of an actin comet: the deformation of the drops is a signature of the elastic normal stress distribution around the drop (Boukellal *et al.* 2004).

Nematode sperm cells do not swim as other sperm cells but crawl. Their cytoskeleton is made of a protein called MSP (major sperm protein), which is very similar to actin but not polar. No molecular motors are involved in the motion of these cells. It has been suggested that the motion of nematode sperm cells is due to an interplay between internal stresses in the cytoskeleton owing to a self-generated pH gradient and adhesion on the substrate (Bottino *et al.* 2002). We have constructed a theoretical elastic model describing the interplay between adhesion at specific sites and cytoskeleton deformation (Joanny *et al.* 2003).

The cytoskeleton of most cells is an active gel and cell motion involves the deformation of the gel under the action of the molecular motors. This has been studied experimentally on several types of cells such as fibroblasts or fish keratocyte cells (Verkhovsky *et al.* 1999), which are among the fastest moving cells. Keratocytes are an important model system for cell motility because small cell fragments that contain mostly the active gel made of actin and myosins and are therefore much simpler than the whole cell can move spontaneously. We have constructed a general hydrodynamic theory for active polar gels using the non-equilibrium formalism used for the hydrodynamics of nematic liquid crystals. The hydrodynamic description takes into account actin polymerisation, actin polarity, the viscoelastic character of the gel and the active behaviour of the molecular motors. A first application of this approach is an extremely simplified model of keratocyte motion based on the existence of two spiral defects in the polarised actin network (Kruse *et al.* 2004, 2005).

Acknowledgements

The theoretical research presented has been done in common with J. Prost and O. Campas (Institut Curie), F. Jülicher and K. Kruse (Dresden) and K. Sekimoto (Strasbourg). I am also grateful to C. Sykes and H. Boukellal (Institut Curie) who performed the experiments on 'Soft Listeria'.

References

Bottino D, Mogilner A, Roberts T, Stewart M, Oster G, 2002, *J Cell Sci* **115** 367.
Boukellal H, Campas O, Joanny J.F, Prost J and Sykes C, 2004, *Phys Rev E* **69** 061906.
Joanny J F, Jülicher F, Prost J, 2003 *Phys Rev Lett* **90** 168102.
Kruse K, Joanny J F, Jülicher F, Prost J and Sekimoto K, 2004, Phys Rev Lett **92** 078101.
Kruse K, Joanny J F, Jülicher F, Prost J and Sekimoto K, 2005, *Eur Phys J E* **16** 5.
Prost J, 2002, The physics of Listeria motion, in *Physics of bio-molecules and cells*, Les Houches Lecture Notes LXXV, eds Flyvberg H, Jülicher F, Ormos P and David F (EDP Sciences, Springer).
Verkhovsky A B, Svitkina T M and Borisy G G, 1999, *Current Biology* **9** 11.

Part III

Experimental Techniques

Single-molecule force spectroscopy of proteins

Matthias Rief and Hendrik Dietz

Ludwigs Maximilians University of Munich
Amalienstr 54 80799, Munich, Germany [1]

Matthias.rief@physik.uni-muenchen.de

1 Introduction

Proteins are wonderful examples of self-organised nano-machines. Proteins are peptide polymer chains built from an amino acid library containing 20 different members. These members have different side chains that differ in basic properties such as electrical charge or hydrophobicity. (For an introduction to proteins, see Poon's chapter in this volume.) One of the most fundamental and challenging problems in molecular biophysics is understanding how these polypeptide chains fold into complex 3-D structures that can act as channels, enzymes and even motor proteins.

The folding process of proteins is generally described as diffusion in a high-dimensional energy landscape. (See McLeish's chapter in this volume for an introduction to diffusion in energy landscapes.) A simplified scheme of such an energy landscape is shown in Figure 1. In its folded conformation the protein occupies an energetic minimum. Unfolded conformations lie energetically higher, and according to the current view of protein folding, the way from the unfolded to the folded conformation is biased by a funnel-shaped energy landscape. Within this energy landscape there may be local minima representing intermediate states. Exposing the protein to heat or chemical denaturants is the classical way of driving a protein out of its folded conformation through this energy landscape to an unfolded conformation (grey pathway, Figure 1). However, the exact pathway or reaction coordinate is not known in such experiments.

---

[1]Current address:
Physics Department, Technical University of Munich
Lehrstuhl für Biophysik E22, D-85748 Garching b. Munich, Germany

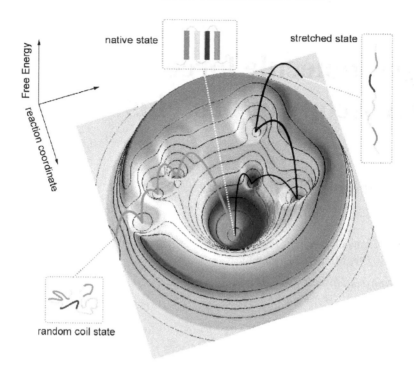

Figure 1. *Schematic of a folding landscape for a protein. Grey trace: hypothetical thermal unfolding pathway. Black trace: hypothetical mechanical unfolding pathway.*

Recent advances in single-molecule force spectroscopy have made it possible to explore the energy landscape of a single protein molecule along a well-defined reaction coordinate by applying a mechanical force. Figure 2 shows how such an experiment works. A protein molecule is tethered at one end to the substrate and at the other end to a sharp tip mounted onto a sensitive cantilever spring. The cantilever spring is then retracted with sub-nanometre precision, and the protein is subject to an increasing force until it unfolds. The force acting can be detected via the deflection of the cantilever spring amplified by a light pointer.

Several aspects make mechanical experiments with single molecules a valuable tool for protein science. A large fraction of the proteins in our body have structural and thus mechanical function. Examples include muscle proteins, cytoskeletal proteins and proteins of the extracellular matrix. Single-molecule force experiments can help to investigate the mechanical function of proteins, an aspect that has not been easily investigated in experiments so far. But, beyond physiology, force as a structural control parameter also offers attractive possibilities for exploring the energy landscape of biomolecule folding.

The pathway of unfolding (e.g., the black line in Figure 1) in a mechanical unfolding experiment is biased along the direction of applied force that is precisely known. This pathway is necessarily different from pathways in chemical or thermal denaturation. We can therefore use single-molecule mechanical experiments to explore the folding energy landscape of proteins along a well-defined reaction coordinate.

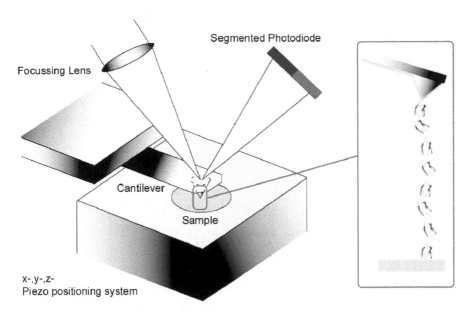

Figure 2. *Schematic of a force spectroscope.*

Proteins can respond very differently to the application of mechanical force. In the simplest case the protein behaves as a two-state folder/unfolder, and owing to the high cooperativity of the weak interactions that stabilise its structure, the transition will be all or none. An example is given in Figure 3(a) showing unfolding of Immunoglobulin (Ig) domains of the muscle protein titin. Each peak in the force-extension curve reflects cooperative unfolding of an individual Ig domain in titin. No intermediate states can be detected. Mechanical unfolding of titin is a process far from thermal equilibrium. Unfolding occurs in discrete steps and refolding rates are negligible under mechanical load. In contrast to the discrete non-equilibrium unfolding behaviour of titin domains, topologically simpler proteins like coiled-coil structures fold and unfold close to equilibrium under the influence of mechanical force. As an example, Figure 3(b) shows unfolding of the myosin coiled-coil. A force plateau at 25 pN marks the unfolding transition. Since the process occurs in equilibrium, relaxation of a mechanically unfolded coiled-coil results in the same force-extension curve, and this protein structure is therefore truly elastic.

The two example proteins demonstrate the variability of mechanical behaviour among different protein structures. However, it is important to note that single-molecule experiments are experimentally challenging, and due to surface effects and non-specific background generally less than 1% of the experimental recordings reflect clean single-molecule unfolding. The curves shown in Figure 3 therefore represent a strongly selected sub-population of all force-extension curves recorded in the experiments. A key issue in single-molecule force spectroscopy is to identify the characteristic mechanical 'fingerprint' of a specific protein. One strategy to obtain a clear selection criterion is the use of modular proteins. As in the case of titin (Figure 3(a)), a chain of identical subunits linked together yields a repetitive and characteristic sawtooth pattern. Only curves exhibiting this pattern are then used for data analysis. However, for many proteins, espea)

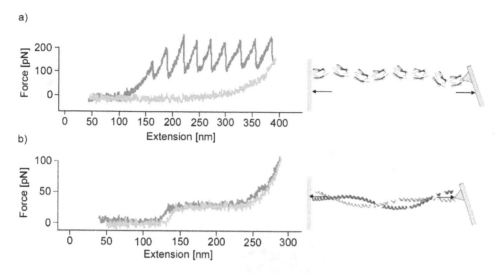

b)

Figure 3. *Two examples of the forced unfolding of proteins in non-equilibrium and equilibrium. (a) Sample trace of titin unfolding. Each peak corresponds to the unfolding of a single Ig domain. A schematic representation of the protein showing a series of individual Ig domains is given on the right. (b) Unfolding and refolding of the myosin II coiled coil. The folding and refolding traces show almost no hysteresis, indicating a process in equilibrium (adapted from Schwaiger et al. 2002). A schematic representation of the protein showing the coiled-coil structure is given on the right.*

cially larger ones, it is difficult to construct such multimeric chains. In the following we introduce a pattern detection algorithm based on a cross-correlation function that allows us to identify a molecular fingerprint for a single protein out of the large experimental background noise.

2 Pattern recognition in force-extension traces

2.1 Subjective vs. objective

The unspecific nature of the protein binding to substrate and cantilever tip leads to a rather low efficiency (below 1%) of a force spectroscopy experiment. The term efficiency refers to the ratio between clear single-molecule force-extension traces and the total number of force curves pulled. The overwhelming majority of force traces is dominated by non-specific and multiple-molecule interactions. The resulting patterns are generally omitted from analysis. But it has to be appreciated that such non-specific interactions, in principle, can assume any form; especially, they could even match the patterns one argues to be representative, for instance, for unfolding of a specific protein domain. An example of such a selected trace reflecting the unfolding of a single green fluorescent protein (GFP) domain is shown in Figure 4. Region A shows the unfolding of a proteins structure leading to a protein elongation of 77 nm and is therefore consistent with the expected contour length of unfolded GFP.

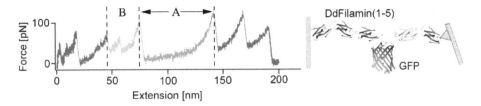

Figure 4. *Sample trace measured on a chimeric protein containing a single GFP domain flanked by domains of the dictyostelium discoideum filamin (DdFilamin). The section marked in dark grey (region A) corresponds to the unfolding of GFP and subsequent stretching of the lengthened protein. The section marked in light grey (region B) corresponds to unfolding of a special DdFilamin exhibiting a stable unfolding intermediate, as it has been described by Schwaiger et al. 2004.*

A question often asked about single-molecule data is: how do we know that this is a single-molecule event, representative and of statistical significance? It could also be argued that the data have been subjectively filtered since most experimentalists never discuss all force-extension traces obtained throughout a single experiment except those showing a certain pattern.

In the following we will introduce objective criteria that can be applied to experimental traces and help to find the specific mechanical fingerprint of a protein. This will result in a pattern recognition method providing a tool to filter data sets of force-extension traces for a certain pattern and assigns a degree of coincidence with the given pattern as a quantitative measure of matching with the pattern.

2.2 Pattern recognition in force-extension traces

Let the function $g(x)$ represent a certain pattern, while the function $f(x)$ shall be tested on partial or full coincidence with the pattern function $g(x)$. We will make use of cross-correlations as functions of a displacement variable u to develop such a testing method:

$$K_{g,f}(u) = \int_0^b g(x) \cdot f(x+u)dx. \tag{1}$$

The number b in Equation 1 denotes the width of the pattern $g(x)$. We are interested in a recognition function $C_{g,f}(u)$, which has value 1 for displacements u, leading to perfect match between pattern $g(x)$ and test function $f(x)$, and otherwise is always smaller than 1. That is, if the function $f(x)$ contains exactly the pattern $g(x)$ at a certain displacement u, the condition given in Equation 2 has to be fulfilled:

$$\lim_{f \to g} \frac{\int_0^b g(x) \cdot f(x+u)dx}{\int_0^b g(x) \cdot g(x)dx} = 1. \tag{2}$$

This condition motivates to introduce a displacement-dependent (u-dependent) normalisation for the cross-correlation function $K_{g,f}$. The Cauchy-Schwarz inequality can be used for this purpose:

$$\frac{\left(\int_0^b g(x) \cdot f(x+u)dx\right)^2}{\int_0^b g^2(x)dx \cdot \int_0^b f^2(x+u)dx} \leq 1. \tag{3}$$

Now, we simply define the recognition function $C_{g,f}(u)$ as the square root of the left-hand side of Inequality 3 and end up with a function fulfilling the desired condition (Equation 2):

$$C_{g,f}(u) = \frac{K_{g,f}(u)}{\sqrt{\int_0^b g^2(x)dx \cdot \int_0^b f^2(x+u)dx}}. \tag{4}$$

By determining the argument $u_{C=max}$ for which the function $C_{g,f}(u)$ has a maximum, we can thus determine the section of width b of the test function $f(x)$ exhibiting best match to the pattern $g(x)$. This 'best matching section' we call $\bar{f}(x)$:

$$\bar{f}(x) = f(x - u_{max}) \; ; \; x \in [0, b]. \tag{5}$$

2.3   Degree of coincidence

The number $C_{g,\bar{f}}(u_{C=max})$ determined in the foregoing section represents, in principle, already a measure to evaluate a *degree of coincidence*, that is, the goodness of the matching between pattern $g(x)$ and the best matching section $\bar{f}(x)$ of the test function $f(x)$. But this measure is still not very suitable to compare the matching of the pattern $g(x)$ to different test functions. This section introduces the necessary refinement.

For this purpose, we consider two different test functions $\bar{f}(x)$ and $w(x) = \eta \cdot \bar{f}(x)$ of the same width as the pattern $g(x)$ (η is a real number). Calculating now the value of the recognition function at zero displacement between the pattern and the test function leads to the same number $C_{g,\bar{f}}(0) = C_{g,w}(0)$ for both test functions, which is definitely undesirable. A necessary scaling correction factor $s_{g,\bar{f}}$ is suitably defined as follows:

$$s_{g,\bar{f}} = \begin{cases} \sqrt{\frac{\langle \bar{f}^2 \rangle}{\langle g^2 \rangle}} & \text{if } \langle \bar{f}^2 \rangle \leq \langle g^2 \rangle \\ \sqrt{\frac{\langle g^2 \rangle}{\langle \bar{f}^2 \rangle}} & \text{otherwise.} \end{cases} \tag{6}$$

A generally valid definition of a *degree of coincidence* Γ can now be given by:

$$\Gamma := C_{g,f}(u_{C=max}) \cdot s_{g,\bar{f}}. \tag{7}$$

This definition of an objective *degree of coincidence* could now, in principle, be used to perform statistical analysis of force-extension data sets for appearance of a certain pattern. Γ assumes any value in the interval $0 \leq \Gamma \leq 1$, where an increasing Γ reflects increasing coincidence between the test function and pattern. $\Gamma = 1$ is fulfilled if the test function and pattern are identical.

However, owing to the nature of the patterns investigated in protein force spectroscopy, another refinement improves the resolution of the degree of coincidence measure.

If the test function $\bar{f}(x)$ matches 'well' the pattern $g(x)$ while a function $z(x) = \bar{f}(x) + \eta$ matches 'less' because of a certain (force) offset η, we would like to have this also expressed in a considerably higher value $C_{g,\bar{f}}(u = 0) > C_{g,z}(u = 0)$ of the recognition function at displacement zero. The current situation instead leads to:

$$\frac{C_{g,\bar{f}}(u = 0)}{C_{g,z}(u = 0)} = \frac{\int_0^b z^2(x)dx}{\int_0^b \bar{f}^2(x)dx} = 1 + \frac{\int_0^b 2\eta\bar{f}(x) + \eta^2 dx}{\int_0^b \bar{f}^2(x)dx} \geq 1 \tag{8}$$

(The average of the pattern $g(x)$ is assumed to be zero, that is, $\langle g(x) \rangle = 0$.) The term $\int_0^b 2\eta f(x)dx$ in Inequality 8 could become negative and therefore even produce an '$= 1$' in Inequality 8. Such a situation can be avoided by replacing the pattern function $g(x)$ and the best matching part of the test function $\bar{f}(x)$ by their average values, that is,

$$\tilde{g}(x) = g(x) - \langle g(x) \rangle \text{ and } \tilde{f}(x) = \bar{f}(x) - \langle \bar{f}(x) \rangle, \tag{9}$$

and then calculating the value of the recognition function at displacement zero $C_{\tilde{g},\tilde{f}}(0)$. We now define a practical *degree of coincidence* γ to measure objectively the matching between patterns as they occur in force-extension traces to be:

$$\gamma := C^2_{g-\langle g \rangle, \bar{f}-\langle \bar{f} \rangle}(0) \cdot s_{g,\bar{f}}. \tag{10}$$

The number γ assumes only positive real values ≤ 1. For $\gamma = 1$, the pattern $g(x)$ matches perfectly to the section $\bar{f}(x)$ of the investigated trace $f(x)$.

By squaring $C^2_{g-\langle g \rangle, \bar{f}-\langle \bar{f} \rangle}(0)$, the recognition method becomes more sensitive on the exact form of the pattern, a strategy suitable for recognition of complex patterns. Other definitions of the degree of coincidence have to be considered to improve the resolution of the method depending on the nature of the pattern. If, for instance, the pattern of interest is rather featureless (e.g., a simple force plateau of a certain length as often seen in polymer desorption experiments), the scaling factor s could be squared instead.

2.4   Application of the recognition method to GFP unfolding

First of all, we have to define a function $g(x)$ containing the pattern we want to check the data sets for. For this purpose, we chose the section marked in black in Figure 5(a) of a simulated force-extension trace. This section corresponds to an idealised force curve for unfolding of a single GFP domain. As a set of test functions $f(x)$ we chose the complete data set of a force spectroscopy experiment on a modular chimeric protein containing a single GFP domain with a size of 1012 single force-extension traces. The pattern recognition algorithms first identifies the best matching sections in each force trace and then calculates the corresponding value of the *degree of coincidence* γ as defined in

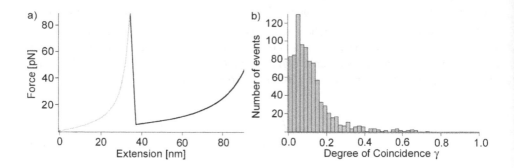

Figure 5. *(a) Simulated force-extension trace. As a pattern function $g(x)$ we chose the section marked in black, corresponding to a calculated GFP unfolding pattern. (b) Distribution of γ-values as they have been calculated for a experimental dataset containing 1012 force-extension traces measured on a modular protein containing a single GFP domain.*

Equation 10 for each force curve. Figure 5(b) shows the distribution of γ values as they have been calculated for this dataset.

From Figure 5(b) we see that high γ-values, $\gamma \geq 0.5$, are rarely observed. But how do the force curves, which have been judged to exhibit a certain degree of coincidence γ with the pattern look like? Figure 6 shows typical force-extension traces with a degree of coincidence γ out of four different intervals.

Figure 6(a) shows four arbitrarily selected traces with a degree of coincidence of $0.0 \leq \gamma \leq 0.01$. The traces do not show any similarity with the given pattern. Especially, we note that they show only few interactions or drift effects.

Figure 6(b) (interval $0.2 \leq \gamma \leq 0.21$) shows traces reflecting more interactions (adhesions, desorptions, etc.) with the cantilever; nevertheless, we note only very poor matching with the given pattern.

Figure 6(c) (interval $0.38 \leq \gamma \leq 0.42$) in turn shows traces exhibiting sections with a certain similarity to the given pattern, but still those structures match poorly to the given pattern.

Finally, Figure 6(d) (interval $0.55 \leq \gamma \leq 1$) shows traces exhibiting sections that reproduce very well the given pattern. A manual analysis of the full data set of 1012 force traces reveals that all force traces exhibiting GFP unfolding (that is, the pattern $g(x)$) have been judged by the pattern recognition routines to have a degree of coincidence of $\gamma \geq 0.55$.

Therefore, we conclude that, first, the pattern recognition method is able to filter big datasets for traces exhibiting a certain pattern and, second, for each trace we can now provide a quantitative degree of coincidence (as defined in Equation 10) with a certain pattern.

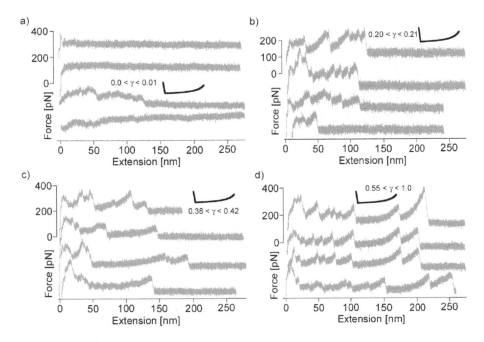

Figure 6. *Typical measured force-extension traces whose degree of coincidence with the pattern (insets) lays within four different intervals.*

2.5 Statistical analysis of datasets

In the preceding section we discussed the proof of principle of the presented pattern recognition method. In this section we demonstrate how the pattern recognition method can be used for a quantitative analysis of different datasets with respect to the occurrence of a certain pattern. For this purpose, we applied the pattern recognition routines on two datasets of similar size. One dataset contains 1012 force curves measured on a chimeric protein of modular quaternary structure containing a single GFP domain, whereas the other dataset contains 934 force curves measured on a different modular protein *not* containing GFP. The pattern we chose for the analysis is a calculated GFP unfolding pattern as it has been used already in the preceding section. Therefore, we expect that higher degrees of coincidence (interval $0.5 \leq \gamma$) should only be observed in the first dataset. Figure 7 shows the result of this analysis.

The superposition and close-up of the two histograms in Figure 7 shows that the degree-of-coincidence distribution of the dataset measured on the sample *not* containing GFP falls rapidly to zero. From a value of $\gamma = 0.2$, only 0.5% of the total number of force-extension traces have been judged to have a better matching of the GFP unfolding pattern. The highest value for γ is 0.46 and was attributed to only one force trace. In contrast, we see that the γ distribution calculated for the data set measured on a sample *containing* GFP falls smoothly to zero. From a value of $\gamma = 0.2$ still 6.4% of the total number of force curves have been judged with better degrees of coincidence. Nine force traces show a matching to the pattern with a quality of $\gamma \geq 0.5$. These degrees of coincidence reflect very good matching to the pattern in the investigated case, as discussed in

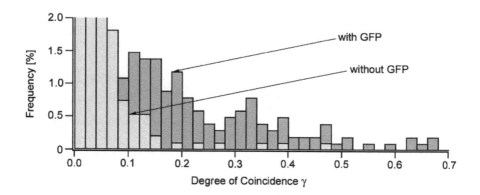

Figure 7. *Comparison of the distribution of the degree of coincidence γ as calculated for two different datasets.*

the foregoing section.

In comparison we conclude that the calculated GFP unfolding pattern does only occur with very good matching in the dataset measured on a sample containing GFP, while this pattern cannot be found in the dataset measured on a sample *not* containing GFP. We also find that the frequency of force traces with partial matching to this pattern is an order of magnitude higher in the experiment containing GFP rather than in the experiment lacking GFP. Hence, we can argue that an unfolding pattern is of such a specifity that it never gets reproduced by unspecific interactions.

The presented method for judging pattern matching of force traces by assigning a certain degree of coincidence γ can therefore be used for quantitative analysis of data sets and is able to show objectively that a certain pattern is in deed a property — a molecular fingerprint — of an investigated sample.

2.6 Conclusion

Single-molecule force spectroscopy is a novel tool for investigating mechanical properties and folding energy landscapes of proteins. It is crucial for noisy single-molecule experiments to develop and apply objective criteria for the selection of data traces. We have presented such an approach based on correlation functions.

References

Carrion-Vazquez M, Oberhauser A F, Fisher T E, Marszalek P E, Li H and Fernandez J M, 2000, *Progress in Biophysics and Molecular Biology* **74** 63.
Clausen-Schaumann H, Seitz M, Krautbauer R and Gaub H E, 2000, *Curr Opin Chem Biol* **4** 524.
Rief M, Gautel M, Oesterhelt F, Fernandez J F and Gaub H E, 1997, *Science* **276** 1109.
Schwaiger I, Sattler C, Hostetter D R and Rief M, 2002, *Nat Mater* **1** 232.
Schwaiger I, Kardinal A, Schleicher M, Noegel A A and Rief M, 2004, *Nat Struct Mol Biol* **11** 81.

A practical guide to optical tweezers

Christoph F Schmidt

Department of Physics and Astronomy, Free University of Amsterdam
De Boelelaan 1081, 1081 HV Amsterdam, The Netherlands

cfs@nat.vu.nl

1 Introduction

This chapter provides a brief review of the physical principles behind the method of trapping microscopic dielectric particles by focused laser light (optical tweezers) and then discusses some of the practical possibilities and limitations encountered when using optical tweezers. Since its invention in the 1980s, the use of optical tweezers has spread far in biology and also in physics and materials science. This chapter is not intended as a review of applications, but rather to highlight some of the technical issues one needs to consider in applications. Several extensive reviews of optical tweezers and experimental uses have been published (Svoboda and Block 1994, Ashkin 1997, Sheetz 1997) and an exhaustive literature overview is available (Lang and Block 2003).

2 Basic principles

Light interacting with matter exerts forces commonly described as radiation pressure. In a particle picture, a photon carries momentum, and it can transfer that momentum entirely or partially to an absorbing or scattering object. In the macroscopic world such forces are usually minute and negligible.

On the scales of cells and macromolecules, however, light forces, especially when imparted by high power lasers, can become important in comparison to other forces present, such as gravitation, hydrodynamic drag or active forces produced by cells. Maximal forces are on the order of 100 pN with laser powers on the order of 1 W, which is enough to overpower molecular motors, deform cells or cytoskeletal protein polymers,

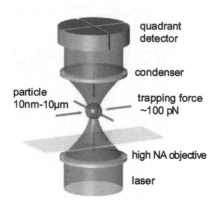

Figure 1. *Schematic diagram of an optical trap. A high-numerical-aperture objective lens forms a focus in which a refractile particle is trapped in 3D with forces of typically up to 100 pN. The condenser collects the transmitted light and a quadrant photodiode serves as displacement and force detector.*

stretch DNA or even unfold proteins. Much of the pioneering work on radiation pressure on small particles and atoms was done by Arthur Ashkin and his co-workers in the 1970s and 1980s (Ashkin 1970, Ashkin 1978, Ashkin 1980). They demonstrated in 1986 a single-beam optical trap consisting of a laser beam focused through a microscope objective (Figure 1) (Ashkin *et al.* 1986). They showed that objects such as polystyrene or silica spheres can be trapped, and also living objects such as virus particles, bacteria, or protozoa (Ashkin and Dziedzic 1987, Ashkin *et al.* 1987). As a non-contact method of micromanipulation, optical tweezers have advantages over classical micromanipulation methods, such as micropipettes. The optics of the microscope do not have to be modified for mechanical access to the sample. Particles can be manipulated within closed compartments, for example cells, and trapped particles can be easily released by shuttering the laser beam. Furthermore, micromanipulation by optical tweezers has recently been combined with high resolution measurements of forces and motions, reaching a single-molecule scale. Thus the method joins the group of techniques (such as atomic force microscopy, near-field scanning optical microscopy or single molecule fluorescence microscopy) that can be used to observe the dynamics of individual biological macromolecules. Such methods probe the functionally important mechanical properties of biomolecules and aggregates up to whole cells and complement high-resolution static imaging by electron microscopy and by x-ray crystallography on the one hand, and the typically very fast dynamic measurements on ensembles by conventional spectroscopy methods on the other hand.

An exact model of optical trapping would involve solving the Maxwell equations with the appropriate boundary conditions, which can become prohibitively complex in typical experimental scenarios. Simplifying approximations can be used depending on the specific situation. Particles that are small compared to the light wavelength can be modelled as point dipoles (Rayleigh scattering). A gradient in light intensity exerts a force on the particle (Ashkin 1978, Ashkin and Gordon 1983, Visscher and Brakenhoff

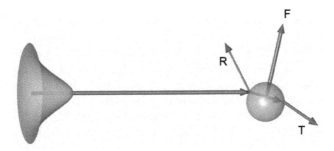

Figure 2. *Ray optics model for trap force. Rays are reflected (R) and refracted (T) by the particle, generating a net force (F) which can be thought of as composed of a gradient force directed towards the focus and a scattering force directed along the optical axis.*

1992a,b) in the same way as an iron nail gets attracted towards the pole of a magnet, while the (isotropic) scattering imparts a force on the particle pointing in the direction of light propagation. In the other extreme, ray optics can be used for particles that are large compared to the light wavelength. In that case, reflection and refraction occur, and the reactive forces on the trapped object can be calculated from the deflection and the bundling of the rays passing through it (Figure 2) (Ashkin 1992, Visscher and Braken-hoff 1992a,b).

In this case, the backward reflection of some light contributes to a forward force and can prevent trapping if the numerical aperture of the beam is not large enough. In both cases there is thus a force pointing towards the focus (gradient force) and a force pointing in the propagation direction of the laser beam (scattering force) (Ashkin 1992). Assuming a Gaussian focus and a particle much smaller than the laser wavelength (Rayleigh limit) the gradient forces in radial and axial direction are (Agayan *et al.* 2002), gradient force:

$$\hat{\mathbf{z}} \cdot \langle \mathbf{F}_g \rangle = -\frac{\epsilon_0}{\pi} \alpha' |\mathbf{E}_0|^2 z \frac{w_0^4}{z_0^2} \left[\frac{1}{w^4(z)} - \frac{2r^2}{w^6(z)} \right] \cdot \exp\left(-\frac{2r^2}{w^2(z)} \right) \tag{1}$$

$$\mathbf{r} \cdot \langle \mathbf{F}_g \rangle = -\frac{2\epsilon_0}{\pi} \alpha' |\mathbf{E}_0|^2 r \frac{w_0^2}{w^4(z)} \cdot \exp\left(-\frac{2r^2}{w^2(z)} \right) \tag{2}$$

and scattering force:

$$\hat{\mathbf{z}} \cdot \langle \mathbf{F}_s \rangle = \frac{\epsilon_0}{\pi} \alpha'' |\mathbf{E}_0|^2 \frac{w_0^2}{w^2(z)}$$
$$\times \left[k_m \left(1 - \frac{r^2}{2} \frac{(z^2 - z_0^2)}{(z^2 + z_0^2)^2} \right) - \frac{w_0^2}{z_0 w^2(z)} \right] \cdot \exp\left(-\frac{2r^2}{w^2(z)} \right) \tag{3}$$

$$\mathbf{r} \cdot \langle \mathbf{F}_s \rangle = -\frac{\epsilon_0}{\pi} \alpha'' |\mathbf{E}_0|^2 \frac{w_0^2}{w^2(z)} \frac{k_m r}{R(z)} \cdot \exp\left(-\frac{2r^2}{w^2(z)} \right), \tag{4}$$

with complex polarisability $\alpha = \alpha' + i\alpha''$, field E; z and r are unit vectors in radial and laser propagation directions, respectively, and w_0 the beam radius in the focus and $w(z) = w_0 \sqrt{1 + (z/z_0)^2}$ the beam radius near the focus; $z_0 = \pi w_0^2 / \lambda_m$ the Rayleigh range, $k_m = 2\pi/\lambda_m$ the wave vector with λ_m the wavelength in the medium with refractive index n_m.

Stable trapping will only occur if the gradient force wins over the trapping force all around the focus. Trap stability thus depends on the geometry of the applied field and on properties of the trapped particle and the surrounding medium. The forces generally depend on particle size and the relative index of refraction $n = n_p/n_m$ where n_p and n_m are the indices of the particle and medium, respectively, which is hidden in the polarisability α in equations (1)–(4). In the geometrical optics regime, maximal trap strength is particle-size independent, but increases with n over some intermediate range until, at larger values of n, the scattering force exceeds the gradient force. The scattering force on a non-absorbing Rayleigh particle of diameter d is proportional to its scattering cross section, thus the scattering force scales with the square of the polarisability (volume) (Jackson 1975), or as d^6. The gradient force scales linearly with polarisability (volume), i.e., it has a d^3-dependence (Ashkin *et al.* 1986, Harada and Asakura 1996).

3   Heating in optical tweezers

In order to obtain forces on the order of tens or hundreds of pN, laser powers of typically tens to hundreds of mW are used, leading to focal intensities exceeding MW/cm^2 (for comparison, the intensity of bright sunlight on the surface of the earth is on the order of 100 mW/cm^2). The potential of thermal and non-thermal damage caused by these high intensities to (biological) samples has been a matter of concern and investigation (Ashkin and Dziedzic 1987, Barkalow and Hartwig 1995, Allen *et al.* 1996, Neuman *et al.* 1999). One way of reducing non-thermal photodamage to many biological materials is to use near-infrared lasers (such as Nd:YAG, Nd:YLF, diode- or Ti:Sapphire lasers) rather than visible lasers (Neuman *et al.* 1999). The temperature increase due to trapping a particle in water has been roughly estimated to be rather low, in the order of 1 K/W (Block 1990).

One can directly observe temperature changes in a focused 1064 nm laser beam by analysing the thermal motion of trapped polystyrene or silica beads, as commonly used in optical tweezers experiments (Figure 3). Results of such experiments were modelled taking into account the entire spatial profile of the focused beam in a low-NA approximation (Peterman *et al.* 2003a). By comparing data taken in water and glycerol, it was shown that light absorption by and dissipation in the solvent is the primary determinant of the temperature change, rather than heat absorbed by the trapped particle. The heat conductivity of glass is substantially higher than that of water, so that the sample chamber acts as a cooling device for the optical tweezers. This cooling effect of the sample cell wall is part of the model and was also demonstrated in the experiments. The model can be used for other solvents (with not too high absorption) as long as heat absorption and conductivity are known. The results are largely independent of trapped particle properties, provided again that the absorption of the particle is not too high. The temperature increase in the focus is:

$$\Delta T(0) = B \cdot P = b \cdot [\ln(2\pi \cdot R/\lambda) - 1] \cdot P \quad , \quad b \equiv \frac{\alpha}{2\pi \cdot C}. \quad (5)$$

P is the laser power, b a factor that depends on absorption coefficient α and heat conductivity C of the solvent, R is the distance of the laser focus to the closest boundary, which

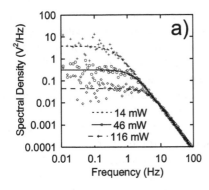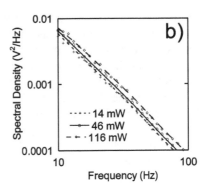

Figure 3. *Power spectra of the Brownian motion of a trapped 502 nm diameter polystyrene bead in glycerol. The laser power was as indicated. The lines represent fits of a Lorentzian to the data. In (a) the whole spectra are shown, (b) is a zoom of the power-law, high-frequency region where temperature effects are most clearly visible. Data from (van Dijk et al. 2004).*

is supposed to be infinitely conductive for heat, and λ is the wavelength of the laser light. The factor b equals 3.8 K/W for water and 12.2 K/W for glycerol. Assuming a distance of 10 μm to the next boundary, the calculated temperature increase is 12 K/W for water and 38 K/W for glycerol. The experimentally measured values were about 8 K/W for water and 40 K/W for glycerol at 10 μm distance from the surface.

Even a small temperature increase can have significant effects because often the viscosity depends strongly on temperature. For example, in an optical trapping experiment in water with a laser power at 1064 nm of 100 mW (500 mW) a temperature increase of about 0.8 K (4 K) occurs in the focus. In many cases the Lorentzian fit to a power spectrum of a trapped bead is used for the calibration of the trap and detector response (Allersma *et al.* 1998). If the heating effect is not taken into consideration, using this calibration method the trap stiffness (which is proportional to the estimated viscosity (η) times the measured corner frequency (f_0)) is overestimated by 2% (10%) when a laser power of 100 mW (500 mW) is used. Here we assume a temperature increase of 8 K/W and a base temperature of 294.55 K. The detector response (in m/V) is proportional to the temperature (T) divided by the viscosity (η) and the zero-frequency intercept of the power spectrum ($S_0 f_0{}^2$) and is in the same circumstances underestimated by 2% (11%). Consequently, heating effects due to laser-light absorption by the solvent in optical trapping experiments even in aqueous solution have a small but measurable effect and should be taken into consideration, especially when laser powers higher than about 100 mW are used.

4 Resonant trapping

A disadvantage of optical tweezers in comparison, for example, to magnetic manipulation is that in a crowded environment, such as the inside of a cell, force is exerted indiscriminately. Another limitation is the maximum force available with optical tweezers while avoiding damage to the specimen and to conventional optics by heating. Optical tweezers operating under 1 W of average laser power can exert forces of up to about 100 pN — enough to stall mechanoenzymes and stretch DNA (Ashkin 1997). Stronger forces are typically needed to irreversibly move organelles such as chloroplasts or nuclei within cells. Somewhat larger forces can be reached using alternative trap geometries or high-index particles. Maximal force scales with particle volume ($\approx d^3$, d = particle diameter) for particles small compared to the laser wavelength, and becomes independent of particle radius in the ray-optics regime. Smaller particles have a faster dynamic response and are therefore often advantageous for use as dynamic probes (Gittes and Schmidt 1998c). Thus, a method for maximising trapping force at a given particle size is desirable. A selective enhancement of trapping strength would also provide a degree of specificity in optical tweezing, i.e., make it possible to preferentially trap probe particles in a crowded environment.

Selective trapping with high force can potentially be achieved when trapping an absorptive particle near its resonant absorption frequency, but not directly on-resonance (Agayan *et al.* 2002). The particle's complex refractive index is strongly increasing, and trapping force as well as absorption are expected to be enhanced. The scattering force on a non-absorbing Rayleigh particle of diameter d is proportional to its scattering cross section, thus the scattering force scales with the square of the polarisability (volume) (Jackson 1975, Harada and Asakura 1996), or as d^6. The gradient force scales linearly with polarisability (volume), i.e., it has a d^3-dependence (Ashkin *et al.* 1986, Harada and Asakura 1996). Since the dependences on polarisability for the scattering and gradient forces are quadratic and linear, respectively, stable single-beam trapping only occurs for particles smaller than some maximum threshold size. Thus in the Rayleigh regime, trap strength will also be maximal for particular values of n and d. So far, optical tweezers have commonly been used at frequencies far from any resonances in the trapped particles, whereas the excitation-frequency dependence of optical forces on atoms has been studied in great depth (Bjorkholm *et al.* 1986). For detuning below resonance, an optical potential well is formed at the focus. Detuning above resonance ejects atoms from the beam focus. Tuning exactly on resonance maximises absorption thus maximising the scattering force while minimising the gradient force. In a dielectric particle or macromolecule, numerous transitions can contribute to the absorption profile and resonant transitions are usually much broader than for atoms. Nevertheless, we expect a small particle, i.e., a collection of interacting dipoles, to behave qualitatively similarly to atoms when the excitation frequency is varied. One can model these effects by considering a Rayleigh particle in an electromagnetic field and approximate the induced dipole as a classical electron oscillator (CEO) (Agayan *et al.* 2002).

Figure 4 shows scaled trap stiffness calculated using the CEO curve fit for the absorption for different convergence angles (θ). At smaller θ, trapping can only occur for wavelengths far from resonance ($\lambda_0 = 0.403\ \mu$m) as the scattering force tends to dominate the gradient force. As θ increases, the gradient of the beam focus increases and trapping can occur closer to resonance. Comparison of Figures 4(a) and 4(b) indicates that for a given $\theta = 45°$ and λ, the radial trap strength k_r is up to an order of magnitude larger than the axial trap strength k_z. Also, for a given θ, the wavelengths corresponding to the maximum radial and axial trap stiffness have similar values, always remaining above

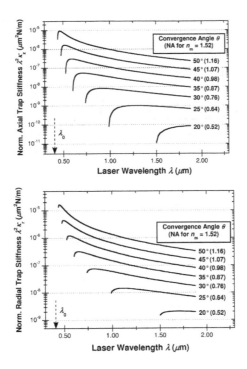

Figure 4. *Normalised axial (a) and radial (b) trap stiffnesses for different numerical apertures of the focusing lens for a small absorbing particle with absorption maximum at λ_0, calculated numerically, using a classical electron oscillator approximation (see text). Data from Agayan et al. (2002).*

resonance, but the wavelength for maximum radial stiffness is shifted towards slightly lower wavelengths compared to that for the axial stiffness. For all three cases, trapping in the radial direction is slightly easier than in the axial direction for a given wavelength since a slightly smaller θ is required for maximum stiffness. On resonance, absorption is greatest and the scattering force is at its maximum, therefore there is no trapping. The overall result is that trapping is only enhanced for wavelengths red-detuned from resonance. It is thus expected that red-detuned optical tweezing near sharp resonance lines in absorptive solids or liquids will afford enhanced trapping opportunities.

5 Photobleaching in optical tweezers

To obtain high-resolution information on position or conformation of a molecule and at the same time apply forces to it, one can combine optical trapping with single-molecule fluorescence microscopy. The technical challenge in such an experiment is to discriminate a minute fluorescence signal from the much larger background signals caused by the trap and the fluorescence excitation laser light. Experiments have shown that this is feasible even when the fluorophore is directly attached to the trapped particle, by using

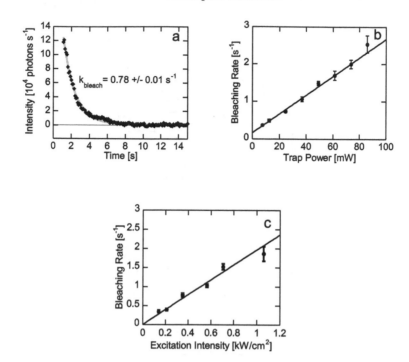

Figure 5. *Effect of the power of the trapping and the fluorescence excitation laser on photobleaching. (a) Photobleaching of many Cy3 molecules attached to a trapped bead with an exponential fit with decay constant $0.78 \pm 0.01\ s^{-1}$ (grey line), 850 nm trapping laser at 25 mW, 532 nm fluorescence excitation laser 350 W/cm². (b) Trap laser power dependence of the bleaching rate. (c) Fluorescence excitation intensity dependence of the bleaching rate. Shown are the fitted rates, and a linear fit to the averaged data (solid line), with slope $1.96 \pm 0.05\ cm^2 s^{-1} kW^{-1}$. The power of the 850 nm trapping laser was 25 mW. Data from van Dijk et al. (2004).*

optimised optical filters (van Dijk *et al.* 2004). It was found, however, that the photo-stability of the fluorophore tested suffered from the presence of the additional laser light used for trapping, i.e., they bleach more rapidly (Figure 5(a)). Bleaching rates increased linearly both with the intensity of the trapping laser and the intensity of the fluorescence excitation light (Figure 5(b), (c)). Photobleaching rates were unaffected by the presence or absence of oxygen, but were nevertheless significantly diminished in the presence of antioxidants (Table 1). Rates are also lower when the polarisations of trapping and fluorescence excitation laser were crossed. The most likely explanation for these observations is that the enhanced photobleaching is caused by the absorption of a visible photon followed by the excited-state absorption of a near-infrared photon. The higher excited singlet states generated in this way then readily form non-fluorescent dye cations. Different dyes were found to suffer to a different extent from the excited-state absorption, with Cy3 being worst and tetramethylrhodamine least affected (Figure 5).

These results are related to enhanced photobleaching in 2-photon excited fluores-

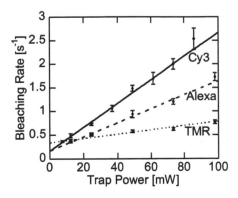

Figure 6. *Bleaching rates for three different dyes: Cy3, Alexa555 and TMR. The intensity of the 532 nm fluorescence excitation laser was kept constant at 350 W/cm². Data from van Dijk et al. (2004).*

condition	relative bleaching rate
untreated	1
degassed under argon	0.97 ± 0.03
oxygen scavenger	0.46 ± 0.03
ascorbic acid	0.25 ± 0.01

Table 1. *Effect of antioxidants and oxygen depletion on the bleaching rates. The data were measured with 89 mW trapping power and 350 W/cm² fluorescence excitation intensity and are represented relative to the values for the untreated sample under the same optical conditions. Data from van Dijk et al. (2004).*

cence experiments, which show a bleaching rate proportional to the third power of the excitation intensity (Patterson and Piston 2000, Dittrich and Schwille 2001). In those experiments in which only a non-resonant near-infrared laser beam is present at much higher power, bleaching takes place from higher excited states produced via a 3-photon process, either by direct 3-photon excitation, or by 2-photon excitation into the lowest singlet state immediately followed by an additional excitation by one photon into the higher excited states.

In summary, combining trapping and single molecule fluorescence is possible, but measures should be taken to reduce the additional photobleaching in the combined fluorescence and trapping experiments: (1) choice of the right fluorophore (TMR is better than Alexa 555, which is better than Cy3); (2) use of low trapping and fluorescence excitation powers; (3) use of antioxidants such as ascorbic acid; (4) use of perpendicular polarisation of the fluorescence excitation and trapping laser beams.

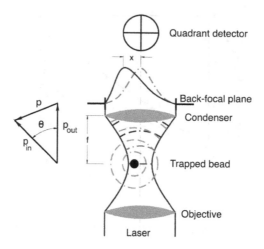

Figure 7. *Schematic illustration of the setup used for displacement detection.*

6   Displacement detection and detection bandwidth

Tied to momentum transfer to the particle in optical tweezers is an equal and opposite change of momentum (i.e., primarily a change of the angular distribution) in the light beam (Figure 7). This can be used to measure the exerted force or the particle displacement (Ghislain and Webb 1993, Smith *et al.* 1996, Allersma *et al.* 1998, Gittes and Schmidt 1998a, Pralle *et al.* 1999, Smith *et al.* 2003). Using segmented photodiodes, displacement and force can be measured with nanometre and piconewton resolution. Some further optical theory is needed to understand such high resolution detection schemes that are not based on simple video imaging. Trapped objects are mostly phase objects, i.e., they do not absorb the laser light. This is desirable, because absorption would cause heating with higher laser powers. Position detection will therefore make use of interference or phase shifts. Position detection in two dimensions can be achieved by imaging not the trapped object itself, but the light intensity distribution in the back-focal plane of the collimating lens (usually the microscope condenser) on a quadrant photodiode, thereby monitoring the angular scattering pattern from the trapped object, which changes with position in the trap (Figure 7). For small particles, the angular intensity shifts can be understood as a first order interference effect between illuminating and scattered light (Gittes and Schmidt 1998a):

$$\frac{\delta I(F,x)}{I_{tot}} = \frac{2k^3\alpha}{\pi r^2}e^{-x^2/w_0^2}\sin(kx\sin\theta\cos\phi)e^{-k^2w_0^2\theta^2/4}, \tag{6}$$

where $\delta I(F,x)$ is the change of intensity at position in the far field, i.e., for large r (in spherical coordinates r, θ, ϕ with the laser focus as the origin) for a given lateral displacement x of the trapped bead, with wave vector k, beam waist radius w_0, and bead polarisability α. The force F on the particle is proportional to the intensity weighted average angular shift $\langle \sin \theta \rangle_I$ of the light:

$$F = \frac{I}{c} \langle \sin \theta \rangle_I = \frac{I}{c} \frac{\langle x_{bfp} \rangle_I}{f}, \tag{7}$$

with the total intensity hitting the trapped bead I, and speed of light c. If the condenser is constructed to fulfil the Abbé sine-condition (Born and Wolf 1989), which is true in most cases, one can directly measure the angular shift from the lateral light shift in the back-focal plane of the microscope condenser (Figure 7) with focal length f.

This technique has in principle a wide bandwidth, from hours (limited by the mechanical drift of the instrument) down to microseconds (limited by shot noise and electronics). When using a near-infrared laser (1064 nm) for trapping, one has to choose the proper photodiodes, however. Common Si-photodiodes have a frequency response that is attenuated above about 5 kHz (Peterman *et al.* 2003b) due to approaching the band-gap energy in the near infrared (≈ 1100 nm). Although well-known, this is not usually quantitatively elaborated on in manufacturers' catalogues. As long as light is absorbed in the depletion layer of the photodiode, the response time is fast (nanoseconds). When, however, charge pairs are created outside this layer in the substrate, they will have to diffuse (microseconds) towards the depletion layer where they then contribute to the photocurrent, unless they have recombined before that. If the charge carrier diffusion time is smaller than the recombination time, the response time will be slowed down, in the opposite case the sensitivity will be decreased, without an effect on the response time.

There are specialised silicon photodiodes on the market that operate at a high reverse bias causing a complete depletion of the substrate, by which they maintain a fast response in the near infrared. Alternatively, standard InGaAs photodiodes have a band gap of ≈ 1700 nm and consequently do not suffer from this problem at 1064 nm. A very accurate test for detection bandwidth is the observation of the Brownian motion of micron-sized particles in an optical trap. This Brownian motion has a well known power spectral density of Lorentzian shape for low frequencies. At higher frequencies inertial solvent effects modify the Lorentzian shape, again in a theoretically well-known manner (Landau and Lifshitz 1959).

Figure 8 shows the frequency-dependent attenuation of the power spectral density (psd) for a Si- diode and the true psd measured with both a specialised high-bandwidth silicon diode and an InGaAs diode. The figure only shows the high-frequency part of the power spectral density, multiplied with the square of the frequency, which should result in a horizontal line for an ideal Lorentzian.

7 Signal-to-noise ratio and resolution

To determine the position of a known object, the classical diffraction limit (Born and Wolf 1989) of a microscope for the distinction of two unknown objects is not relevant. The fundamental limitation for the amplitude of motion that can be detected in a given time comes from the 'shot noise' of photon counting, i.e., the standard deviation of $N^{1/2}$ associated with counting N random events in a sampling time interval (Reif 1965). With a few milliwatts of laser power focused on a micron-sized bead in water, displacement sensitivities of on the order of 10^{-3}Å can theoretically be reached at 100 kHz band-

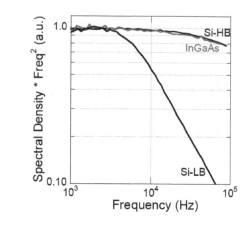

Figure 8. *Power spectral densities of the Brownian motion of 900 nm silica beads in water, trapped in the focus of a 1064 nm laser beam, multiplied by the square of the frequency. The three spectra were measured with three different quadrant photodiodes common Silicon: UDT SPOT 9-DMI, InGaAS: Hamamatsu G6849, and High bandwidth Silicon: Perkin Elmer YAG444-4A. Data from Peterman et al. (2003b).*

width (Denk and Webb 1990). Practical limitations usually stem from other sources of noise, such as electronic noise in the preamplifier stages, or mechanical instabilities in the microscope as well as Brownian motion superimposed on processes to be measured (Gittes and Schmidt 1998b,c).

Thermal motions of microscopic probes limit the possibilities of experiments that are designed to resolve, for example, single-macromolecule dynamics in aqueous conditions. In the following text, I will describe strategies for maximising signal-to-noise ratios or resolution in typical experimental situations (see Gittes and Schmidt 1998b,c for more details). It turns out that the viscous drag on a micromechanical probe is more important than the compliance of the probe.

An optically trapped bead is typically used to measure either a displacement or a force (or both) generated by an object of interest, e.g., a motor protein. One primarily monitors the position $\Delta x(t) = x_p(t) - x_0(t)$ of the probe with respect to the centre of the trap in which it is elastically suspended as it interacts with the target object (Figure 9). The suspension force can be inferred from the 'probe strain' $\Delta x(t)$ and the stiffness K_p of the optical trap by:

$$F(t) = K_p \Delta x(t). \tag{8}$$

The position detection can be used to construct a feedback loop, to perform either a pure force measurement by preventing probe displacement (position clamp) or a pure displacement measurement by keeping the force constant (force clamp). I will first discuss these two prototypical feedback experiments and then the case without feedback. Acousto-optic modulators can be used to move the laser beam of optical tweezers rapidly so that ideal feedback can be approximated quite well.

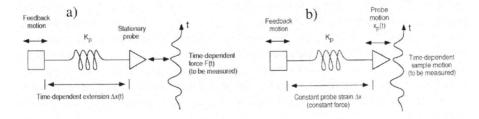

Figure 9. *(a) Position-clamp experiment to measure (time dependent) forces by fixing the probe position by feedback on the trap position. (b) Force-clamp experiment to measure (time dependent) displacements by fixing the force exerted by the trap by feedback on the trap position.*

7.1 Position-clamp experiments

Position-clamp experiments measure a time-varying force $F_{sig}(t)$ exerted on a probe held stationary by using feedback to move the trap centre position $x_0(t)$ (Figure 9(a)). The probe position has to be measured absolutely in this case and not relatively to the (moving) trap centre. This can, for example, be done by superimposing a second weak laser beam on the trapping beam. As a changing force begins to displace the probe, feedback changes the probe strain $\Delta x(t) = x_p - x_0(t)$ to keep the probe position x_p constant. The time-dependent force is found from $\Delta x(t)$:

$$F_{tot}(x) = K_p \Delta x(t). \tag{9}$$

With perfect feedback control, the fundamental limitation in measuring the force exerted by the target object comes from the white-noise thermal force that also acts on the probe. The power spectrum of the thermal force is given by (Gittes and Schmidt 1998b):

$$S_F(f) = 4\gamma k_B T, \tag{10}$$

where k_B is Boltzmann's constant and T is absolute temperature and γ is the frictional drag coefficient of the probe in the fluid. Because the thermal noise power extends to very high frequencies, low-pass filtering the strain signal $\Delta x(t)$ will increase the signal-to-noise ratio (the cut-off frequency f_s must of course lie above frequencies of interest in the signal). The remaining uncertainty $F(t) = (F_{tot}(t) - F_{sig}(t))_{rms}$ is the integrated noise power below f_s,

$$\delta F_{rms} = \sqrt{\overline{\Delta F^2}} = \sqrt{4\gamma k_B T f_s}. \tag{11}$$

The force uncertainty is minimised by either: (1) reducing the drag γ on the probe or (2) if possible, slowing down the force signal to be measured and making f_s as low as possible. In the case of a bead of diameter 1 μm, optically trapped in water far from a surface, the drag coefficient is about $\gamma = 8 \times 10^{-9}$ kg/s. Assuming a reasonable bandwidth of 1 kHz, this implies a force uncertainty of $F_{rms} \approx 0.4$ pN. This situation implies a trade-off between temporal and force resolution, which must be considered carefully in a given application. Static forces can thus, in principle, be measured to arbitrary precision, with very small f_s and very long measurement times, but, in practice, drift in the apparatus becomes limiting. Equation 11 does not contain the trap stiffness and thus shows that it is not relevant in principle. In practice, electronic noise in the photo-diode amplifier can limit how small a strain can be detected, in which case a softer trap allows measurement of a smaller force change and of a smaller absolute force. On the other hand, electronic noise and laser noise can be controlled fairly well so that detector noise is usually not the limiting factor.

7.2   Force-clamp experiments

If the probe strain Δx is held constant by feedback as the probe moves, the constant suspension force $F_{set} = K_p \delta x$ is balanced by a constant force of interaction with the target (Figure 9(b)). The probe position $x_p(t)$ is then monitored and its changes are assumed to reflect precisely the changes in position of the target because under a constant force, any elastic coupling between probe and target would not change in length. We ignore here direct dynamical effects of viscous drag and probe mass; thus, in the absence of thermal forces, the probe would always exert *exactly* the chosen force on the sample. In reality, the suspension force actually balances the constant sum of the force of interaction with the target and a fluctuating thermal force on the probe. The probe position $x_p(t)$ is then only approximately following the true constant-load position of the target.

Thermal noise now imposes distinct limitations in two experimental situations: (1) it limits the accuracy with which a high-force spatial response can be determined and (2) it sets a minimum force at which a spatial response can be obtained at all. By a 'large' force I mean that F_{set} is large compared to the root-mean-square thermal force on the probe: $F_{set} \gg F_{rms}$ (see Equation 11). The position uncertainty in the experiment is caused by the force uncertainty ΔF_{rms}. If the local stiffness of the probe-sample interaction is K_s, then

$$\Delta x_{rms} = \frac{\sqrt{4\gamma k_B T f_s}}{K_s}. \tag{12}$$

As with the position clamp, the stiffness of the trap does not enter directly, but now the details of the probe-target interaction are determining the error. In measuring the displacement $x_p(t)$ caused by molecular motor action with optical tweezers, for example, the uncertainty Δx_{rms} can be very small if the stiffness K_s of the bead-motor linkage is high. Thermal noise can be further reduced, as before, by reducing the drag on the probe, or by slowing the target motion if possible.

At the low-force limit, the thermal force noise determines a minimum set force or load at which a displacement signal can still be recorded. In order to keep the feedback loop stable F_{set} must stay larger than ΔF_{rms}:

$$F_{set(min)} = \Delta F_{rms} = \sqrt{4\gamma k_B T f_s}. \tag{13}$$

For noise reduction, again the primary strategies are to reduce both the drag on the probe and the filter frequency f_s. Surprisingly, probe stiffness is again not a direct consideration in avoiding large forces on the sample (assuming the detector resolution not to be limiting).

7.3 Displacement measurement without feedback

With optical tweezers, displacements of a single active molecule, such as a motor protein, against the trap compliance are often measured without using feedback. The molecule is tethered by a compliant link (stiffness K_s) to the probe held in the trap (stiffness K_p). Due to the compliant attachment, the molecular displacement, δx_m, will result in an attenuated probe displacement, δx_p:

$$\delta x_p = \delta x_m \frac{K_s}{K_s + K_p}. \tag{14}$$

Due to thermal forces, the position uncertainty of the probe will be (analogous to Equation 12):

$$\Delta(\delta x_p)_{rms} = \frac{\sqrt{4\gamma k_B T f_s}}{K_s + K_p}. \tag{15}$$

The signal-to-noise ratio can then be defined as:

$$\frac{\delta x_p}{\Delta(\delta x_p)_{rms}} = \frac{K_s \delta x_m}{\sqrt{4\gamma k_B T f_s}}. \tag{16}$$

The signal-to-noise ratio is therefore independent of trap stiffness and again dependent on filter frequency and drag coefficient. The signal-to-noise ratio also becomes larger with increasing stiffness in the molecule-to-probe connection; this has been observed (Svoboda *et al.* 1993).

8 Conclusions

Optical tweezers have rapidly become a popular investigative tool in a variety of research fields from physics and materials science all the way to molecular and cell biophysics. There are many ways to construct them and a variety of techniques to combine them with. The basic principles are well understood although many aspects of their detailed function remain to be explored. I have in this chapter attempted to give a synopsis of a number of technical issues of practical relevance that we have encountered and explored in my laboratory in recent years. Although surely not exhaustive, I have aimed to provide an impression of both limitations and capabilities.

Acknowledgements

I thank Erwin Peterman for a critical reading and discussion of the manuscript.

References

Agayan R R, Gittes F, Kopelman R and Schmidt C F, 2002, *Appl Opt* **41** 2318.
Allen P G, Laham L E, Way M and Janmey P A, 1996, *J Biol Chem* **271** 4665.
Allersma M W, Gittes F, deCastro M J, Stewart R J and Schmidt C F, 1998, *Biophys J* **74** 1074.

Ashkin A, 1970, *Phys Rev Lett* **24** 156.
Ashkin A, 1978, *Phys Rev Lett* **40** 729.
Ashkin A, 1980, *Science* **210** 1081.
Ashkin A, 1992, *Biophys J* **61** 569.
Ashkin A, 1997, *Proc Natl Acad Sci U S A* **94** 4853.
Ashkin A and Dziedzic J M, 1987, *Science* **235** 1517.
Ashkin A, Dziedzic J M, Bjorkholm J E and Chu S, 1986, *Opt Lett* **11** 288.
Ashkin A, Dziedzic J M and Yamane T, 1987, *Nature* **330** 769.
Ashkin A and Gordon J P, 1983, *Opt Lett* **8** 511.
Barkalow K and Hartwig J H, 1995, *Biochem Soc Trans* **23** 451.
Bjorkholm J E, Chu S, Ashkin A and Cable A, 1986, Laser cooling and trapping of atoms Advances in Laser Science-II, *Second International Laser Science Conference, Seattle, WA, USA.*
Block S M, 1990, Optical tweezers: a new tool for biophysics, in *Noninvasive Techniques in Cell Biology* volume 9 375–402, eds Foskett J K and Grinstein S (Wiley-Liss, New York).
Born M and Wolf E, 1989, Principles of Optics (Pergamon Press, Oxford).
Denk W and Webb W W, 1990, *Appl Opt* **29** 2382.
Dittrich P S and Schwille P, 2001, *Appl Phys B-Lasers and Optics* **73** 829.
Ghislain L P and Webb W W, 1993, *Opt Lett* **18** 1678.
Gittes F and Schmidt C F, 1998a, *Opt Lett* **23** 7.
Gittes F and Schmidt C F, 1998b, *Methods in Cell Biology* **55** 129.
Gittes F and Schmidt C F, 1998c, *E Biophys J Biophys Lett* **27** 75.
Harada Y and Asakura T, 1996, *Opt Commun* **124** 529.
Jackson J D, 1975, Classical Electrodynamics (Wiley, New York).
Landau L D and Lifshitz E M, 1959, Fluid Mechanics (Pergamon Press and Addison-Wesley Pub Co, London, Reading, MA).
Lang M J and Block S M, 2003, *American J Phys* **71** 201.
Neuman K C, Chadd E H, Liou G F, Bergman K and Block S M, 1999, *Biophys J* **77** 2856.
Patterson G H and Piston D W, 2000, *Biophys J* **78** 2159.
Peterman E J G, Gittes F and Schmidt C F, 2003a, *Biophys J* **84** 1308.
Peterman E J G, van Dijk M A, Kapitein L C and Schmidt C F, 2003b, *Review of Scientific Instruments* **74** 3246.
Pralle A, Prummer M, Florin E L, Stelzer E H K and Horber J K H, 1999, *Microscopy Research and Technique* **44** 378.
Reif F, 1965, Fundamentals of Statistical and Thermal Physics (McGraw-Hill, New York).
Sheetz M P, 1997, Laser Tweezers in Cell Biology, in *Methods in Cell Biology* (Academic press, London).
Smith S B, Cui Y J and Bustamante C, 1996, *Science* **271** 795.
Smith S B, Cui Y J and Bustamante C, 2003, *Biophotonics* **361** 134.
Svoboda K and Block S M, 1994, *Annu Rev Biophys Biomol Struct* **23** 247.
Svoboda K, Schmidt C F, Schnapp B J and Block S M, 1993, *Nature* **365** 721.
van Dijk M A, Kapitein L C, van Mameren J, Schmidt C F and Peterman E J G, 2004, *J Phys Chem B* **108** 6479.
Visscher K and Brakenhoff G J, 1992a, *Optik* **90** 57.
Visscher K and Brakenhoff G J, 1992b, *Optik* **89** 174.

Solution Scattering

Stefan U Egelhaaf

School of Physics and School of Chemistry, The University of Edinburgh
Mayfield Road, Edinburgh EH9 3JZ, UK

egelhaaf@ph.ed.ac.uk

1   Introduction

In a scattering experiment, radiation is incident on a sample and the fraction scattered by the sample is recorded. Solution scattering is concerned with the scattering from samples containing objects in a solvent rather than the diffraction from crystalline material. Scattering is a key technique to study the structure and dynamics of biomolecules and assemblies of biomolecules. Here we are concerned with two classes of scattering experiments: static and dynamic scattering. In a static scattering experiment, the dependence of the (time-averaged) scattered intensity on the direction of observation is determined. This depends on the distribution of scattering centres and thus on the structure and arrangement of the objects. Therefore, we can obtain information on the shape, structure, size and molar mass of the individual objects as well as on their spatial arrangement. In a dynamic scattering experiment, the time dependence of the scattered intensity is analysed. This can provide information on the dynamic properties of the objects, such as their centre of mass motion or, in the case of flexible particles, fluctuations of their shape. Static and dynamic scattering experiments thus provide complementary information.

We will consider light, x-ray and neutron scattering. The typical wavelengths of these radiations are different. They determine the length scales that can be probed in a scattering experiment; for a light scattering experiment, typical length scales are about 100 nm – 10 μm and for x-rays or neutrons about 0.1 – 100 nm. Furthermore, different radiations interact with matter in specific ways, which has some important implications. Still, most effects are independent of the type of radiation and only rely on the interference of secondary scattered waves. The theoretical background is thus presented as generally as possible. Because a detailed, quantitative knowledge of the background is necessary for a proper understanding of many phenomena observed with scattering methods, the presentation is quite mathematical. However, an attempt has been made to explain qualitatively the equations and their significance.

Based on scattering data, it is not possible to determine *directly* the size, shape or structure of scattering objects or their arrangement. Scattering methods are 'indirect' techniques; a structural model of the object always has to be assumed to make progress in analysing scattering data. Based on this model, the expected scattering is calculated and compared to the experimentally determined scattering data. It can thus only be decided whether a specific model agrees or disagrees with the data, leaving open the possibility that other models fit the data equally well. This approach is prone to controversies: in principle, all possible models have to be tested. Nevertheless, scattering methods are powerful techniques offering very useful and often unique possibilities, e.g., to perform *in situ*, non-destructive experiments, which usually yield data with very good statistics.

There are several books on scattering techniques. A general overview is provided by Lindner and Zemb (2002), while the other books given in the references concentrate on different aspects of scattering methods, as indicated by their titles.

2   Static scattering

In a static, or time-averaged, scattering experiment, the sample is irradiated and the intensity of the scattered radiation observed in different directions, which are characterised by the scattering angle θ (Figure 1(A)). Here we are only interested in the scattered intensity averaged over some time interval, typically at least a few seconds. The radiation can either be light, x-rays or neutrons. For the different radiations, the interaction with matter is based on different processes. Light is sensitive to the dielectric properties of the sample, while x-rays and neutrons are scattered by the electrons and nuclei, respectively (Section 2.7). However, we are mainly interested in the angular dependence of the scattered intensity, which is due to the wave characteristics. The theoretical background is hence identical for all radiations, except when we are interested in the absolute scattered intensity, where the specific scattering processes become important. This chapter is thus kept as general as possible, but to avoid being too abstract, we will occasionally refer to the scattering of light, since this is closest to our everyday experience.

The main aim of this section is to understand what causes the angular dependence of the (time-averaged) scattered intensity for a sample containing many particles, such as proteins, DNA or lipid assemblies, suspended in a solvent and what information on the sample can be extracted. We will decompose the calculation into several steps. We start by considering the scattering of a small volume element (Equation 5), which will then be used as a 'building block' for an extended particle (Equation 7), before we consider an ensemble of particles (Equation 9). Up to this point, these considerations are also the basis for dynamic scattering, which will be discussed in the following section (Section 3). Here we continue by calculating the time-averaged intensity (Equation 12), which is the most important result in this section. It is determined by three terms. First, the form factor, which depends on intra-particle interference effects and thus on the size, shape and structure of the individual objects (Section 2.5). Second, the structure factor, which depends on inter-particle interference effects and hence contains information on the arrangement of the objects in the sample (Section 2.6). Third, a direction (or scattering angle θ) independent 'prefactor', which determines the absolute level of the scattered intensity and depends on the specific radiation, i.e., light, x-rays or neutrons (Section 2.7). Finally, we will briefly comment on the main methods used for data analysis (Section 2.8).

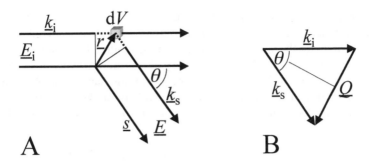

Figure 1. *(A) Scattering geometry. (B) Definition of the scattering vector $\underline{Q}$.*

2.1 Scattering by a small volume element

From our everyday experience we know that pure water appears transparent. In contrast, we can observe soil particles suspended in river water or small dust particles as they scatter light in an intense sunbeam. To scatter light, particles need to have a refractive index (or dielectric constant) different from the liquid in which they are dispersed. More generally, scattering is caused by fluctuations in the (scattering) density of the sample. Such fluctuations can, for example, be caused by particles with a refractive index different from the solvent; this is what we are interested in here. They can also result from density fluctuations of otherwise homogeneous material; the liquid itself has spontaneous density fluctuations or, if it is a mixture of liquids, density and concentration fluctuations, which also lead to scattering. The contribution of this 'background' scattering can be measured and subtracted from the total scattering (Section 2.7). However, here we assume for simplicity that this contribution is negligible.

We consider the scattered field $d\underline{E}(\underline{r}, t)$ of a very small volume element dV at position $\underline{r}$ (where $\underline{r}$ is measured relative to an arbitrary origin), which is illuminated by a plane wave with wavelength λ_0. (Then the wavelength in the suspension is $\lambda = \lambda_0/n$, with n being the refractive index of the suspension.) The incident field is hence $\underline{E}_i(\underline{r}, t) = \underline{E}_0 \exp\{i(\underline{k}_i \cdot \underline{r} - \omega t)\}$, where $\underline{E}_0$ is the amplitude of the incident radiation, $\underline{k}_i$ its wavevector with magnitude $k_i = 2\pi/\lambda_0$, ω its angular frequency and t the time. Each 'point-like' volume element $dV \ll \lambda^3$ acts as an oscillating dipole which re-radiates or scatters light in all directions, i.e., leads to a spherical wave, proportional to $(1/s)\exp\{i(k_s s - \omega t)\}$, with s the distance from the volume element and $\underline{k}_s$ the wavevector of the scattered beam. The amplitude of the spherical wave depends on the amplitude of the incident field $\underline{E}_0$, the size of the small volume element, $dV(\underline{r})$, and the scattering length density $\rho(\underline{r})$ (the 'scattering ability') of the material that fills $dV(\underline{r})$. (We will come back to how $\rho(\underline{r})$ depends on the material and radiation in Section 2.7). Including all these factors, one obtains the field $d\underline{E}(\underline{r}, t)$ scattered by a small volume element dV centred at $\underline{r}$ and detected at a large distance s, i.e., in the far field:[1]

$$d\underline{E}(\underline{r}, t) = \underline{E}_0 \, \rho(\underline{r}) \, \frac{1}{s} e^{i(k_s s - \omega t)} \, e^{i\delta\phi(t)} \, dV. \tag{1}$$

---

[1] The detector distance s is typically between a tenth (light scattering) and several (x-ray and neutron scattering) metres and thus to a good approximation the same for all $dV(\underline{r})$ in the scattering volume.

The factor $\exp\{i\delta\phi(t)\}$ contains information on the phase of the scattered field or, more precisely, on the phase difference $\delta\phi(t)$ compared to a field scattered at the origin. This factor is crucial for the angular dependence of the scattered intensity of an extended particle. To derive $\delta\phi(t)$, we look at the geometry of a scattering experiment as shown in Figure 1(A). (We only consider scattering in the scattering plane.) The phase difference $\delta\phi(t)$ is caused by the extra distance $\delta L(t)$ travelled relative to the path through the origin (dotted line). The extra distance $\delta L(t)$ is the sum of the projections of $\underline{r}(t)$ on $\underline{k}_i$ and $\underline{k}_s$, resulting in

$$\delta\phi(t) = \frac{2\pi}{\lambda}\delta L(t) = \underline{k}_i \cdot \underline{r}(t) - \underline{k}_s \cdot \underline{r}(t) = -\underline{Q} \cdot \underline{r}(t), \tag{2}$$

where we have defined the scattering vector $\underline{Q} = \underline{k}_s - \underline{k}_i$.

In static (and dynamic) light scattering we consider scattering processes that are (quasi) elastic. The wavelength is thus assumed not to change during the scattering process and the magnitudes of the incident and scattered wave vectors are identical: $k_i = k_s = k$. The magnitude of $\underline{Q}$ is hence related to the scattering angle θ by (Figure 1(B))

$$Q = |\underline{Q}| = 2k\sin\frac{\theta}{2} = \frac{4\pi}{\lambda}\sin\frac{\theta}{2}. \tag{3}$$

If we return to the phase factor $\exp\{i\delta\phi(t)\} = \exp\{-i\underline{Q}\cdot\underline{r}(t)\}$ (Equation 2), we realise that $\underline{Q}$ is the wave vector of a wave that defines positions $\underline{r}$ with an identical phase factor. Two points with an identical phase factor have a distance along $\underline{Q}$ equal to the wavelength of this wave, $2\pi/Q$ (or an integer multiple of this). This sets the length scale $\mathcal{L}$ probed in a scattering experiment:

$$\mathcal{L} \approx \frac{2\pi}{Q} = \frac{\lambda}{2\sin(\theta/2)}. \tag{4}$$

This length $\mathcal{L}$ is the length of the 'ruler' used to investigate the scattering object. Since the size of this ruler should be adequate for the object to be studied, we have to consider the range of length scales, $\mathcal{L}$, which can be covered by the different scattering techniques. The wavelength of visible light is in the range 400 nm$<\lambda<$800 nm. Typical light-scattering equipment allows to access an angular range of $10°<\theta<160°$, while specifically designed small-angle light-scattering (SALS) instruments allow for scattering angles down to $\theta \approx 1°$. This results in Q values from about 0.2×10^{-3} nm^{-1} to 0.04 nm^{-1} (assuming the solvent is water with $n_s = 1.33$). The accessible length scales thus cover a range 150 nm$<\mathcal{L}<$30 μm.

The wavelength of x-rays and neutrons is much shorter, typically 0.1 nm$<\lambda<$1 nm. If an angular range similar to light scattering would be covered, i.e., $1°<\theta<160°$, this would correspond to 0.1 nm$^{-1}<Q<$100 nm^{-1} and 0.05 nm$<\mathcal{L}<$50 nm ($n \approx 1$ for most materials). In particular, the high Q and thus low-$\mathcal{L}$ range is not very useful for the investigation of biomolecules in solution. It is, however, advantageous to expand the range toward smaller Q to obtain an overlap with the Q range accessible with light scattering. Small-angle x-ray scattering (SAXS) and small-angle neutron scattering (SANS) instruments fulfil this requirement. They typically cover an angular range $0.1°<\theta<15°$, although some instruments go significantly beyond this range. This angular range results in 0.01 nm$^{-1}<Q<$15 nm^{-1} and 0.4 nm$<\mathcal{L}<$500 nm, which covers the typical length scales of biomolecules and overlaps with the Q range of light scattering.

In summary, in this section we looked at the parameters that determine the field $dE(r, t)$ scattered by a small volume element dV (Equation 1). Keeping only the parameters characteristic for a certain scattering volume dV located at r, i.e., neglecting all dependencies on the experimental set-up, we obtain

$$dE(Q, t) \sim \rho(r)e^{-iQ \cdot r(t)} \, dV, \qquad (5)$$

with the scattering vector Q and the scattering length density $\rho(r)$ of the material (and thus the scattering length $\rho(r)dV$ of the whole volume element). The scattering vector Q is an important parameter as it determines the accessible length scales $\mathcal{L}$ (Equation 4) and thus which structures can be observed. Its magnitude is determined by the wavelength λ and the scattering angle θ (Equation 3).

2.2 Scattering by an extended particle

Having derived the scattered field $dE(Q, t)$ of a small volume element $dV(r)$, we now examine the scattering of an extended particle. We will consider the particle as consisting of small volume elements dV as we discussed them in the preceding text. To obtain the scattering of a particle, we thus have to add the contributions from all the small volume elements that make up the extended particle. We have to consider two important consequences: First, the scattered waves from the different volume elements have different path lengths or, equivalently, phases, which we have to add correctly. Second, the incident and scattered waves have to pass through more material (other small volume elements), which might affect the amplitudes and phases of the waves.

Depending on the size and refractive index of the dissolved particles, different approaches must be used to describe the scattering. Particles small compared to the wavelength and/or with a low refractive index difference relative to the solvent can be described by the so-called Rayleigh-Gans-Debye theory. For larger particles the Lorenz-Mie theory and for very big particles Fraunhofer diffraction applies. Because biomolecules do not usually fall into this size range, we will not discuss these regimes and only refer to the literature (Kerker 1969, van de Hulst 1981).

Hence, in the following we will only consider the Rayleigh-Gans-Debye regime. This is based on the assumption that the incident and scattered fields are not altered by the presence of other small volume elements, i.e., the wavefronts are not distorted on passing through the particles. This is the first Born approximation. It requires that most of the incident field passes through the sample unscattered and that the particles adsorb only very little and introduce only a negligible phase shift, i.e., $(4\pi R/\lambda)|n_p/n_s - 1| \ll 1$ with R being the particle radius and n_p and n_s the refractive indices of the particle and solvent, respectively. Based on the approximation that the amplitude and phase of the waves are not affected by the material, we can assume that each small volume element dV of the particle acts independently as a Rayleigh scatterer, as discussed in the previous section.

To calculate the scattered field of an extended particle j, $E_j(Q, t)$, we thus consider the particle as consisting of many small volume elements $dV(r)$, which independently act as Rayleigh scatterers. The contributions $dE(Q, t)$ of all volume elements $dV(r)$ are added or, more precisely, integrated to give the total scattered field $E_j(Q, t)$. This is done by considering the position r of the volume element $dV(r)$ relative to a specific

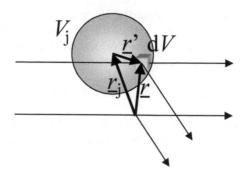

Figure 2. *Decomposition of the position $\underline{r}$ of a volume element dV into the position of the centre of mass, $\underline{r}_j$, of the extended particle (or any other fixed position with respect to the particle) and the distance $\underline{r}'$ within the particle.*

point, $\underline{r}_j$ (Figure 2):[2]

$$\underline{r}(t) = \underline{r}_j(t) + \underline{r}'. \tag{6}$$

For flexible particles $\underline{r}' = \underline{r}'(t)$, we suppress this here for brevity, but will come back to it in Section 2.5.3, particularly Equation 19, where we consider flexible objects.

Based on the contribution $d\underline{E}(Q, t)$ of a small volume element $dV(\underline{r})$ (Equation 5), the integral over the whole volume V_j of the extended particle then gives the total scattered field $\underline{E}_j(\underline{Q}, t)$:

$$
\begin{aligned}
\underline{E}_j(\underline{Q}, t) &= \int_{V_j} d\underline{E}(\underline{Q}, t) \sim \int_{V_j} \rho(\underline{r}) e^{-i\underline{Q}\cdot\underline{r}(t)} \, dV \\
&= \int_{V_j} \rho(\underline{r}') e^{-i\underline{Q}\cdot\underline{r}'} \, dV \, e^{-i\underline{Q}\cdot\underline{r}_j(t)} = b_j(\underline{Q}) \, e^{-i\underline{Q}\cdot\underline{r}_j(t)}.
\end{aligned} \tag{7}
$$

The contribution of a single extended particle, irrespective of its position $\underline{r}_j$, is summarised in $b_j(\underline{Q})$ with

$$b_j(\underline{Q}) = \int_{V_j} \rho(\underline{r}') e^{-i\underline{Q}\cdot\underline{r}'} \, dV. \tag{8}$$

Thus, $b_j(\underline{Q})$ is essentially a Fourier transform of $\rho(\underline{r}')$, which describes the (scattering) mass distribution within the particle. Furthermore, $\underline{Q}$ is the conjugate parameter to $\underline{r}'$ and thus determines the length scale $\mathcal{L}$ studied as already discussed above (Equation 4).

To summarise, we have calculated the field scattered by an *extended* particle situated at $\underline{r}_j$ (Equation 7). This relation, $\underline{E}_j(\underline{Q}, t) \sim b_j(\underline{Q}) \exp\{-i\underline{Q}\cdot\underline{r}_j(t)\}$, is similar to the relation we obtained for a small volume element, $d\underline{E}(\underline{Q}, t) \sim \rho(\underline{r}) \exp\{-i\underline{Q}\cdot\underline{r}(t)\} \, dV$ (Equation 5), except that the scattering length of a small volume element, $\rho(\underline{r})dV$, is now replaced by the $\underline{Q}$-dependent 'scattering length' of an extended particle, $b_j(\underline{Q})$, with $b_j(\underline{Q})$ being the Fourier transform of $\rho(\underline{r}')$.

---

[2] Although this could be any point, the particle's centre of mass is a typical choice.

2.3   Scattering by an ensemble of particles

In the last section we derived the scattered field of an extended particle based on the scattering of many small volume elements. The contributions from all volume elements forming the particle were added to obtain the scattered field of the extended particle (Equation 7). Now, we go a step further and calculate the scattered field of an ensemble of particles based on the scattering of an individual particle. Nevertheless, a similar procedure is applied. The contributions from all particles in the scattering volume are added to obtain the scattered field of an ensemble of particles. In this case, we have to consider the *sum* of the discrete contributions of all particles, while before an *integral* over all the small volume elements resulted in the contribution of the whole (continuous) particle. Apart from this difference, the procedure is the same.

The total field $\underline{E}(Q, t)$ scattered by N particles is given by

$$
\begin{aligned}
\underline{E}(Q,t) &= \sum_{j=1}^{N} \underline{E}_j(Q,t) \sim \sum_{j=1}^{N} b_j(Q)\, e^{-i\underline{Q}\cdot\underline{r}_j(t)} \\
&= \sum_{j=1}^{N} \left(\int_{V_j} \rho(\underline{r}')e^{-i\underline{Q}\cdot\underline{r}'}\, dV \right) e^{-i\underline{Q}\cdot\underline{r}_j(t)},
\end{aligned}
\tag{9}
$$

where we used $b_j(Q)$ as given in Equation 8. This equation gives the scattered field $\underline{E}(Q, t)$ of an ensemble of particles. It illustrates that the contributions from *all* volume elements dV within particles in the scattering volume have to be added. (As mentioned earlier, we neglect the contribution from the solvent.) It is only for our convenience that we first add (integrate) the contributions within an individual particle and then add (sum) these contributions to obtain the scattering of the whole ensemble of particles. Although this is the appropriate procedure in the majority of experiments, the best choice depends on the specific sample and the desired information. In principle, the sample can be divided in small volume elements in any way. It can, for example, be advantageous to consider a domain of a particle instead of a whole particle and all the domains of one particle, i.e., an 'ensemble' of domains, instead of an ensemble of particles. Furthermore, rather than divide the sample into different subunits, one can integrate over it in one step.

2.4   Average scattered intensity

In the last section we derived the instantaneous scattered field $\underline{E}(Q, t)$ with particles at positions $\underline{r}_j(t)$ (Equation 9). Since static light-scattering experiments take a finite time, we calculate the average $\langle \underline{E}(Q) \rangle$ over a (macroscopic) time interval. As the particles execute Brownian motion, the positions $\underline{r}_j(t)$ change, and hence the phase angles $\delta\phi_j(t) = -Q\cdot\underline{r}_j(t)$ fluctuate randomly. (In the case of flexible particles, we also have to consider the time-dependence of $\underline{r}'(t)$, i.e., 'internal modes'.) Thus, the scattered fields $\underline{E}_j(Q,t) \sim b_j(Q) \exp\{-iQ \cdot \underline{r}_j(t)\}$ (Equation 7) correspond to steps forming a random walk in the two-dimensional complex plane. For large enough Q, $\delta\phi(t) = -\underline{Q}\cdot\underline{r}_j(t)$ can, in time, take on any value ranging over many times 2π. For the time-averaged scattered field $\langle \underline{E}(Q) \rangle$, we thus obtain the typical result for a random walk that is symmetrical about the origin; the mean is zero:

$$\langle \underline{E}(\underline{Q}) \rangle \sim \sum_{j=1}^{N} \left\langle b_j(\underline{Q}) \, e^{-i\underline{Q}\cdot\underline{r}_j(t)} \right\rangle = 0. \tag{10}$$

The average scattered field $\langle \underline{E}(\underline{Q}) \rangle$ therefore gives no information on the sample.

However, in most scattering experiments the average scattered intensity $\langle I(\underline{Q}) \rangle$ and not the average scattered field $\langle \underline{E}(\underline{Q}) \rangle$ is measured, for which we get

$$\begin{aligned} \langle I(\underline{Q}) \rangle &= \sqrt{\frac{\epsilon}{\mu}} \, \langle \underline{E}(\underline{Q}) \cdot \underline{E}^*(\underline{Q}) \rangle \\ &\sim \left\langle \sum_{j=1}^{N}\sum_{k=1}^{N} b_j(\underline{Q}) \, b_k(\underline{Q}) \, e^{-i\underline{Q}\cdot[\underline{r}_j(t)-\underline{r}_k(t)]} \right\rangle, \end{aligned} \tag{11}$$

with the dielectric constant ϵ and the permeability μ. If the particle properties and thus $b_j(\underline{Q})$ are not correlated with the particle positions $\underline{r}_j(t)$, the average can be separated and we obtain

$$\begin{aligned} \langle I(\underline{Q}) \rangle &\sim \sum_{j=1}^{N}\sum_{k=1}^{N} \langle b_j(\underline{Q}) \, b_k(\underline{Q}) \rangle \left\langle e^{-i\underline{Q}\cdot[\underline{r}_j(t)-\underline{r}_k(t)]} \right\rangle \\ &= \langle b^2(\underline{Q}) \rangle \sum_{j=1}^{N}\sum_{k=1}^{N} \left\langle e^{-i\underline{Q}\cdot[\underline{r}_j(t)-\underline{r}_k(t)]} \right\rangle \\ &= N\langle b^2(0) \rangle \frac{\langle b^2(\underline{Q}) \rangle}{\langle b^2(0) \rangle} \frac{1}{N} \sum_{j=1}^{N}\sum_{k=1}^{N} \left\langle e^{-i\underline{Q}\cdot[\underline{r}_j(t)-\underline{r}_k(t)]} \right\rangle \\ &= N\langle b^2(0) \rangle \, P(\underline{Q})S(\underline{Q}), \end{aligned} \tag{12}$$

where we have introduced the following (normalised) functions

$$P(\underline{Q}) = \frac{\langle b^2(\underline{Q}) \rangle}{\langle b^2(0) \rangle}, \tag{13}$$

$$S(\underline{Q}) = \frac{1}{N} \sum_{j=1}^{N}\sum_{k=1}^{N} \left\langle e^{-i\underline{Q}\cdot[\underline{r}_j(t)-\underline{r}_k(t)]} \right\rangle. \tag{14}$$

The contributions from intra-particle and inter-particles interferences have been separated and are accounted for by the form factor $P(\underline{Q})$ and structure factor $S(\underline{Q})$, respectively. This is, as mentioned, only possible if the particle positions $\underline{r}_j$ are not correlated with the particle properties, especially $b_j(\underline{Q})$. This is in particular the case for identical particles, i.e., $b_j(\underline{Q}) = b_k(\underline{Q}) = b(\underline{Q})$.

According to Equation 12, the average scattered intensity depends on three main factors. First, the form factor $P(\underline{Q})$; since $b(\underline{Q})$ contains an integral over the particle volume (Equation 8), it depends on intra-particle interferences and thus on the size, shape and structure of the individual particles. Second, the structure factor $S(\underline{Q})$ only depends on the particle positions $\underline{r}_j$ and not their individual structure. $S(\underline{Q})$ is therefore determined by inter-particle interferences and hence by the arrangement of the particles. Third, the prefactor (which is partially contained in the proportionality sign) determines the absolute level of the scattered intensity, in particular $I(Q = 0)$. In the following we will discuss these three contributions in turn.

2.5 Form factor

As detailed earlier, the form factor $P(Q) = \langle b^2(Q)\rangle/\langle b^2(0)\rangle$ (Equation 13) depends on intra-particle interferences and is thus determined by the size, shape and structure of the particles. It is normalised such that $P(Q\to 0)\to 1$. Furthermore, $b(Q)$ is essentially the Fourier transform of the scattering mass distribution $\rho(\underline{r}')$ within an individual particle (Equation 8). To illustrate the characteristics of the form factor $P(Q)$, we now discuss a few examples with relevance to biological samples. For an extensive collection of form factors, see Pedersen's article in Lindner and Zemb (2002).

2.5.1 Power law

Even if a particle has a very complex structure, it is sometimes possible to 'decompose' its structure into simpler structures, for example, cylinders, discs or spheres, on limited length scales $\mathcal{L}$. The different length scales $\mathcal{L}$ can individually be probed by an appropriate choice of the magnitude of the scattering vector $Q \approx 2\pi/\mathcal{L}$ (Equations 3 and 4). (For an example, see the polymer form factor, Section 2.5.3.) Even for the investigation of complex structures it is thus very useful to know the form factors of 'simple' objects, such as cylinders, discs or, more generally, fractals. For these objects, the mass M increases with a typical length L according to $M \sim L^{d_f}$, where d_f is the object dimension ($d_f = 1$ for a cylinder and $d_f = 2$ for a disc).

We have seen earlier that, loosely speaking, the intensity $I(Q)$ is proportional to the scattering material in a sphere of radius $1/Q$. If, for example, we consider a cylinder with cross-sectional area A and length L, then L inside a sphere of radius $1/Q$ increases as $L \sim 1/Q$ and we obtain $I(Q) \sim AL = A/Q \sim Q^{-1}$, i.e., we expect the scattered intensity of a cylinder to decay as $I(Q) \sim Q^{-1} = Q^{-d_f}$. Similarly, for a disc with thickness d and radius L, inside a sphere of radius $1/Q$ we have $L \sim 1/Q$ and thus $I(Q) \sim dL^2 = d/Q^2 \sim Q^{-2} = Q^{-d_f}$. More general, for objects with dimension d_f we get

$$P(Q) \sim L^{d_f} \sim Q^{-d_f}. \qquad (15)$$

It is important to note that, since real objects are of finite extent, this scaling will only hold over a limited range of length scales $\mathcal{L}$, corresponding to a limited Q range.

2.5.2 Homogeneous sphere

To calculate the form factor $P(Q)$ of a homogeneous sphere with radius R, we start with the definition of $b(Q)$ (Equation 8), consider that $\rho(\underline{r}') = \rho$ for a homogeneous sphere, use spherical coordinates for which $\mathrm{d}V = r'^2 \sin\vartheta \, \mathrm{d}\varphi \, \mathrm{d}\vartheta \, \mathrm{d}r'$, take $\underline{Q}$ to be along the z axis and write out the dot product in the exponent accordingly, and substitute

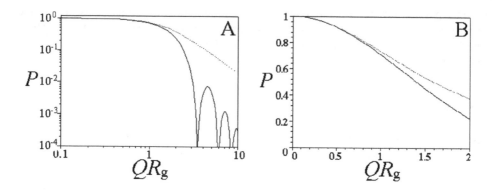

Figure 3. *Form factor $P(Q)$ of a sphere (Equation 17, solid line) and flexible polymer (Equation 23, dashed line) with the same radius of gyration R_g as a function of the dimensionless scattering vector QR_g. (A) log-log representation. (B) Linear representation for small QR_g.*

$u = Qr' \cos \vartheta$, so that

$$
\begin{aligned}
b(\underline{Q}) &= \int_V \rho(\underline{r}') e^{-i\underline{Q}\cdot\underline{r}'} dV = \rho \int_0^{2\pi} d\varphi \int_0^R \int_0^\pi e^{-iQr' \cos \vartheta} \, r'^{\,2} \sin \vartheta d\vartheta \, dr' \\
&= \rho \frac{4\pi R^3}{(QR)^3} \left[\sin (QR) - QR \cos (QR) \right].
\end{aligned}
\tag{16}
$$

From this we obtain

$$
P(\underline{Q}) = \frac{\langle b^2(\underline{Q}) \rangle}{\langle b^2(0) \rangle} = \left(\frac{3}{(QR)^3} \left[\sin (QR) - QR \cos (QR) \right] \right)^2.
\tag{17}
$$

The form factor $P(Q)$ only depends on the magnitude and not the direction of $\underline{Q}$ (Figure 3). Moreover, it is only a function of the dimensionless variable QR. For a first estimate of the radius R of a sphere, it is useful to know that the first minimum is located at $QR \approx 4.49$. Note that Figure 3 shows $P(Q)$ as a function of QR_g with the radius of gyration R_g of a sphere given by

$$
\begin{aligned}
R_g^2 &= \frac{1}{V} \int_V r'^{\,2} dV = \frac{1}{V} \int_0^{2\pi} d\varphi \int_0^R r'^{\,2} \int_0^\pi r'^{\,2} \sin \vartheta d\vartheta \, dr' \\
&= \frac{4\pi}{(4\pi/3)R^3} \int_0^R r'^{\,4} \, dr' = \frac{3}{5} R^2.
\end{aligned}
\tag{18}
$$

2.5.3 Polymer

The 'freely jointed chain' is the simplest model of a polymer. (For an introduction to polymers, see the chapter by Warren in this volume.) It describes the polymer as a linear chain of N beads separated by a bond or segment of length l. Adjacent bonds can take any angle, and thus the conformation of the polymer represents a random walk. If the polymer is flexible enough, l is much smaller than the wavelength λ (or $Ql \ll 1$), i.e., all volume elements of one segment scatter in phase. Then the N beads can be regarded as point scatterers at positions $\underline{r}'_j$ with $j = 1..N$. The scattered field of the whole polymer is calculated by adding the contributions of the N beads, similar to adding all the small volume elements that form an extended particle in Section 2.2 (Equations 7 and 8). However, here we add discrete beads and calculate the sum (as opposed to the integral) of the individual contributions. In this respect, the procedure is similar to the calculation of the scattered field of an ensemble of (discrete) particles (Equation 9), with the fact that here we deal with locations within an individual particle (indicated by a prime). Furthermore, to allow for flexibility, we introduce a time dependence:

$$b(\underline{Q}, t) \sim \sum_{j=1}^{N} e^{-i\underline{Q} \cdot \underline{r}'_j(t)}. \tag{19}$$

The corresponding intensity $\langle I(\underline{Q}, t) \rangle \sim \langle b_j(\underline{Q}, t) b_k^*(\underline{Q}, t) \rangle$ averaged over all configurations of the polymer is

$$\langle I(\underline{Q}, t) \rangle \sim \sum_{j=1}^{N} \sum_{k=1}^{N} \left\langle e^{-i\underline{Q} \cdot [\underline{r}'_j(t) - \underline{r}'_k(t)]} \right\rangle. \tag{20}$$

For a large polymer with $N \gg 1$, there will be many segments between beads j and k for most j and k. Thus, $\underline{r}'_j - \underline{r}'_k$ is a three-dimensional Gaussian variable with distribution $\mathcal{P}(\underline{r}'_j - \underline{r}'_k)$. The calculation of the average is similar to the calculation leading to the intermediate scattering function, which will be discussed later (Section 3.2). With steps corresponding to Equations 50 and 51, we obtain

$$\langle I(\underline{Q}) \rangle \sim \sum_{j=1}^{N} \sum_{k=1}^{N} e^{-Q^2 \langle |\underline{r}'_j - \underline{r}'_k|^2 \rangle / 6} = \sum_{j=1}^{N} \sum_{k=1}^{N} e^{-Q^2 |j-k| l^2 / 6}, \tag{21}$$

where we have used the statistics of a random walk, namely, $\langle |\underline{r}'_j - \underline{r}'_k|^2 \rangle = |j-k| l^2$. For

$Ql \ll 1$, we can transform to continuous variables:

$$\langle I(\underline{Q}) \rangle \sim \int_0^N dj \int_0^N dk \, e^{-Q^2|j-k|l^2/6} = 2 \int_0^N dj \int_0^j dk \, e^{-Q^2(j-k)l^2/6}$$

$$= \frac{2}{Q^2 l^2/6} \int_0^N dj \, e^{-Q^2 j l^2/6} \left(e^{Q^2 j l^2/6} - 1 \right)$$

$$= \frac{2}{(Q^2 l^2/6)^2} \left(\frac{Q^2 N l^2}{6} + e^{-Q^2 N l^2/6} - 1 \right)$$

$$= \frac{2N^2}{(Q^2 \langle R_g^2 \rangle)^2} \left(Q^2 \langle R_g^2 \rangle + e^{-Q^2 \langle R_g^2 \rangle} - 1 \right). \tag{22}$$

In the second step we considered j and k to be equivalent and twice we took $j \geq k$. Furthermore, the mean-square radius of gyration of a freely jointed chain is $\langle R_g^2 \rangle = Nl^2/6$.

To normalise $P(Q)$, we need to consider the limit $\langle I(Q \to 0) \rangle \sim N^2$, which indicates that the scattered light from all N beads adds in phase. The form factor $P(Q)$ of a Gaussian polymer, the 'Debye function' (Figure 3), is hence given by

$$P(Q) = \frac{\langle I(Q) \rangle}{\langle I(Q \to 0) \rangle} = \frac{2}{(Q^2 \langle R_g^2 \rangle)^2} \left(e^{-Q^2 \langle R_g^2 \rangle} + Q^2 \langle R_g^2 \rangle - 1 \right). \tag{23}$$

At very large Q, $P(Q)$ becomes

$$P(Q \to \infty) = \frac{2}{Q^2 \langle R_g^2 \rangle} \sim \frac{1}{Q^2}. \tag{24}$$

This is as expected (Equation 15): The radius of gyration of a polymer, $\langle R_g^2 \rangle = Nl^2/6$ implies that the polymer's mass $M \sim N \sim \langle R_g^2 \rangle$ and hence $d_f = 2$.

The preceding results are based on the model of a freely jointed chain. We now want to extend this to more realistic models. Rather than performing rigorous calculations, we apply the scaling concepts introduced above (Equation 15). The configuration of a freely jointed chain resembles a random walk of the segments. While a random walk can visit positions twice, this is not possible for a real polymer, which has a finite volume. Hence, space occupied by one segment is no longer accessible to other segments. This results in excluded volume interactions between individual segments and the configuration resembling a so-called 'self-avoiding random walk'. As a result, the polymer swells and the radius of gyration becomes $R_g \sim N^{3/5}$ and thus $M \sim R_g^{5/3}$. We thus expect $P(Q) \sim Q^{-5/3}$.

So far, we have considered a polymer with or without excluded volume interactions that is assumed to be completely flexible. Below a certain length, however, a real polymer shows some stiffness; it is only 'semi-flexible'. We can deal with this effect by regrouping N_K segments into a new segment. With an appropriate number N_K (which has to increases with stiffness), the conformations of the new chain with N/N_K segments of length $l_K = N_K l$ can again be described as a (self-avoiding) random walk. The Kuhn length l_K characterises the flexibility (or rather stiffness) of the polymer and

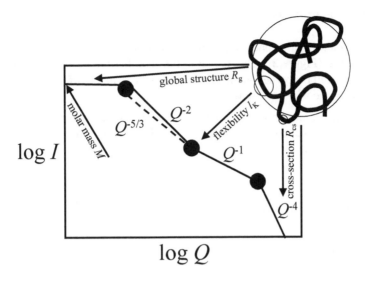

Figure 4. *Schematic representation of the scattered intensity $I(Q)$ of a polymer with the different asymptotic regimes, whose relations to the corresponding structural features in real space are indicated.*

corresponds to the length over which the polymer behaves as a rigid rod. We thus expect for $Q \lesssim 1/l_K$ the scattering behaviour typical for a flexible polymer with $I(Q) \sim Q^{-2}$ or $I(Q) \sim Q^{-5/3}$, and for $Q \gtrsim 1/l_K$ the scattering behaviour typical for a rigid rod with $I(Q) \sim Q^{-1}$. (Detailed calculations show that the transition region is actually located around $Q \approx 3.8/l_K$).

For even larger Q and thus smaller length scales, we expect the finite cross section to dominate. The local monomer structure typically leads to a much steeper decay of $I(Q)$ due to the cross-sectional scattering function.

In summary, the scattering of a polymer shows different asymptotic regimes with distinct scattering behaviour (Figure 4). They are due to the specific characteristics of a polymer on the corresponding length scales $\mathcal{L}$. Beginning at small Q and thus large length scales $\mathcal{L}$ (Equation 4), we observe the following sequence of behaviour. At $Q \to 0$ the intensity is proportional to the molar mass M of the polymer (Section 2.7); in the low-Q or the so-called Guinier regime (Section 2.5.4) with $Q \lesssim \langle R_g \rangle^{-1}$ the intensity depends on the overall size of the polymer and is independent of its structure; in the range $\langle R_g \rangle^{-1} \lesssim Q \lesssim l_K^{-1}$ there is an asymptotic regime with $I(Q) \sim Q^{-d_f}$, where $d_f = 2$ for a random walk (no excluded volume effects) and $d_f = 5/3$ for a self-avoiding random walk (excluded volume effects); in the range $l_K^{-1} \lesssim Q \lesssim R_{cs}^{-1}$ the slope is decreased with an asymptotic regime $I(Q) \sim Q^{-1}$ characteristic for rigid rods; for large $Q \gtrsim R_{cs}^{-1}$ we finally have a much steeper decay of the intensity due to the limited cross section with a typical length R_{cs}. However, often the length scales involved, i.e., R_g, l_K and R_{cs}, are not sufficiently separated and the transitions between regimes are fairly broad. An example can be found in Section 4.1.

2.5.4 Guinier approximation

The low-Q scattering of a sphere and a polymer of the same radius of gyration R_g are virtually identical (Figure 3). This is due to the fact that the low-Q scattering is determined by the overall properties and is insensitive to the structure of the object. This holds for any shape, as we will show now.

Any particle, and not only a polymer molecule, can be considered as being made up of N volume elements at positions $\underline{r}'_j$ (Section 2.5.3). The form factor $P(\underline{Q})$ (Equations 8 and 13) of a homogeneous particle $(\rho(\underline{r}') = \rho)$ then is

$$P(\underline{Q}) = \frac{\langle b^2(\underline{Q})\rangle}{\langle b^2(0)\rangle} = \frac{1}{N^2}\sum_{j=1}^{N}\sum_{k=1}^{N}\left\langle e^{-i\underline{Q}\cdot(\underline{r}'_j-\underline{r}'_k)}\right\rangle = \frac{1}{N^2}\sum_{j=1}^{N}\sum_{k=1}^{N}\frac{\sin Q\left|r'_j-r'_k\right|}{Q\left|r'_j-r'_k\right|}, \quad (25)$$

where the average is over all orientations:

$$\left\langle e^{-i\underline{Q}\cdot(\underline{r}'_j-\underline{r}'_k)}\right\rangle = \frac{1}{4\pi}\int_0^{2\pi}d\varphi\int_0^{\pi}e^{-iQ(r'_j-r'_k)\cos\vartheta}\sin\vartheta\,d\vartheta = \frac{\sin Q|r'_j-r'_k|}{Q|r'_j-r'_k|}. \quad (26)$$

The low-Q behaviour is obtained by expanding Equation 25 using $\sin\xi \approx \xi-\xi^3/3!$ and with the definition of the mean-square radius of gyration $\langle R_g^2\rangle=(1/2N^2)\sum\sum|\underline{r}'_j-\underline{r}'_k|^2$:

$$P(Q) = 1 - \frac{1}{3}Q^2\langle R_g^2\rangle + \mathcal{O}(Q^4). \quad (27)$$

This is known as the Guinier approximation. Independent of the shape of the object, it provides, based on the low-Q scattering, a well-defined measure of the size of the object: the mean-square radius of gyration $\langle R_g^2\rangle$. The average in $\langle R_g^2\rangle$ refers, for flexible objects, to the average over all configurations and, for polydisperse samples, to the intensity-weighted average over the distribution of sizes, shapes or any other (polydisperse) parameter.

2.6 Structure factor

The structure factor $S(\underline{Q})$ (Equation 14) depends on $\underline{r}_j-\underline{r}_k$, i.e., the relative particle positions. It thus contains information on the spatial arrangement of particles. The particle arrangement can be quantitatively described by the distribution function $g(\underline{r})$ with $\underline{r} = \underline{r}_j-\underline{r}_k$. Given a particle at the origin, $(N/V)g(\underline{r})d^3\underline{r}$ gives the number of particles in the volume element $d^3\underline{r}$ at position $\underline{r}$, where for large distances r, $g(r\rightarrow\infty) = 1$ and thus the bulk number density N/V is reproduced. We now want to derive a quantitative relation between $S(\underline{Q})$ and $g(\underline{r})$.

The sums in Equation 14 can be separated into terms concerning the same particle

$(j = k$, with N terms) and different particles $(j \neq k$, with $N^2 - N$ terms):

$$
\begin{aligned}
S(\underline{Q}) &= \frac{1}{N} \sum_{j=1}^{N} \sum_{k=1}^{N} \left\langle e^{-i\underline{Q}\cdot(\underline{r}_j - \underline{r}_k)} \right\rangle \\
&= \frac{1}{N} \left(\sum_{j=1}^{N} 1 + \sum_{j=1}^{N} \sum_{j \neq k=1}^{N} \left\langle e^{-i\underline{Q}\cdot\underline{r}} \right\rangle \right) \\
&= 1 + \frac{N}{V} \int_{V} (g(\underline{r}) - 1)\, e^{-i\underline{Q}\cdot\underline{r}}\, \mathrm{d}^3\underline{r},
\end{aligned}
\tag{28}
$$

where we have used $g(\underline{r})$ to calculate the average and transformed to a continuous variable, i.e., the sum into an integral. This equation again represents a Fourier transform: $(S(\underline{Q})-1)$ is essentially the Fourier transform of $(g(\underline{r})-1)$ and vice versa. This is analogous to the pair $b(\underline{Q})$ and $\rho(\underline{r})$ (Equation 8), which describes the distribution of small volume elements in an extended particle, while we now consider the distribution of particles in the scattering volume. Owing to the Fourier transform relationship between the structure factor $S(Q)$ and the radial distribution function $g(r)$, the positions, Q_{max} and r_{max} of the main peaks in $S(Q)$ and $g(r)$, respectively, satisfy $Q_{max}r_{max} \approx 2\pi$.

In the case of isotropic scattering, which is usually encountered with biomolecular solutions, we obtain (similar to Equation 26)

$$
S(Q) = 1 + 4\pi \frac{N}{V} \int (g(r) - 1)\, r^2\, \frac{\sin Qr}{Qr}\, dr.
\tag{29}
$$

A small Q expansion gives

$$
S(Q) = 1 - 2\frac{N}{V} B_2 + \frac{2}{3}\pi \frac{N}{V} Q^2 C_2 + \mathcal{O}(Q^4),
\tag{30}
$$

where $B_2 = -2\pi \int (g(r)-1) r^2 dr$ and $C_2 = -\int (g(r)-1) r^4 dr$. At low concentrations, $g(r) \approx \exp\{-V(r)/k_B T\}$, with $V(r)$ being the interaction potential of two particles, and B_2 reduces to the second virial coefficient. Furthermore, for very dilute suspensions, $N/V \to 0$ and thus $S(Q) \to 1$ (Equation 29), as expected. The same result is obtained by considering that, in dilute suspensions, the positions of different particles $(j \neq k)$ are uncorrelated and thus $\langle \exp\{-i\underline{Q}\cdot[\underline{r}_j(t) - \underline{r}_k(t)]\} \rangle = \langle \exp\{-i\underline{Q}\cdot\underline{r}_j(t)\} \rangle \langle \exp\{i\underline{Q}\cdot\underline{r}_k(t)\} \rangle$ and that these averages are zero (for the same reason as discussed for $\langle \underline{E}(\underline{Q}) \rangle$, Equation 10) and we are only left with the first term in Equations 28 and 29, i.e., 1, due to $j = k$.

2.7 Absolute scattered intensity

The typical result of a static scattering experiment is the time-averaged scattered intensity $\langle I(Q) \rangle$. Its absolute magnitude depends on exact experimental conditions. It is more convenient to consider a normalised quantity, which only depends on the sample and thus allows for an easier comparison of data obtained with different equipments. This requires us to take into account several factors: (i) The intensity of the incident beam $I_0 \sim |\underline{E}_0|^2$. (ii) The observed solid angle $\Omega = A/s^2$ (with the covered detector area

A and detector distance *s*), which represents a segment of the spherical scattered wave. (In contrast, the incident beam is assumed to be a plane wave, whose intensity does hence only depend on *A* and not *s*.) (iii) The scattering volume *V*, which determines the number of objects contributing to the scattering. The requirement to take these factors into account is partially fulfilled by the differential scattering cross section per unit solid angle, $d\sigma/d\Omega$, and, in its entirety, by the differential scattering cross section per unit solid angle and unit sample volume, $d\Sigma/d\Omega$:

$$\frac{d\Sigma}{d\Omega}(\underline{Q}) = \frac{1}{V}\frac{d\sigma}{d\Omega}(\underline{Q}) = \frac{1}{V}\frac{\langle I(\underline{Q})\rangle}{I_0}s^2. \tag{31}$$

$d\Sigma/d\Omega$ is the probability that a photon or neutron of the incident beam is scattered into the solid angle $d\Omega$. It is identical to the Rayleigh ratio, $\Delta\mathcal{R}$, which is commonly used in light scattering. With Equations 1 and 12 we obtain

$$\frac{d\Sigma}{d\Omega}(\underline{Q}) = \frac{N}{V}\langle b^2(0)\rangle P(\underline{Q}) S(\underline{Q}). \tag{32}$$

The *Q* dependence due to intra-particle and inter-particle interference effects, $P(\underline{Q})$ and $S(\underline{Q})$, respectively, was discussed earlier (Sections 2.5 and 2.6, respectively). The *Q*-independent 'prefactor', which determines the absolute level of the intensity, corresponds to $Q = 0$ and dilute suspensions where $P(0) = 1$ and $S(0) = 1$. We obtain $[d\Sigma/d\Omega](0) = (N/V)\langle b^2(0)\rangle$. Since *N* is the number of independent scatterers, each of which contributes the amplitude $b(0)$, $[d\Sigma/d\Omega](0)$ is essentially the mean-square displacement of a random walk in the complex plane with *N* steps of length $b(0)$. This result is hence not unexpected (and consistent with the fact that the mean displacement of a random walk, i.e., the scattered field $\langle\underline{E}(\underline{Q})\rangle$, is zero, Equation 10).

From Equation 8 we see that $b(0)$ is the scattering length of a whole particle:

$$b(0) = \int_V \rho(\underline{r}')\,dV. \tag{33}$$

Since we are concerned with objects suspended in a solvent, such as proteins, DNA or lipid assemblies in buffer, we are less interested in the scattering length density of a small volume element at position $\underline{r}'$ inside a particle, $\rho(\underline{r}')$, but rather in its scattering length density with respect to the scattering length density of the surrounding solvent, ρ_s, i.e., its scattering length density contrast $\Delta\rho(\underline{r}') = \rho(\underline{r}')-\rho_s$. For ρ_s constant throughout the scattering volume *V*, i.e., a 'featureless' solvent, and $V\gg\lambda$, this corresponds to an additional contribution to the scattering that is strongly peaked around $Q = 0$ and can be neglected. However, due to spontaneous density fluctuations in the solvent and, in the case of a solvent mixture, concentration fluctuations, $\rho_s(\underline{r}')$ depends on the position and is hence not constant. This leads to a background scattering, which in practice can be determined and subtracted but will be neglected here for simplicity. We will thus replace $\rho(\underline{r}')$ by $\Delta\rho(\underline{r}')$, which reflects the fact that we are interested in objects suspended in a

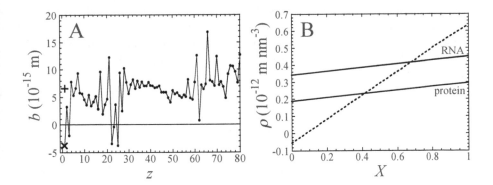

Figure 5. *(A) Dependence of the scattering length* b *on the atomic number z. All values are averages according to the natural occurrence of isotopes, except for hydrogen where* b *of* ^{1}H *(cross) and* ^{2}H *(deuterium D, plus) is indicated separately. (B) Dependence of the scattering length density* ρ *on the* D_2O *content, X, of water for a typical protein and RNA. The dotted line indicates the scattering length density for an* H_2O-D_2O *mixture of composition X.*

solvent. We thus get

$$
\frac{d\Sigma}{d\Omega}(\underline{Q}) = \frac{N}{V}\left(\int_V \Delta\rho(\underline{r}')\,dV\right)^2 P(\underline{Q})\,S(\underline{Q})
$$

$$
= \frac{1}{N_A}\bar{v}^2\Delta\rho^2\,c\,M\,P(\underline{Q})\,S(\underline{Q}) = K\,c\,M\,P(\underline{Q})\,S(\underline{Q}), \qquad (34)
$$

where N_A is Avogardo's number, M the molar mass of the object, $c = NM/VN_A$ its concentration (in mass/volume), $\bar{v} = VN_A/M$ its specific volume, $\Delta\rho$ its (average) scattering length density contrast and $K = (\bar{v}\Delta\rho)^2/N_A$ a factor which depends on the sample and kind of radiation but is independent of Q (Section 2.7.3).

Equation 34 is given in terms of the concentration c and molar mass M of the particles and thus reflects the fact that we are interested in objects suspended in a solvent. It also indicates that scattering allows us to determine the molar mass M from $[d\Sigma/d\Omega](Q{\rightarrow}0)$, if we can independently determine $\Delta\rho$.

We will now discuss the scattering length contrast density $\Delta\rho$, which depends on the scattering process and thus the interaction of the specific radiation (neutrons, x-rays or light) with matter. This is different from the Q dependence of $[d\Sigma/d\Omega](\underline{Q})$, which results from the interference of secondary waves and thus relies on the wave character of the applied radiation and is independent of the specific radiation.

2.7.1 Neutron scattering

Neutrons are scattered by the nuclei of atoms. Their scattering ability is characterised by the scattering length, which is identical to the $b(0)$ introduced earlier (Equation 33), except that it refers to a nucleus as opposed to a whole particle. Owing to this difference and for brevity we use the symbol $b = b(0)$ for the scattering length. The neutron scattering length b varies in an unsystematic way with the type of nucleus (Figure 5(A)). The scattering length density contrast $\Delta\rho$ is obtained from the scattering lengths b (similar to Equation 33) by

$$\Delta\rho = \frac{1}{v_p}\sum_j b_{p,j} - \frac{1}{v_s}b_s = \frac{1}{v_p}\sum_j \left(b_{p,j} - \frac{v_p}{v_s}b_s\right) \tag{35}$$

with b as the scattering length and v the volumes, with the subscripts referring to particle (p) and solvent (s), respectively. The index j includes all atoms in the volume v_p, which often is chosen to be the particle volume, but may, for example, also be a specific part of the particles only. While values for b are available (Sears 1992), values for the volumes, v_p and v_s, are usually inferred from the densities and molar masses.

Owing to the dependence of b on the properties of the nucleus, $\Delta\rho$ not only depends on the kind of atom but also on the specific isotope and the nuclear spin. This has several important consequences. First, there are two contributions to the scattering: coherent and incoherent scattering, which provide different information on the sample. Coherent scattering contains the entire structural information, in particular, on collective properties such as the spatial arrangement of the particles. In contrast, incoherent scattering is caused by the irregularity of the isotope distribution and contains no phase relation or interference terms. It provides no structural information, but because of the missing phase relation can be exploited to study the behaviour of individual particles, such as their self-diffusion. Second, the dependence on the specific isotopes allows for contrast variation, which is discussed in Section 2.7.4.

2.7.2   X-ray scattering

X-rays are scattered by the electrons in atoms, which, except for wavelengths near an electronic transition, scatter as if they were free. An electron has a scattering length $b_0 = 2.8 \times 10^{-15}$ m, the electron radius. Thus, an atom or molecule with z electrons has an x-ray scattering length $b = zb_0$; in contrast to neutrons, the x-ray scattering lengths of atoms follow a simple relationship, they are proportional to the atomic number z, and thus (Equation 35)

$$\Delta\rho = \frac{1}{v_p}\sum_j \left(b_{p,j} - \frac{v_p}{v_s}b_s\right) = \frac{b_0}{v_p}\sum_j \left(z_{p,j} - \frac{v_p}{v_s}z_s\right). \tag{36}$$

If the wavelength is close to the absorption edge of an atom, the scattering length b of this atom depends on the wavelength. This so-called anomalous scattering can be used to change the scattering length of a specific kind of atom and hence allows for contrast variation (Section 2.7.4). For biological samples, it is particularly interesting to change the contrast of light atoms, such as hydrogen, carbon, nitrogen or oxygen. Their absorption edges, however, are at very high wavelengths, which are not (yet) accessible.

2.7.3 Light scattering

As mentioned in Section 2.1, the scattering of light is caused by differences in the dielectric properties or, equivalently, the refractive indices of the objects and the solvent. With ϵ_p and ϵ_s the (average) dielectric constants of the particle and solvent, respectively, and the relationship $\epsilon = n^2$, we obtain

$$\Delta\rho = \frac{k^2}{4\pi}\frac{\epsilon_p - \epsilon_s}{\epsilon} = \frac{k^2}{4\pi}\frac{n_p^2 - n_s^2}{n^2} \approx \frac{k^2}{2\pi n}(n_p - n_s), \tag{37}$$

where ϵ and n are the average values for the whole suspension. The approximation is valid for small differences between n_p and n_s. The prefactor $k^2/4\pi$ can be obtained by solving Maxwell's equations. It indicates a dependence on the magnitude of the wave vector $k = 2\pi/\lambda$ or, equivalently, the wavelength λ. Scattering is stronger for shorter wavelength; blue light is scattered more than red light. A consequence is the blue colour of the sky and the red sun observed during sunsets, which are due to the importance of scattered and unscattered light, respectively.

Often, the refractive index of the suspended particles, n_p, is not known. However, the dependence of the refractive index of the suspension, n, on particle concentration, c, can be measured. This yields the refractive index increment dn/dc. Assuming additivity of refractive indices by volume, namely, $n = c\bar{v}n_p + (1 - c\bar{v})n_s$, we obtain for the refractive index increment $dn/dc = (n_p - n_s)\bar{v}$. (For proteins, typically $dn/dc \approx 0.2$ ml/g.) Equation 37 hence becomes

$$\Delta\rho \approx \frac{k^2}{2\pi\bar{v}n}\frac{dn}{dc} = \frac{2\pi n}{\bar{v}\lambda_0^2}\frac{dn}{dc}. \tag{38}$$

This results in a constant K, as defined in Equation 34:

$$K = \frac{(\bar{v}\Delta\rho)^2}{N_A} = \frac{4\pi^2 n^2}{N_A \lambda_0^4}\left(\frac{dn}{dc}\right)^2. \tag{39}$$

2.7.4 Contrast variation

Contrast variation can be applied to highlight different parts of the sample, for example, domains in composite particles or a single species in a mixture. It is thus particularly useful for complex samples. It involves changing the contrast of the part of the sample that is of interest, which can be achieved by changing the scattering length either of that part or the solvent. It is usually much easier to change the scattering length of the solvent, but highlighting different parts of a particle might require modification of the scattering length of a part of the particle. It is thus necessary to be able to vary the scattering length of a specific kind of atom; the methods available to achieve this depend on the radiation used. (A pictorial illustration of contrast variation can be found in Paul and Thomas (1989).)

In neutron scattering, the scattering length b depends on the kind of nucleus, and thus $\Delta\rho$ is not only determined by the kind of atom but on the specific isotope (Section 2.7.1). The scattering length of parts of the sample can thus be changed by substituting one isotope with another isotope of the same kind of atom. It is assumed that this does not

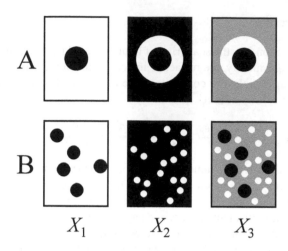

Figure 6. *Schematic illustration of contrast variation for different isotopic compositions of the solvent, X_1, X_2 and X_3. Solvent contrast variation is used to highlight (A) different parts in an inhomogeneous particle or (B) different species of particles in a mixture.*

change the properties, in particular the structure, of the sample. (This assumption can be validated by using x-ray scattering, which is not sensitive to isotope substitution.) In x-ray scattering, anomalous scattering (Section 2.7.2) can be used to change the scattering length and thus the contrast of a specific kind of atom, even without interfering with the sample. However, the absorption edges of light atoms, which are particularly important for biological samples, are not accessible to date. Also, with light scattering, only limited contrast variation is possible; changing the refractive index usually significantly changes the properties of the sample. Therefore, here we focus on contrast variation using neutron scattering, but the general principles are also valid for x-ray or light scattering.

Since hydrogen is ubiquitous in most biomolecules and solvents, it is fortunate that hydrogen shows a strong dependence of the neutron scattering length b on the specific isotope; $b_H = -3.74{\times}10^{-15}$ m for a proton vs. $b_D = 6.67{\times}10^{-15}$ m for deuterium. Thus, a variation of the H_2O-D_2O ratio of water is a convenient way to change the contrast. Figure 5(B) shows the scattering length density $\rho(X)$ of biomolecules as a function of the D_2O content of water, X. The lines are usually not horizontal because some hydrogens of the biomolecules, mainly acidic hydrogens accessible to the solvent, exchange with deuterium in the solvent. The intersection of these lines with the water line, which represents the scattering length density $\rho_w(X) = b_w(X)/v_w = [b_H + X(b_D - b_H)]/v_w$ of the particular H_2O-D_2O mixture (dotted line), indicates the so-called 'match point' ($v_w = 0.03$ nm³ is the volume of a water molecule). In a H_2O-D_2O mixture of this composition, the biomolecules are 'invisible'. For example, the match point of proteins is about 40% D_2O and for nucleic acids about 70% D_2O (Figure 5(B)).

For an inhomogeneous particle, for example, a virus with a 'core' of nucleic acid and a 'shell' of protein (Section 4.3), solvent contrast variation can reveal details of the two parts. In about 40% D_2O (composition X_1 in Figure 6(A)) the protein is matched and the intensity reflects the size and shape of the nucleic acid, while at about 70% D_2O (composition X_2 in Figure 6(A)) the nucleic acid is invisible and the scattering is only due to the protein. All other solvent compositions (for example, composition X_3 in Figure 6(A)) lead to scattering, which is, to varying degrees, determined by both. A similar procedure can be applied to study the arrangement of different particles (Figure 6(B)). For example, in Section 4.4 we briefly discuss the distribution of ions around DNA. With solvent contrast variation we can determine the arrangement of the DNA (with the ions matched), the ions (with the DNA matched) and their relative positions (from other compositions).

Solvent contrast variation can be used to obtain information on the structure of inhomogeneous particles or the arrangement of different particles. However, to study a part of a homogeneous particle or a subset of particles, which are all made of the same material, one needs to selectively deuterate small domains or a particular species, for example, one of two populations of particles. With the selectively deuterated samples, the same procedure as described in the preceding text can then be followed.

2.8 Data analysis

In a first step, the raw data has to be reduced, i.e., all the instrument-specific contributions have to be accounted for. This depends on the specific experiment, but typically includes correcting for the scattering from the sample cell and solvent (Section 2.7), the sample transmission and the electronic background as well as normalising to a reference to obtain data on an absolute scale. This procedure results in data that depend only on the sample. They can thus easily be shared or compared with other data, and reduced data are typically the form in which scattering data are published.

Once raw data is reduced, the reduced data needs to be analysed and interpreted. The extraction of the particle structure or arrangement from the reduced data, the 'inverse scattering problem', is not unambiguous. In contrast, the calculation of the scattered intensity for a given model structure, the 'scattering problem', is unambiguous and is done along the lines of the derivations presented in this chapter. However, this requires the assumption of a model. There are thus two main methods to obtain structural information from scattering data: Either the scattering problem is solved based on a structural model ('model-fitting' approach), or the ill-posed inverse scattering problem is attempted with the reduced data as starting point ('model-independent' approach). Both methods have their advantages and shortcomings, which we will briefly outline in the following. Which method is better suited also depends on the available *a priori* information and the complexity of the sample, for example, whether the particles are polydisperse in size, shape or interactions, whether the particles are homogeneous, or whether strong interactions are present between the particles. However, due to the complementarity of the approaches, a combination of both methods is often desirable.

In the model-fitting approach, one starts with a 'guess' or model of the particle structure and arrangement, which might be based on information that is independently available. For this model, the scattering problem is then solved and the theoretical model function compared to the experimentally determined scattered intensity. This can lead to modifications to the model or the determination of model parameters. After several iterations, one might arrive at a model that describes the data adequately. However, with this approach, one can only decide whether a specific model agrees or disagrees with the data, leaving open the possibility that other models fit the data equally well. This approach is thus prone to controversies because, in principle, all possible models have to be tested.

The 'model-independent' method starts with the (reduced) data. The measured data essentially represent a Fourier transform of the scattering mass distribution or density profile within the particle (Equations 8 and 28). Using a Fourier transformation, we can obtain the pair distance distribution function, which provides information on the probability to find scattering material within an object in a certain distance. This function not only provides real-space information but also suggestions for the particle structure, because it has characteristic shapes for some structural models, such as spheres, cylinders or discs. The pair distance distribution function can further be deconvolved for some structural models, in particular spherical geometries, to obtain the density profile. The Fourier transformation and deconvolution are, however, not always unambiguous. This is due to the limited number of data points and the limited Q range covered. The basic idea of the indirect Fourier transformation (IFT) is to perform calculations and a corresponding fit from real space into Q space, thus creating an optimised function system. Hence, this method does not really start in Q space (it is an *indirect* Fourier transform) and is therefore, strictly-speaking, also a model-fitting approach.

Light, x-ray and neutron scattering are indirect techniques for the determination of size, shape and structure of biomolecules. Ideally they are combined with direct methods, such as optical or electron microscopy, which can provide real-space information. These direct techniques, however, have also disadvantages; they are often prone to preparation artifacts and suffer from poor statistics. In contrast, scattering techniques allow us to study biomolecules in solution, i.e., without any special sample preparation, and usually have excellent statistics owing to the large number of particles in the scattering volume. A combination of these complementary methods is thus a very powerful tool to investigate biomolecules.

3   Dynamic scattering

Above we have argued that the interference between the scattered waves of two particles depends on the particles' relative positions, $\underline{r}_j(t) - \underline{r}_k(t)$ (Equation 12). If the particles (or parts of the particles) move, their relative positions and thus their interference will continuously change. The scattered intensity hence fluctuates in time with the typical timescale of these fluctuations depending on the particles' motion. In Section 2 we derived the total field $\underline{E}(Q, t)$ scattered by an ensemble of particles (Equation 9) and subsequently calculated the time-averaged scattered intensity (Equation 12); information contained in the fluctuations of the scattered intensity was thus disregarded. Now we will discuss *dynamic* scattering experiments and hence return to Equation 9 and study the fluctuations. An analysis of the intensity fluctuations provides information on the particles' dynamics, for example, their Brownian motion; the faster the intensity fluctuates, the faster the particles are expected to move.

To observe these fluctuations, the scattering from different particles needs to be coherent, i.e., the incident radiation has to be in phase over the scattering volume. Dynamic light scattering (DLS; or photon correlation spectroscopy, PCS) is hence performed using lasers. It is one of the most important tools to study the dynamics of soft condensed matter, including the motion of biomolecules in solution or the motion of parts of biomolecules relative to their centre of mass ('internal modes'). DLS is also frequently used to determine the size of particles.

Owing to the high intensity of modern synchrotron x-ray beams, it has become possible to obtain sufficiently intense, *coherent* x-ray beams by using spatial and frequency filtering. This allows for the x-ray analogue of dynamic light scattering: x-ray photon correlation spectroscopy (Thurn-Albrecht *et al.* 1996). In principle, neutron beams could be filtered similarly, but current neutron sources do not provide a high enough flux to obtain a reasonably intense beam of coherent neutrons. However, to obtain dynamic information with neutrons, one can use so-called inelastic neutron-scattering instruments, such as neutron spin-echo, time-of-flight or backscattering spectrometers. We will not cover these techniques in this chapter, but refer, for example, to Zorn's article in Lindner and Zemb (2002) or the book by Higgins and Benoit (1994).

3.1 Correlation functions

With the characteristic timescale of fluctuations, τ_c, the intensities at times t_1 and t_2 are expected to be 'similar' for short delay times, $\tau = t_2 - t_1 \ll \tau_c$, and the intensities $I(Q, t_1)$ and $I(Q, t_2)$ are uncorrelated, i.e., we cannot know whether they are similar or not, for very long delay times, $\tau \gg \tau_c$. The 'similarity' of intensities $I(Q, t)$ and $I(Q, t+\tau)$ for different delay times τ is quantified by the time correlation function. The time correlation function of the intensity, $G^{(2)}$, is defined as the product of the intensities at times t_1 and t_2; $G^{(2)}(Q, t_1, t_2) = I(Q, t_1)I(Q, t_2)$. If the fluctuations are independent of the start time, then they are called stationary and $G^{(2)}$ only depends on the delay time $\tau = t_2 - t_1$ (usually one arbitrarily chooses $t_1 = 0$). Furthermore, to get sufficiently good statistics, we have to average $G^{(2)}$ over many configurations of the sample. If the sample assumes many different configurations in time, this average can be replaced by the time average; $G^{(2)}(Q, \tau) = \langle I(Q, 0)I(Q, \tau)\rangle$. To avoid a dependence on the incident intensity (which depends on the specific equipment used), the *normalised* time correlation function of the intensity, $g^{(2)}(Q, \tau)$, is used:

$$g^{(2)}(Q, \tau) = \frac{\langle I(Q, 0)I(Q, \tau)\rangle}{\langle I(Q)\rangle^2}. \tag{40}$$

(Note that the symbol for the normalised time correlation function, $g^{(2)}$, always has superscript '(2)', while the symbol for the (unrelated) distribution function g (Section 2.6) does not have a superscript.) The time correlation function $g^{(2)}(Q, \tau)$ hence compares the delayed intensity $I(Q, \tau)$ with the 'initial' intensity $I(Q, 0)$ with the independent variables being the scattering vector Q and the delay time τ. It can be used to measure the characteristic time τ_c of fluctuations in a random process and how the random signal becomes uncorrelated and thus how it 'loses memory'.

For very long delay times $\tau \to \infty$, the intensities $I(Q, 0)$ and $I(Q, \tau)$ will be independent (or uncorrelated) and thus

$$\lim_{\tau \to \infty} g^{(2)}(Q, \tau) = \lim_{\tau \to \infty} \frac{\langle I(Q, 0)I(Q, \tau)\rangle}{\langle I(Q)\rangle^2} = \lim_{\tau \to \infty} \frac{\langle I(Q, 0)\rangle\,\langle I(Q, \tau)\rangle}{\langle I(Q)\rangle^2} = 1, \tag{41}$$

where for a stationary signal $\langle I(\underline{Q})\rangle = \langle I(\underline{Q}, 0)\rangle = \langle I(\underline{Q}, \tau)\rangle$.

For very short delay times $\tau \to 0$, we obtain

$$\lim_{\tau \to 0} g^{(2)}(\underline{Q}, \tau) = \lim_{\tau \to 0} \frac{\langle I(\underline{Q}, 0) I(\underline{Q}, \tau)\rangle}{\langle I(\underline{Q})\rangle^2} = \frac{\langle I(\underline{Q}, 0)^2\rangle}{\langle I(\underline{Q})\rangle^2} = 2. \tag{42}$$

The last step, i.e., that this limit is equal to 2, follows from the following argument (Equation 9):

$$\langle I^2(\underline{Q}, t)\rangle \quad \sim \quad \langle |\underline{E}(\underline{Q}, t)|^2 |\underline{E}(\underline{Q}, t)|^2\rangle$$

$$\sim \quad \sum_{j=1}^{N} \sum_{k=1}^{N} \sum_{l=1}^{N} \sum_{m=1}^{N} \langle e^{-i\underline{Q}\cdot[\underline{r}_j(t)-\underline{r}_k(t)+\underline{r}_l(t)-\underline{r}_m(t)]}\rangle. \tag{43}$$

If one of j, k, l and m is different from all the others, it factors out in the average. Since the average over the contribution of a single particle is zero (this is a random walk as in the calculation of $\langle \underline{E}(\underline{Q})\rangle$, Equation 10), it will not contribute. For example, if j is independent we obtain

$$\langle e^{-i\underline{Q}\cdot\underline{r}_j(t)}\rangle \langle e^{-i\underline{Q}\cdot(-\underline{r}_k(t)+\underline{r}_l(t)-\underline{r}_m(t))}\rangle = 0. \tag{44}$$

We therefore only need to consider pairs of j, k, l and m that are equal. Furthermore, if, e.g., $j=l\neq k=m$ we obtain a factor $\langle \exp\{-i\underline{Q}\cdot[\underline{r}_j(t)+\underline{r}_l(t)]\}\rangle = \langle \exp\{-i\underline{Q}\cdot 2\underline{r}_j(t)\}\rangle = 0$. Non-vanishing contributions to the quadruple sum thus only come from

$$j = k = l = m: \quad \sum_{j=1}^{N} \langle 1\rangle = N$$

$$j = k \neq l = m: \quad \sum_{j=1}^{N} \sum_{j\neq l=1}^{N} \langle 1\rangle\langle 1\rangle = N^2 - N$$

$$j = m \neq k = l: \quad \sum_{j=1}^{N} \sum_{j\neq k=1}^{N} \langle 1\rangle\langle 1\rangle = N^2 - N, \tag{45}$$

and thus for $N \to \infty$:

$$\langle I^2(\underline{Q})\rangle \sim 2N^2 - N \approx 2N^2. \tag{46}$$

Using $\langle I(\underline{Q})\rangle \sim N$ (Equation 12) we finally get $\langle I^2(\underline{Q})\rangle = 2\langle I(\underline{Q})\rangle^2$ as used in Equation 42.

In summary, the motion of particles in solution leads to continuously changing relative particle positions and thus varying interference between the scattered waves. This results in fluctuations in the scattered intensity (Figure 7(A)), which are analysed using the normalised time correlation function $g^{(2)}(\tau)$. With increasing delay time, this function decays from 2 to 1 with a characteristic time τ_c (Figure 7(B)). In the following section we will deduce a relation between $g^{(2)}$ and the underlying particle motion for the case of particles undergoing Brownian motion. This will allow us to analyse quantitatively the fluctuations in the scattered intensity to obtain the diffusion coefficient D of the particles.

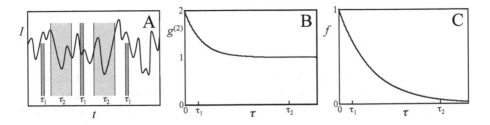

Figure 7. *(A) Schematic representation of a typical time dependence of the scattered intensity $I(t)$. Two different delay times ($\tau_1 < \tau_2$) are indicated as examples. (B and C) Schematic representations of a typical normalised time correlation function $g^{(2)}(\tau)$ (B) and intermediate scattering function $f(\tau)$ (C) as a function of delay time τ. Note that the horizontal axes in (B) and (C) are stretched by a factor of about 5 compared to the horizontal axis in (A).*

3.2 Correlation function of Brownian particles

We now proceed to calculate the normalised time correlation function $g^{(2)}(\underline{Q}, \tau)$ (Equation 40) for particles undergoing Brownian motion. We have already derived the normalisation factor $\langle I(\underline{Q}) \rangle^2 \sim N^2$ (Equation 12) and will thus first concentrate on the time correlation function $G^{(2)}(\underline{Q}, \tau)$:

$$
\begin{aligned}
G^{(2)}(\underline{Q}, \tau) &= \langle I(\underline{Q}, 0) I(\underline{Q}, \tau) \rangle \\
&\sim \sum_{j=1}^{N} \sum_{k=1}^{N} \sum_{l=1}^{N} \sum_{m=1}^{N} \left\langle e^{-i\underline{Q} \cdot [\underline{r}_j(0) - \underline{r}_k(0) + \underline{r}_l(\tau) - \underline{r}_m(\tau)]} \right\rangle.
\end{aligned} \tag{47}
$$

We assume that the particles are independent. Then, as in the preceding text (Equation 43), considering only pairs of j, k, l and m that are equal, we obtain

$$
j = k = l = m : \quad \sum_{j=1}^{N} \langle 1 \rangle = N
$$

$$
j = k \neq l = m : \quad \sum_{j=1}^{N} \sum_{j \neq l = 1}^{N} \langle 1 \rangle \langle 1 \rangle = N^2 - N
$$

$$
\begin{aligned}
j = m \neq k = l : \quad &\sum_{j=1}^{N} \left\langle e^{-i\underline{Q} \cdot [\underline{r}_j(0) - \underline{r}_j(\tau)]} \right\rangle \sum_{j \neq k = 1}^{N} \left\langle e^{-i\underline{Q} \cdot [\underline{r}_k(0) - \underline{r}_k(\tau)]} \right\rangle \\
&= N^2 \left| \left\langle e^{-i\underline{Q} \cdot [\underline{r}_j(0) - \underline{r}_j(\tau)]} \right\rangle \right|^2,
\end{aligned} \tag{48}
$$

where the last step follows from the fact that identical particles all have the same average behaviour. Hence, collecting all contributions and using $\langle I(\underline{Q}) \rangle^2 \sim N^2$ for the normalisation (Equation 12), we obtain

$$g^{(2)}(Q,\tau) = 1 + \left| \left\langle e^{-i\underline{Q}\cdot[\underline{r}(0)-\underline{r}(\tau)]} \right\rangle \right|^2 = 1 + f^2(Q,\tau), \tag{49}$$

with the so-called intermediate scattering function $f(Q,\tau)$ (Figure 7(C)), which we now calculate for independent, monodisperse spheres undergoing Brownian motion. The exponent depends on the (random) displacement of the particle from an arbitrary initial point, $\Delta\underline{r}(\tau) = \underline{r}(\tau) - \underline{r}(0)$. For the random walk of Brownian particles, the distribution $\mathcal{P}(\Delta\underline{r})$ is Gaussian with

$$\mathcal{P}(\Delta\underline{r}) = \left(\frac{3}{2\pi\langle\Delta\underline{r}^2\rangle} \right)^{3/2} e^{-\frac{3(\Delta\underline{r})^2}{2\langle\Delta\underline{r}^2\rangle}}. \tag{50}$$

The average $\langle\exp\{iQ\cdot\Delta\underline{r}(\tau)\}\rangle$ (Equation 49) has to be taken over this distribution of displacements:

$$
\begin{aligned}
f(Q,\tau) &= \left\langle e^{i\underline{Q}\cdot\Delta\underline{r}(\tau)} \right\rangle = \int e^{i\underline{Q}\cdot\Delta\underline{r}}\, \mathcal{P}(\Delta\underline{r})\, d^3\Delta\underline{r} \\
&= \left(\frac{3}{2\pi\langle\Delta\underline{r}^2\rangle} \right)^{3/2} \int e^{i\underline{Q}\cdot\Delta\underline{r}}\, e^{-\frac{3(\Delta\underline{r})^2}{2\langle\Delta\underline{r}^2\rangle}}\, d^3\Delta\underline{r} \\
&= \left(\frac{1}{\pi\alpha} \right)^{3/2} \int e^{i\underline{Q}\cdot\Delta\underline{r}}\, e^{-(\Delta\underline{r})^2/\alpha}\, d^3\Delta\underline{r} \\
&= \left(\frac{1}{\pi\alpha} \right)^{3/2} \int e^{-(\Delta\underline{r}-i\alpha\underline{Q}/2)^2/\alpha\, -\, \alpha\underline{Q}^2/4}\, d^3\Delta\underline{r} \\
&= \left[\left(\frac{1}{\pi\alpha} \right)^{3/2} \int e^{-\underline{u}^2/\alpha}\, d^3\underline{u} \right] \times e^{-\alpha Q^2/4} \\
&= e^{-\alpha Q^2/4} = e^{-\langle\Delta r^2\rangle Q^2/6} = e^{-DQ^2\tau},
\end{aligned}
\tag{51}
$$

where the last steps involve setting $\alpha = 2\langle\Delta\underline{r}^2\rangle/3$, completing the square, making a linear change in variable $\underline{u} = \Delta\underline{r} - i\alpha\underline{Q}/2$ with $d^3\underline{u} = d^3\Delta\underline{r}$, realising that the term in squared brackets represents the integral of a normalised three-dimensional Gaussian and is thus unity, and, finally, using the fact that for a particle undergoing Brownian motion $\langle\Delta\underline{r}^2\rangle = 6D\tau$. We thus finally get (Equations 49 and 51)

$$g^{(2)}(Q,\tau) = 1 + f^2(Q,\tau) = 1 + \left(e^{-DQ^2\tau} \right)^2. \tag{52}$$

The correlation function thus decays with a time constant $\tau_c = 1/DQ^2$, the characteristic time of the fluctuations. With Equation 3 we obtain

$$\tau_c = \frac{1}{DQ^2} = \frac{\lambda^2}{D(4\pi\sin(\theta/2))^2} \approx \frac{\lambda^2}{D}. \tag{53}$$

This is the time taken by a particle to explore a distance of about the wavelength λ (note that $\langle\Delta r^2\rangle \approx D\tau$) and therefore to change the phase of the scattered light by about 2π. The characteristic time τ_c is usually microseconds to milliseconds. Depending on the equipment, typically delay times τ with 10^{-8} s $\leq \tau \leq 1000$ s are accessible. A log-linear plot of the measured intermediate scattering function $f(Q,\tau) = \exp\{-DQ^2\tau\}$ vs. delay time τ should result in a straight line with slope $-DQ^2$. Even more appropriate is a log-linear plot of $f(Q,\tau)$ vs. $Q^2\tau$, in which the results from different Q should superimpose into a single line with slope D (for an example, see Figure 12).

3.3 Data analysis

In a typical experiment the time correlation function $G^{(2)}(Q,\tau)$ is measured and $f(Q,\tau)$ deduced (Equations 40, 47 and 52). The intermediate scattering function $f(Q,\tau)$ is fitted to give the characteristic time $\tau_c = 1/DQ^2$ (if appropriate) from which (using Equation 3), the particles' diffusion coefficient D can be obtained. The hydrodynamic radius R_h can then be determined using the Stokes-Einstein relation

$$D = \frac{k_B T}{6\pi\eta R_h},$$ (54)

with Boltzmann's constant k_B, temperature T and viscosity η.

In the case of monodisperse spherical particles with radii R, the hydrodynamic radius R_h is identical to the radius, $R = R_h$. In the case of non-spherical particles, the dependence of R_h on the particle parameters is more complicated but is known for a number of different shapes (for an example, see Section 4.5). This method thus allows us to determine the size of particles.

For samples containing polydisperse particles or flexible particles that undergo centre-of-mass and, simultaneously, internal motions, the measured intermediate scattering function $\langle f(Q,\tau)\rangle$ represents an average over individual intermediate scattering functions, which are weighted by the corresponding scattered intensity of the particle, i.e., the intensity-weighted distribution of diffusion coefficients, $P(D)$:

$$\langle f(Q,\tau)\rangle = \int f(Q,\tau)\, P(D)\, \mathrm{d}D = \int e^{-DQ^2\tau}\, P(D)\, \mathrm{d}D.$$ (55)

The average intermediate scattering function $\langle f(Q,\tau)\rangle$ thus represents a Laplace transform of $P(D)$. An inverse Laplace transform of $\langle f(Q,\tau)\rangle$ could thus be expected to yield $P(D)$. However, this is a so-called 'ill-conditioned' problem; even a small statistical or systematic error in the measurement translates into a large uncertainty in $P(D)$. Nevertheless, there are methods and programmes, for example, CONTIN, which were developed to extract the maximum possible information on $P(D)$ from $\langle f(Q,\tau)\rangle$. Furthermore, for particles with a relatively narrow size distribution, the method of cumulants can provide an indication on the width of the distribution. It treats polydispersity as a perturbation to monodisperse single-exponential behaviour and determines the moments (or cumulants) of $P(D)$. Often it is sufficient to consider the first two terms in the expansion: The mean $\langle D\rangle$ and the square of the relative standard deviation, $\sigma_D^2 = (\langle D^2\rangle - \langle D\rangle^2)/\langle D\rangle^2$. They are related to the measured $\langle f(Q,\tau)\rangle$ by

$$\ln\langle f(Q,\tau)\rangle = -\langle D\rangle Q^2\tau + \frac{\sigma_D^2}{2}\left(\langle D\rangle Q^2\tau\right)^2.$$ (56)

4   Examples

We now present a few biologically relevant examples to illustrate the concepts explained in the previous sections. The first two examples are concerned with different techniques to analyse and interpret static data: First, we present data analysis based on model fitting, i.e., the calculation of the scattering function for an (assumed) model (Sections 4.1 and 4.2). In the second example, a 'model-independent' approach is used, which uses the indirect Fourier transform (IFT), i.e., it starts with the experimentally determined $I(Q)$ and converts this Q space information into real space (Section 4.2). Then, we describe how the structure of a virus can be determined; this is a classical example of contrast variation (Section 4.3). Having determined the structures of different particles, we are subsequently concerned with the structure of solutions, in particular, the arrangement of highly charged DNA in aqueous salt solutions (Section 4.4). Finally, the last example deals with the use of dynamic light scattering to determine the shape of a protein (Section 4.5).

4.1   Gluten — data analysis (model fitting)

This example illustrates how model fitting can be used to obtain detailed, quantitative information on the solution structure of biomolecules. Here we are interested in the structure of the protein gluten, which consists of three domains: a central, elongated domain flanked by two small, globular domains. (See Poon's chapter in this volume for an introduction to protein structure.) Wheat gluten proteins are of considerable interest due to their functionality in bread. They form extensive, insoluble protein networks in dough, which are stabilised by intermolecular disulfide bonds. These networks contribute to the biomechanical properties, such as strength and elasticity. The precise molecular basis for the elastic properties is, however, still under investigation. It is suggested that the central domain contributes to the elastic behaviour. In this example we present small-angle neutron-scattering experiments on the solution structure of two proteins, dB4 and dB1, which represent the whole central domain (dB4) and about a quarter of the central domain (dB1). They have molar masses of 63.8 kDa (dB4) and 16.9 kDa (dB1), respectively (Egelhaaf *et al.* 2003). Prior to this study, they were thought to have a rodlike structure.

Visual inspection of the scattered intensity $I(Q)$ shows distinctive behaviour in the different Q regions (Figure 8(A)). The intercept, $I(Q \to 0)$, is proportional to the molar mass and indicates a ratio of about 4 between the two proteins. At small Q the intensity decays slowly. This is characteristic for the Guinier regime, which is associated with the overall size of the protein. A Guinier fit (Equation 27) gives radii of gyration, R_g, of 6.7 nm and 2.8 nm, respectively. At about $Q \sim 1/R_g$ the scattered intensity crosses over to a steep decay following a power law $I(Q) \sim Q^{-5/3}$ in the intermediate Q range (Figure 8(A), dashed line). An exponent of $-5/3$ is characteristic for polymers in a good solvent, where good solvent conditions imply excluded volume effects (Section 2.5.3). It is important to note that this 'polymer' does not necessarily correspond to the protein backbone but is a more general model of a flexible cylinder describing the overall structure of the proteins. (The size of the cross section, which will be discussed below, will shed more light on this issue.) These cylinders, however, are not completely flexible but have a tendency to persist in some initial direction (Section 2.5.3). Below the Kuhn

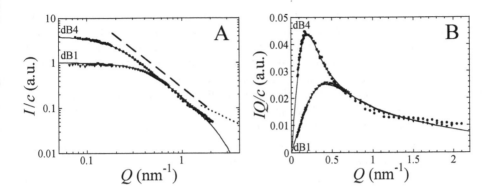

Figure 8. *Small-angle neutron-scattering data of solutions containing representative parts of the central domain of the protein gluten, dB1 and dB4. (A) Normalised scattered intensities $I(Q)/c$ as a function of the magnitude of the scattering vector Q in a log-log representation. Asymptotic $I(Q) \sim Q^{-5/3}$ and $I(Q) \sim Q^{-1}$ behaviours are indicated as the dashed and dotted lines, respectively. (B) $I(Q)Q$ as a function of Q ('Holtzer plot'). The solid lines are fits based on a semi-flexible polymer model (Egelhaaf et al. 2003).*

length l_K, and thus beyond $Q \sim 1/l_K$, they behave like stiff cylinders and an asymptotic $I(Q) \sim Q^{-1}$ dependence is expected (Figure 8(A), dotted line). This transition is located at $Ql_K \approx 3.8$ and is found at a similar value for both proteins; $Q \approx 1.3$ nm^{-1}, which results in $l_K \approx 1.5$ nm. However, the transition region is fairly broad, and hence an accurate determination difficult. A particularly sensitive way to represent scattering data for semi-flexible polymers is the so-called 'Holtzer plot' (or 'bending rod plot') shown in Figure 8(B), in which $I(Q)Q$ is plotted vs. Q. In this plot the $I(Q) \sim Q^{-1}$ dependence for an infinitely thin cylinder results in a plateau and the position of the transition to this plateau depends on l_K. The Holtzer plot also shows a maximum towards smaller values of Q. The position of the maximum only depends on the radius of gyration R_g (and not the flexibility) and the height of the maximum on the number of Kuhn lengths N_K along the contour length L; $N_K = L/l_K$. As expected, the maximum and thus N_K is higher for the longer dB4. The observed maxima are due to the stronger Q dependence of a flexible coil ($Q^{-5/3}$) when compared to a stiff cylinder (Q^{-1}). In particular, no maximum is found for stiff cylinders. These data are thus in clear disagreement with the structure of a stiff cylinder, which was proposed in earlier studies.

Up to now we have been considering the scattering behaviour in terms of asymptotic expressions only. In order to take full advantage of the information content in the data over the entire Q range, the crossover regions have to be considered also. A complete, quantitative characterisation of the scattered intensity $I(Q)$ over the entire Q range of a semi-flexible, self-avoiding cylinder with a circular cross section can be obtained by a decoupling approximation, which assumes that l_K, the shortest length scale of the semi-flexible, self-avoiding cylinder, is much larger than the cross-sectional radius R_{cs}. It is given by

$$I(Q, L, l_K, R_{cs}) \sim P_{wc}(Q, L, l_K)P_{cs}(Q, R_{cs}), \qquad (57)$$

where $P_{wc}(Q, L, l_K)$ and $P_{cs}(Q, R_{cs})$ are the form factors of an infinitely thin self-avoiding polymer (Section 2.5.3) and a circular cross section, respectively. The resulting fit yields good agreement with the data over the entire Q range (Figure 8, solid lines). The fitted values, especially L and R_{cs}, indicate that the cylinder is associated with the super-secondary structure of the protein, a β-spiral, and not the protein backbone. Although the data over the entire Q range very nicely agree with the model of a semi-flexible cylinder, it has to be stressed that, as with all scattering experiments, it cannot be ruled out that another model might explain the data equally well or even better (Section 2.8).

The structure of a flexible cylinder is an extension of the stiff rod model, which was assumed in earlier studies. On one hand, the cylindrical, rodlike shape is retained, while, on the other hand, the data clearly indicate a high degree of flexibility of the cylinder, which is most evident in the Holtzer plot. This adds a new component to the elasticity of the molecule and thus the network it forms. Having quantitatively determined the structure of the protein, concepts from polymer physics readily provide the theoretical background to predict its mechanical properties, in particular, the elasticity of an individual protein as well as a protein network.

4.2   PCNA — data analysis (indirect Fourier transform)

Proliferating cell nuclear antigen (PCNA, 29 kDa for the monomer) plays an important role in DNA replication and repair. In solution it exists as a trimer with a ring-like structure, which lines a hole through which double-stranded DNA can thread (Figure 9(A), inset). It acts as a moving platform or 'sliding clamp'. The aim of this study (Schurtenberger *et al.* 1998) was to determine the solution structure of human PCNA. Human and yeast PCNA are highly conserved at a structural and functional level; human PCNA is thus expected to be similar to yeast PCNA, for which a crystal structure exists.

Human PCNA was studied in solution by small-angle neutron scattering (Figure 9(A)). The extrapolation $[d\Sigma/d\Omega](Q{\to}0)$ yields the molar mass of the objects, which is consistent with the existence of trimers. To obtain further insight into the structure of PCNA, the experimentally determined $[d\Sigma/d\Omega](Q)$ can be compared with the scattering expected from the crystal structure of yeast PCNA, which had been crystallised. Based on this structure, each C^α-backbone atom can be represented by a 'bead' (Figure 9(A), inset and dashed lines) and the expected scattering calculated (Equation 25). Although the agreement with the data is reasonable, there are significant differences. It is very difficult to assign these differences in $[d\Sigma/d\Omega](Q)$ to certain structural characteristics. To obtain information in real space, an indirect Fourier transform (IFT) of the experimental and calculated $[d\Sigma/d\Omega](Q)$ was performed, which yields the corresponding pair distance distribution functions $p(r)$ (Figure 9(B)). Again, the general features are well represented, but differences are visible. However, even in real space it is very difficult to associate these differences with structural features; a trimer of PCNA consists of almost 800 amino acids, making it difficult to find the amino acids responsible for the discrepancy. Furthermore, any detailed interpretation is hampered by the fact that the discrepancies could either be due to differences between human and yeast PCNA or between solution (human PCNA) and crystal (yeast PCNA) structures.

To gain further insight into the scattering data, it is helpful to consider the simplest model, which can still describe the main characteristics of the trimer. Each monomer

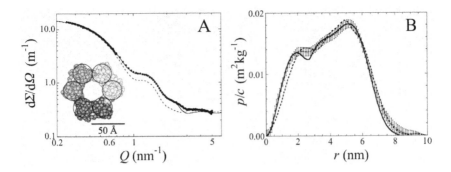

Figure 9. *Small-angle neutron-scattering data of solutions containing human PCNA. (A) Normalised scattered intensity* $d\Sigma/d\Omega(Q)$ *as a function of the magnitude of the scattering vector Q. (B) Normalised distance distribution function* $p(r)/c$ *as obtained by indirect Fourier transform. The model calculations based on the crystal structure of yeast PCNA (dashed lines) and the 'hexamer model' (solid lines) are also shown. The inset shows the* C^{α}*-backbone atoms as used in the crystal structure model and the six homogeneous spheres used in the simple 'hexamer model' (Schurtenberger et al. 1998).*

consists of two domains, thus resulting in a trimer with, on a coarse grained level, sixfold symmetry. The most basic model thus consists of six spheres forming a hexameric ring (Figure 9(A), inset). Surprisingly, this very simple model approximates the experimental $[d\Sigma/d\Omega](Q)$ and $p(r)$ extremely well (Figure 9). There is only a small, but significant, deviation at about $Q_{\Delta} = 3$ nm^{-1}, whose origin is difficult to establish based on $[d\Sigma/d\Omega](Q)$ alone. Here the corresponding $p(r)$ are very helpful. They also show astonishing agreement, except for $r_{\Delta} \approx 2.5$ nm. (Note that $Q_{\Delta}r_{\Delta} \approx 2\pi$.) In contrast to the information in Q space, we can associate this value with a distance in the hexamer model: It is the distance between the contact points of two spheres. Such a contact *point* is indeed not realistic; in a protein we rather expect an extended 'neck' between two spheres. It is thus not surprising that the measured probabilities of distances around 2.5 nm is higher than in the simple model with its contact points. This illustrates how powerful the indirect Fourier transform method can be by providing real-space information, which is usually much more accessible than information in Q space.

4.3 Structure of a virus — contrast variation

The structures of viruses are one of the prime examples how contrast variation can be applied to obtain detailed structural information, which is not available otherwise. Contrast variation allows us to distinguish the different virus components, nucleic acid (RNA or DNA), protein and sometimes lipid, due to their different scattering-length densities ρ (Figure 5(B)). Here we consider an RNA virus, the Southern bean mottle virus (Jacrot 1976, Jacrot *et al.* 1977).

Owing to their different ρ, the weights of the different contributions to the scattering

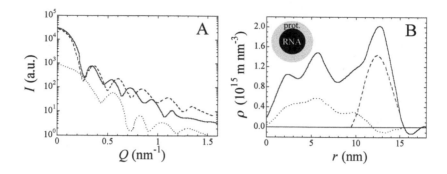

Figure 10. *Contrast variation series obtained by small-angle neutron scattering of aqueous solutions containing a virus. (A) Scattered intensities $I(Q)$ as a function of the magnitude of the scattering vector Q. (B) Radial scattering length density distribution $\rho(r)$ for different solvent compositions (0% D_2O, solid line; 42% D_2O, dotted line; 69% D_2O, dashed line). The inset is a schematic representation of the virus model with an RNA core and a protein shell (Jacrot 1976, Jacrot et al. 1977).*

change upon a change in the solvent scattering length density, i.e., exchange of H_2O with D_2O (Figure 10). At 42% D_2O, the protein is 'invisible' and the scattering is only due to the nucleic acid, whose molar mass and size can thus be determined from this scattering curve. On the other hand, at 69% D_2O, the scattering of the nucleic acid vanishes and the contribution of the protein dominates.

From a visual inspection of these two scattering curves, in particular, the pronounced maxima and minima, we can conclude that both the nucleic acid component and the protein component have spherical symmetry. Furthermore, the first minima of the scattered intensity due to the nucleic acid (42% D_2O; Figure 10, dotted line), is located at larger Q than the first minima due to the protein (69% D_2O, dashed line). This implies that the nucleic acid is more compact than the protein and indicates that the nucleic acid forms a spherical core inside the protein shell. The radial scattering length density distribution, which was obtained by indirect Fourier transform and spherical deconvolution (Figure 10(B)), supports this picture. The compact RNA core (dotted line) is surrounded by a protein shell (dashed line), and a contrast where both RNA and proteins are 'visible' yields contributions from the core and shell (solid line). Another confirmation is obtained by fitting the whole set of scattering data with a model describing the virus as composed of spherical shells with different scattering length densities (Figure 10, inset).

4.4 Concentrated DNA solutions — structure factor as a measure of particle arrangement

DNA is a highly charged polymer. Solutions containing even modest DNA concentrations (and low added salt concentrations) are thus dominated by long-range electrostatic interactions, which lead to a characteristic arrangement of DNA in solution. The organisation of DNA in these solutions is reflected in the structure factor $S(Q)$ (Section 2.6),

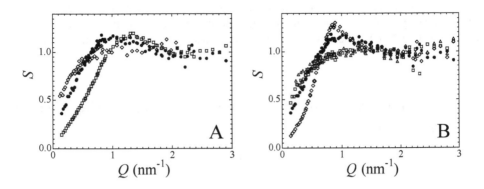

Figure 11. *Structure factor $S(Q)$ as a function of scattering vector Q of aqueous solutions of DNA obtained by small-angle neutron scattering. (A) $S(Q)$ for different DNA concentrations (◇, 0.05 M; •, 0.1 M; □, 0.2 M) and a constant salt concentration of 0.2 M KBr. (B) $S(Q)$ for different excess salt concentrations (◇, 0.04 M KBr; •, 0.2 M KBr; □, 1 M KBr; △, 2 M KBr) and a constant DNA concentration of 0.1 M (van der Maarel and Kassapidou 1998).*

which is obtained by dividing the measured scattered intensity $I(Q)$ by the intra-chain form factor $P(Q)$ (Equation 12). Here we report small-angle neutron-scattering experiments (van der Maarel and Kassapidou 1998) on solutions of short DNA fragments with about 160 base pairs, corresponding to a molar mass of 104 kDa and a contour length of 54 nm. The contour length is hence approximately half a Kuhn length l_K, and the form factor $P(Q)$ of an individual DNA fragment is thus similar to the scattering of a stiff rod.

The arrangement of these DNA fragments in aqueous solution is studied as a function of DNA and salt concentration (Figure 11). (The DNA concentrations investigated in this study imply that the solutions are in the semi-dilute regime.) For high DNA and/or low salt concentration, the structure factor $S(Q)$ oscillates about unity and shows a peak at finite Q. This peak in $S(Q)$ evolves upon an increase in DNA concentration; its height increases and its position shifts to larger Q. This reflects the increasing effect of interactions and the decreasing average DNA distance (Figure 11(A)). A similar increase in peak height as well as a peak sharpening is observed upon a decrease in salt concentration (Figure 11(B)), which implies an increasing effect of the interactions because of a decreased screening. However, the peaks remain at about the same Q, since the average DNA distance is constant. For very high salt concentrations, exceeding about 1 M, the correlation peak is almost completely suppressed and the behaviour of neutral polymers is recovered.

In this example the scattering is dominated by the DNA fragments, because the DNA scattering length contrast density exceeds the contrast of the counter-ions and added salt ions significantly. However, by choosing the appropriate scattering length of the solvent, i.e., adjusting the H_2O–D_2O ratio, the contributions of counter-ions and/or added salt ions can be increased. Using contrast variation, it is thus not only possible to determine the organisation of DNA fragments in solution but also to study the distribution of counter-ions and added salt ions around the charged DNA. Owing to strong interactions, small ions are expected to accumulate around the DNA, leading to the formation of a double layer. The double layer structure and the extent to which particular ions can approach the DNA has been successfully determined using this technique (Zakharova *et al.* 1999).

4.5   Solution structure of ocr — shape determination using dynamic light scattering

Upon infection of an *Escherichia coli* host cell by bacteriophage T7, ocr (overcome classical restriction) is the first protein to be produced. Once produced, ocr inhibits the type I restriction/modification (R/M) enzymes of the host cell. R/M enzymes can detect and destroy foreign DNA. Their inhibition is thus crucial to prevent the destruction of the invading DNA. This allows the remaining bacteriophage DNA to be transcribed and thus ensures a successful infection of the host by the phage. Recently the structure of ocr was solved by x-ray crystallography to a resolution of 0.18 nm. This confirmed a previous structural model, which had a lower resolution and was obtained using static (SLS) and dynamic (DLS) light scattering (Blackstock *et al.* 2001).

In the SLS experiments no systematic dependence of the scattered intensity on Q is observed, as expected for a protein that is small compared to the length scale probed in a light-scattering experiment (Equation 4). However, SLS allows the molar mass to be determined (Section 2.7) and for the present solutions yields a molar mass of about 28 kDa. A comparison with the molar mass of an ocr monomer as calculated from the amino acid sequence, about 14 kDa, indicates that ocr exists in solution as dimers.

DLS can be used to obtain information on the shape of the dimer. The correlation functions $f(Q, \tau)$ were determined at different Q. They become superimposed if plotted as a function of $Q^2\tau$ (Figure 12). The results are thus consistent with diffusion (Section 3.2). Furthermore, the linearity of $f(Q, \tau)$ over the time scale of interest indicates that the decay is dominated by one species. This is supported by the second-order cumulant fit giving a small polydispersity (Section 3.3). (At very small $Q^2\tau$, a faster decay is observed, which is mainly due to the rotational diffusion of ocr, but has a rotational decay time that is too small for an accurate determination in the present experiments.) From the characteristic times τ_c, which can be fitted or estimated from a plot of $\ln f(Q, \tau)$ as a function of $Q^2\tau$ (Figure 12), an average diffusion constant D can be calculated (Equation 53); $D = 79.1 \ \mu\text{m}^2\text{s}$. Using the Stokes-Einstein relation (Equation 54), D can be related to the hydrodynamic radius R_h, resulting in a value of $R_h = 2.6$ nm.

Assuming a spherical shape of the ocr dimer and taking into account its volume (calculated from the molar mass $M = 28$ kDa and typical specific volume of proteins $\bar{\nu} = 0.72 \ \text{cm}^3/\text{g}$), the dimer radius should be about 2.0 nm. This is significantly smaller than the observed R_h, which indicates a non-spherical shape of the dimer and significant hydration effects. With (reasonable) hydration alone, this discrepancy cannot be explained, which strongly suggests a non-spherical shape of the ocr dimer.

Non-spherical proteins are often modelled as oblate or prolate ellipsoids of revolution. Their axial ratio $p = a/b$ is defined as the ratio of the end-to-end length of the axis of revolution, a, and the equatorial diameter of the ellipsoid, b. Thus, for oblate proteins $p<1$, for prolate proteins $p>1$ and for spheres $p = 1$. Combined knowledge of the ro-

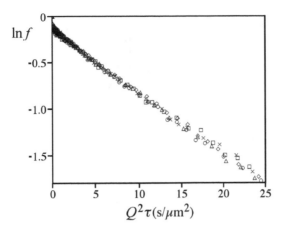

Figure 12. *Intermediate scattering function $f(Q, \tau)$ obtained by dynamic light scattering as a function of $Q^2\tau$ for different scattering vectors Q ($\diamond$, 13.8 μm^{-1}; $\square$, 15.0 μm^{-1}; $\triangle$, 16.3 μm^{-1}; $\times$, 17.5 μm^{-1}; $\circ$, 18.7 μm^{-1}) (Blackstock et al. 2001).*

tational and translational diffusion of non-spherical proteins can be used to determine p directly. Here, however, the rotation is too fast to be accurately determined and one has to rely on D only. To relate p to D, one usually writes

$$D(M, \bar{\nu}, p) = H(M, \bar{\nu})\mathcal{G}(p), \tag{58}$$

where $H(M, \bar{\nu})$ absorbs all the constants specific to the protein (except its shape). $\mathcal{G}(p)$ is then a function of the axial ratio p only (Figure 13). Since $\mathcal{G}(p)$ is independent of protein-specific factors, such as M or $\bar{\nu}$, Figure 13 shows the generic dependence of $\mathcal{G}(p)$ for any protein on its axial ratio p for both prolate and oblate shapes. Having determined D, p can be determined assuming either an oblate or a prolate shape of the ocr dimer. (Using the protein volume also a and b can be determined.) Assuming an oblate shape, its long and short axes are 6.7 nm and 1.4 nm, respectively. This implies that the maximum thickness (the middle of the oblate) is only 1.4 nm. Typical elements of secondary structures, such as α helices or β strands, could thus only be accommodated at the very centre of the ellipsoid. Moreover, a thin oblate ellipsoid would be very flexible, which makes it difficult to maintain a stable tertiary structure and thus an active conformation. These considerations render an oblate shape highly unlikely. For a prolate shape, the long and short axes are found to be 10.4 nm and 2.6 nm, respectively. This size and shape are strikingly similar to that of DNA to which type I R/M enzymes bind. It is furthermore supported by the fact that, as with DNA, ocr is strongly acidic with 28 negative charges per monomer at neutral pH. The prolate model thus suggests that ocr mimics the structure of DNA to which type I R/M enzymes bind. It is therefore consistent with the observation that ocr competes for the DNA binding site of type I R/M systems.

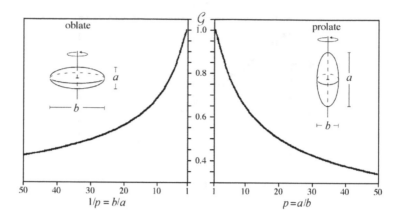

Figure 13. *The function $\mathcal{G}(p)$ as a function of $1/p$ (left, oblate) and p (right, prolate).*

Acknowledgements

It is a pleasure to thank Roland P May, Jan Skov Pedersen, Peter N Pusey and Peter Schurtenberger for introducing me to scattering techniques and many interesting discussions, Johan R C van der Maarel for the data presented in Figure 11, and Louisa Reissig for very helpful comments on this chapter.

References

Blackstock J J, Egelhaaf S U, Atanasiu C, Dryden D T F and Poon W C K, 2001, *Biochemistry* **40** 9944–9949.

Egelhaaf S U, van Swieten E, Bosma T, de Boef E, van Dijk A A and Robillard G T, 2003, *Biopolymers* **69** 311–324.

Higgins J S and Benoit H C, 1994, *Polymers and Neutron Scattering* (Clarendon Press, Oxford).

Jacrot B, 1976, *Rep Prog Phys* **39** 911–953.

Jacrot B, Chauvin C and Witz J, 1977, *Nature* **266** 417–421.

Kerker M, 1969, *The Scattering of Light and Other Electromagnetic Radiation* (Academic Press, New York).

Lindner P and Zemb T, eds, 2002, *Neutrons, X-Rays and Light: Scattering Methods Applied to Soft Condensed Matter* (Elsevier, Amsterdam).

Paul K and Thomas V, 1989, *Winnie the Witch* (Oxford University Press, Oxford).

Schurtenberger P, Egelhaaf S U, Hindges R, Maga G, Jónsson Z O, May R P, Glatter O and Hübscher U, 1998, *J Mol Biol* **275** 123–132.

Sears VF, 1992, *Neutron News* **3** 26–38.

Thurn-Albrecht T, Steffen W, Patkowski A, Meier G, Fischer E W, Grübel G and Abernathy D L, 1996, *Phys Rev Lett* **77** 5437–5440.

van de Hulst H C, 1981 *Light Scattering by Small Particles* (Dover, New York).

van der Maarel J R C and Kassapidou K, 1998, *Macromolecules* **31** 5734–5739.

Zakharova S S, Egelhaaf S U, Bhuiyan L B, Outhwaite C W, Bratko D and van der Maarel J R C, 1999, *J Chem Phys* **111** 10706–10716.

Participants

- August Andersson
  Stockholm University
  Arrhenius vag 12
  10691 Stockholm
  Sweden

- Stefan Auer
  Cambridge University
  Department of Chemistry
  Lensfield Rd
  Cambridge CB2 1EW
  UK

- Michal Avidan
  Tel Aviv University
  School of Physics
  69978 Tel Aviv
  Israel

- Vera Bajutina
  Uzbek Academy of Science
  28 Katartal Street, 700135
  Tashkent
  Uzbekistan

- David Biron
  Weizmann Institute of Science
  Rehovot 76100
  Israel

- Joshua Bloustine
  Brandeis University
  415 South Street
  Waltham, MA 02135
  USA

- Klemen Bohinc
  University of Ljubljana
  Poljanska 26a
  1000 Ljubljana
  Slovenia

- Georgios Boulougouris
  FOM-AMOLF
  Kruislaan 407, 1098 SJ
  Amsterdam
  Netherlands

- Giulio Caracciolo
  University of Rome La Sapienza
  P.le A. Moro 5
  00 185 Rome
  Italy

- Pietro Cicuta
  Cambridge University
  Nanoscience Centre
  JJ Thomsom Av
  Cambridge CB3 0FF
  UK

- Cyrille Claudet
  Laboratoire de Spectrometrie physique CNRS
  140 rue de la physique
  38402 St martin d'Heres, BP 87
  France

- Ivan Coluzza
  FOM Institute
  Kruislaan 407, 1098 SJ
  Amsterdam
  Netherlands

- Ira Cooke
  Max-Planck Institute for Polymer Research
  Ackermannweg 10
  D-55128 Mainz
  Germany

- Jens Danielsson
  Stockholm University
  SE-106 91 Stockholm
  Sweden

- Markus Deserno
  Max-Planck Institute for Polymer Research
  PO Box 3148
  55021 Mainz
  Germany

- Fabiana Diotallevi
  FOM-AMOLF
  Kruislaan 407, 1098 SJ
  Amsterdam
  Netherlands

- Jure Dobnikar
  University Of Konstanz
  PF 5560
  D78957 Konstanz
  Germany

- Arti Dua
  Max-Planck Institute for Polymer Research
  Ackermannweg 10
  D-55128 Mainz
  Germany

- Evgeniy Dubrovin
  Moscow Lomonosov
  Leninskiye Gory
  Moscow, 119992
  Russia

- Michael Maurice Dupin
  Sheffield Hallam University
  Pond Street
  Sheffield, S1 1WB
  UK

- Richard Elliot
  University of Washington
  Department of Physics
  Seattle
  WA 98115
  USA

- Olga Ermak
  Moscow State University
  Moscow 119992
  Russia

- Suzanne Fielding
  University of Leeds
  Department of Physics & Astronomy
  Leeds LS2 9JT
  UK

- Miha Fosnaric
  University of Ljubljana
  Trzaska 25
  Sl-1000 Ljubljana
  Slovenia

- Marat Olegovich Gallyamov
  University of Ulm
  Alber Einstein Allee 11
  D-89069 Ulm
  Germany

- Guillermo Ivan Guerrero-Garcia
  Universidad Autonoma de SanLuis Potosi
  Alvaro Obregon No 64
  Mexico

- Peter Hagedorn
  Riso National Laboratory
  Plant Research department
  PRD-330 Frederiksborgvej 399
  DK-4000 Roskilde
  Denmark

- Poul Martin Hansen
  University of Copenhagen
  Blegdamsvej 17
  2100 Copenhagen
  Denmark

- Imran Hasnain
  Cavendish Laboratory
  Madingley Road
  Cambridge CB3 0HE
  UK

- Rhoda Joy Hawkins
  University of Leeds
  IRC in Polymer Science & Technology
  Leeds LS2 9JT
  UK

- David Head
  Vrije Universiteit,
  Department of Physics & Astronomy
  De Boelelaan 1081, NL-1081 HV Amsterdam
  Netherlands

- Dion Houtman
  Amsterdam University
  Nieuwe Achtergracht 166
  1018 WV Amsterdam
  Netherlands

- Jozef Hritz
  Department of Biophysics PF-UP JS
  Jesenna 5
  04154 Kosice
  Slovak Rep

- Bhavin Khatri
  University of Leeds
  Dept Physics & Astronomy
  Leeds LS2 9JT
  UK

- Joanna Kwiecinsua
  Polish Academy of Sciences
  Al. Lotnikow 32/46
  02–668 Warsaw
  Poland

- Mikael Lund
  University of Lund
  Chemical Centre
  S-22100 Lund
  Denmark

- David Michel
  Sheffield Hallam University
  Materials Research Institute
  City Campus
  Sheffield S1 1WB
  UK

- Mark Miller
  Cambridge University
  Lensfield Road
  Cambridge CB2 1EW
  UK

- Graham Milne
  University of St. Andrews
  North Haugh
  St. Andrews
  UK

- Dmitri Miroshnychenko
  University of Glasgow
  University Gardens
  Glasgow G12 8QW
  UK

- Daisuke Mizuno
  Vrije Universiteit
  De Boelelaan 1081
  1081 HV Amsterdam
  Netherlands

- Christian Nowak
  Max-Planck Institute for Polymer Research
  55021 Mainz
  Germany

- Etay Mar Or
  Tel Aviv University
  School of Physics
  69978 Tel Aviv
  Israel

- Arsteidis Papagiannopoulos
  University of Leeds
  Woodhouse Lane
  LS2 9JT Leeds
  UK

- Angelo Perla
  University of Rome La Sapienza
  Piazzale Aldomoro 2
  00185 Rome
  Italy

- Daniela Pozzi
  University of Rome La Sapienza
  P.le A. Moro 5
  00 185 Roma
  Italy

- Matejn Praprotnik
  National Institute of Chemistry
  Hajdrihova 19
  Sl-1001 Ljubljana
  Slovenia

- Elena Puchianu
  Inst Cellular Biology and Pathology
  8, B.P. Hasdeu St
  Bucharest
  Romania

- Filippo Pullara
  University of Palermo
  Via Archirafi
  36-90123 Palermo
  Italy

- Prashant Purohit
  California Institute of Technology
  1200 East California Blvd
  Pasadena
  CA 91125
  USA

- Shay Rappaport
  Bar-llan University
  Ramat-Gan 52900
  Israel

- Louisa Reissig
  Warsaw University
  ul. Ludna 10/30
  00-414 Warszawa
  Poland

- Salman Rogers
  Cambridge University
  Madingley Road
  Cambridge CB3 0HE
  UK

- Mehmet Sayar
  Max-Planck Institute for Polymer Research
  Ackermannweg 10
  D-55128 Mainz
  Germany

- Simona Sennato
  University of Rome La Sapienza
  Pz. Aldo Moro 2
  00 185 Rome
  Italy

- Tom Shemesh
  Tel Aviv University
  School of Physics
  Tel Aviv 69978
  Israel

- Barak Shenhave
  Weizmann Institute of Science
  Rehovot 76100
  Israel

- Amena Siddiqi
  Cornell University
  223 Eddy S
  Number 2
  Ithaca, NY 14850
  USA

- Aditi Simha
  Max-Planck Institute for Polymer Research
  Noethnltzer Strasse 38
  01187 Dresden
  Germany

- Natalia Sitnikova
  Amsterdam University
  Valckenierstraat 65
  1018 XE Amsterdam
  Netherlands

- Hamutal Soroker
  Tel Aviv University
  School of Physics
  Tel Aviv 69978
  Israel

- Sathish Sukumaran
  Max-Planck Institute for Polymer Research
  Ackermannweg 10
  D-55128 Mainz
  Germany

- Mikhail Vladimirovich Tamm
  Moscow State University
  Vorobyevy Gory
  119992 Moscow
  Russia

- Valdimir Teif
  Institute of Bioorganic Chemistry
  Kuprevich St, 5/2, 220141
  Minsk
  Belarus

- Rochish Thaokar
  Max-Planck Institute for Polymer Research
  Ackermannweg 10
  D-55128 Mainz
  Germany

- Ajay Tharakan
  Birmingham University
  & Unilever Corporate Research
  Unilever R&D
  Colworth house, Sharnbrook
  Bedfordshire MK44 1LQ
  UK

- Chantal Valeriani
  FOM-AMOLF
  Kruislaan 407
  1098 SJ
  Amsterdam
  Netherlands

- Sarah Veatch
  University of Washington
  Department of Physics
  Seattle, WA 98195
  USA

- Thomas Vilmin
  College de France
  11 Place Marcelin Berthelot
  75005 Paris
  France

Index